D1788378

partial differential equations on multistructures

PURE AND APPLIED MATHEMATICS

A Program of Monographs, Textbooks, and Lecture Notes

EXECUTIVE EDITORS

Earl J. Taft
Rutgers University
New Brunswick, New Jersey

Zuhair Nashed
University of Delaware
Newark, Delaware

EDITORIAL BOARD

M. S. Baouendi
University of California, San Diego

Jane Cronin
Rutgers University

Jack K. Hale
Georgia Institute of Technology

S. Kobayashi
University of California, Berkeley

Marvin Marcus
University of California, Santa Barbara

W. S. Massey
Yale University

Anil Nerode
Cornell University

Donald Passman
University of Wisconsin, Madison

Fred S. Roberts
Rutgers University

David L. Russell
Virginia Polytechnic Institute and State University

Walter Schempp
Universität Siegen

Mark Teply
University of Wisconsin, Milwaukee

LECTURE NOTES IN PURE AND APPLIED MATHEMATICS

1. *N. Jacobson,* Exceptional Lie Algebras
2. *L.-Å. Lindahl and F. Poulsen,* Thin Sets in Harmonic Analysis
3. *I. Satake,* Classification Theory of Semi-Simple Algebraic Groups
4. *F. Hirzebruch et al., Differentiable Manifolds and Quadratic Forms*
5. *I. Chavel,* Riemannian Symmetric Spaces of Rank One
6. *R. B. Burckel,* Characterization of C(X) Among Its Subalgebras
7. *B. R. McDonald et al.,* Ring Theory
8. *Y.-T. Siu,* Techniques of Extension on Analytic Objects
9. *S. R. Caradus et al.,* Calkin Algebras and Algebras of Operators on Banach Spaces
10. *E. O. Roxin et al.,* Differential Games and Control Theory
11. *M. Orzech and C. Small,* The Brauer Group of Commutative Rings
12. *S. Thomier,* Topology and Its Applications
13. *J. M. Lopez and K. A. Ross,* Sidon Sets
14. *W. W. Comfort and S. Negrepontis,* Continuous Pseudometrics
15. *K. McKennon and J. M. Robertson,* Locally Convex Spaces
16. *M. Carmeli and S. Malin,* Representations of the Rotation and Lorentz Groups
17. *G. B. Seligman,* Rational Methods in Lie Algebras
18. *D. G. de Figueiredo,* Functional Analysis
19. *L. Cesari et al.,* Nonlinear Functional Analysis and Differential Equations
20. *J. J. Schäffer,* Geometry of Spheres in Normed Spaces
21. *K. Yano and M. Kon,* Anti-Invariant Submanifolds
22. *W. V. Vasconcelos,* The Rings of Dimension Two
23. *R. E. Chandler,* Hausdorff Compactifications
24. *S. P. Franklin and B. V. S. Thomas,* Topology
25. *S. K. Jain,* Ring Theory
26. *B. R. McDonald and R. A. Morris,* Ring Theory II
27. *R. B. Mura and A. Rhemtulla,* Orderable Groups
28. *J. R. Graef,* Stability of Dynamical Systems
29. *H.-C. Wang,* Homogeneous Branch Algebras
30. *E. O. Roxin et al.,* Differential Games and Control Theory II
31. *R. D. Porter,* Introduction to Fibre Bundles
32. *M. Altman,* Contractors and Contractor Directions Theory and Applications
33. *J. S. Golan,* Decomposition and Dimension in Module Categories
34. *G. Fairweather,* Finite Element Galerkin Methods for Differential Equations
35. *J. D. Sally,* Numbers of Generators of Ideals in Local Rings
36. *S. S. Miller,* Complex Analysis
37. *R. Gordon,* Representation Theory of Algebras
38. *M. Goto and F. D. Grosshans,* Semisimple Lie Algebras
39. *A. I. Arruda et al.,* Mathematical Logic
40. *F. Van Oystaeyen,* Ring Theory
41. *F. Van Oystaeyen and A. Verschoren,* Reflectors and Localization
42. *M. Satyanarayana,* Positively Ordered Semigroups
43. *D. L Russell,* Mathematics of Finite-Dimensional Control Systems
44. *P.-T. Liu and E. Roxin,* Differential Games and Control Theory III
45. *A. Geramita and J. Seberry,* Orthogonal Designs
46. *J. Cigler, V. Losert, and P. Michor,* Banach Modules and Functors on Categories of Banach Spaces
47. *P.-T. Liu and J. G. Sutinen,* Control Theory in Mathematical Economics
48. *C. Byrnes,* Partial Differential Equations and Geometry
49. *G. Klambauer,* Problems and Propositions in Analysis
50. *J. Knopfmacher,* Analytic Arithmetic of Algebraic Function Fields
51. *F. Van Oystaeyen,* Ring Theory
52. *B. Kadem,* Binary Time Series
53. *J. Barros-Neto and R. A. Artino,* Hypoelliptic Boundary-Value Problems
54. *R. L. Sternberg et al.,* Nonlinear Partial Differential Equations in Engineering and Applied Science
55. *B. R. McDonald,* Ring Theory and Algebra III
56. *J. S. Golan,* Structure Sheaves Over a Noncommutative Ring
57. *T. V. Narayana et al.,* Combinatorics, Representation Theory and Statistical Methods in Groups
58. *T. A. Burton,* Modeling and Differential Equations in Biology
59. *K. H. Kim and F. W. Roush,* Introduction to Mathematical Consensus Theory

60. *J. Banas and K. Goebel,* Measures of Noncompactness in Banach Spaces
61. *O. A. Nielson,* Direct Integral Theory
62. *J. E. Smith et al.,* Ordered Groups
63. *J. Cronin,* Mathematics of Cell Electrophysiology
64. *J. W. Brewer,* Power Series Over Commutative Rings
65. *P. K. Kamthan and M. Gupta,* Sequence Spaces and Series
66. *T. G. McLaughlin,* Regressive Sets and the Theory of Isols
67. *T. L. Herdman et al.,* Integral and Functional Differential Equations
68. *R. Draper,* Commutative Algebra
69. *W. G. McKay and J. Patera,* Tables of Dimensions, Indices, and Branching Rules for Representations of Simple Lie Algebras
70. *R. L. Devaney and Z. H. Nitecki,* Classical Mechanics and Dynamical Systems
71. *J. Van Geel,* Places and Valuations in Noncommutative Ring Theory
72. *C. Faith,* Injective Modules and Injective Quotient Rings
73. *A. Fiacco,* Mathematical Programming with Data Perturbations I
74. *P. Schultz et al.,* Algebraic Structures and Applications
75. *L Bican et al.,* Rings, Modules, and Preradicals
76. *D. C. Kay and M. Breen,* Convexity and Related Combinatorial Geometry
77. *P. Fletcher and W. F. Lindgren,* Quasi-Uniform Spaces
78. *C.-C. Yang,* Factorization Theory of Meromorphic Functions
79. *O. Taussky,* Ternary Quadratic Forms and Norms
80. *S. P. Singh and J. H. Burry,* Nonlinear Analysis and Applications
81. *K. B. Hannsgen et al.,* Volterra and Functional Differential Equations
82. *N. L. Johnson et al.,* Finite Geometries
83. *G. I. Zapata,* Functional Analysis, Holomorphy, and Approximation Theory
84. *S. Greco and G. Valla,* Commutative Algebra
85. *A. V. Fiacco,* Mathematical Programming with Data Perturbations II
86. *J.-B. Hiriart-Urruty et al.,* Optimization
87. *A. Figa Talamanca and M. A. Picardello,* Harmonic Analysis on Free Groups
88. *M. Harada,* Factor Categories with Applications to Direct Decomposition of Modules
89. *V. I. Istrătescu,* Strict Convexity and Complex Strict Convexity
90. *V. Lakshmikantham,* Trends in Theory and Practice of Nonlinear Differential Equations
91. *H. L. Manocha and J. B. Srivastava,* Algebra and Its Applications
92. *D. V. Chudnovsky and G. V. Chudnovsky,* Classical and Quantum Models and Arithmetic Problems
93. *J. W. Longley,* Least Squares Computations Using Orthogonalization Methods
94. *L. P. de Alcantara,* Mathematical Logic and Formal Systems
95. *C. E. Aull,* Rings of Continuous Functions
96. *R. Chuaqui,* Analysis, Geometry, and Probability
97. *L. Fuchs and L. Salce,* Modules Over Valuation Domains
98. *P. Fischer and W. R. Smith,* Chaos, Fractals, and Dynamics
99. *W. B. Powell and C. Tsinakis,* Ordered Algebraic Structures
100. *G. M. Rassias and T. M. Rassias,* Differential Geometry, Calculus of Variations, and Their Applications
101. *R.-E. Hoffmann and K. H. Hofmann,* Continuous Lattices and Their Applications
102. *J. H. Lightbourne III and S. M. Rankin III,* Physical Mathematics and Nonlinear Partial Differential Equations
103. *C. A. Baker and L. M. Batten,* Finite Geometrics
104. *J. W. Brewer et al.,* Linear Systems Over Commutative Rings
105. *C. McCrory and T. Shifrin,* Geometry and Topology
106. *D. W. Kueke et al.,* Mathematical Logic and Theoretical Computer Science
107. *B.-L. Lin and S. Simons,* Nonlinear and Convex Analysis
108. *S. J. Lee,* Operator Methods for Optimal Control Problems
109. *V. Lakshmikantham,* Nonlinear Analysis and Applications
110. *S. F. McCormick,* Multigrid Methods
111. *M. C. Tangora,* Computers in Algebra
112. *D. V. Chudnovsky and G. V. Chudnovsky,* Search Theory
113. *D. V. Chudnovsky and R. D. Jenks,* Computer Algebra
114. *M. C. Tangora,* Computers in Geometry and Topology
115. *P. Nelson et al.,* Transport Theory, Invariant Imbedding, and Integral Equations
116. *P. Clément et al.,* Semigroup Theory and Applications
117. *J. Vinuesa,* Orthogonal Polynomials and Their Applications
118. *C. M. Dafermos et al.,* Differential Equations
119. *E. O. Roxin,* Modern Optimal Control
120. *J. C. Díaz,* Mathematics for Large Scale Computing

121. *P. S. Milojević* Nonlinear Functional Analysis
122. *C. Sadosky,* Analysis and Partial Differential Equations
123. *R. M. Shortt,* General Topology and Applications
124. *R. Wong,* Asymptotic and Computational Analysis
125. *D. V. Chudnovsky and R. D. Jenks,* Computers in Mathematics
126. *W. D. Wallis et al.,* Combinatorial Designs and Applications
127. *S. Elaydi,* Differential Equations
128. *G. Chen et al.,* Distributed Parameter Control Systems
129. *W. N. Everitt,* Inequalities
130. *H. G. Kaper and M. Garbey,* Asymptotic Analysis and the Numerical Solution of Partial Differential Equations
131. *O. Arino et al.,* Mathematical Population Dynamics
132. *S. Coen,* Geometry and Complex Variables
133. *J. A. Goldstein et al.,* Differential Equations with Applications in Biology, Physics, and Engineering
134. *S. J. Andima et al.,* General Topology and Applications
135. *P Clément et al.,* Semigroup Theory and Evolution Equations
136. *K. Jarosz,* Function Spaces
137. *J. M. Bayod et al.,* *p*-adic Functional Analysis
138. *G. A. Anastassiou,* Approximation Theory
139. *R. S. Rees,* Graphs, Matrices, and Designs
140. *G. Abrams et al.,* Methods in Module Theory
141. *G. L. Mullen and P. J.-S. Shiue,* Finite Fields, Coding Theory, and Advances in Communications and Computing
142. *M. C. Joshi and A. V. Balakrishnan,* Mathematical Theory of Control
143. *G. Komatsu and Y. Sakane,* Complex Geometry
144. *I. J. Bakelman,* Geometric Analysis and Nonlinear Partial Differential Equations
145. *T. Mabuchi and S. Mukai,* Einstein Metrics and Yang–Mills Connections
146. *L. Fuchs and R. Göbel,* Abelian Groups
147. *A. D. Pollington and W. Moran,* Number Theory with an Emphasis on the Markoff Spectrum
148. *G. Dore et al.,* Differential Equations in Banach Spaces
149. *T. West,* Continuum Theory and Dynamical Systems
150. *K. D. Bierstedt et al.,* Functional Analysis
151. *K. G. Fischer et al.,* Computational Algebra
152. *K. D. Elworthy et al.,* Differential Equations, Dynamical Systems, and Control Science
153. *P.-J. Cahen, et al.,* Commutative Ring Theory
154. *S. C. Cooper and W. J. Thron,* Continued Fractions and Orthogonal Functions
155. *P. Clément and G. Lumer,* Evolution Equations, Control Theory, and Biomathematics
156. *M. Gyllenberg and L. Persson,* Analysis, Algebra, and Computers in Mathematical Research
157. *W. O. Bray et al.,* Fourier Analysis
158. *J. Bergen and S. Montgomery,* Advances in Hopf Algebras
159. *A. R. Magid,* Rings, Extensions, and Cohomology
160. *N. H. Pavel,* Optimal Control of Differential Equations
161. *M. Ikawa,* Spectral and Scattering Theory
162. *X. Liu and D. Siegel,* Comparison Methods and Stability Theory
163. *J.-P. Zolésio,* Boundary Control and Variation
164. *M. Křížek et al.,* Finite Element Methods
165. *G. Da Prato and L. Tubaro,* Control of Partial Differential Equations
166. *E. Ballico,* Projective Geometry with Applications
167. *M. Costabel et al.,* Boundary Value Problems and Integral Equations in Nonsmooth Domains
168. *G. Ferreyra, G. R. Goldstein, and F. Neubrander,* Evolution Equations
169. *S. Huggett,* Twistor Theory
170. *H. Cook et al.,* Continua
171. *D. F. Anderson and D. E. Dobbs,* Zero-Dimensional Commutative Rings
172. *K. Jarosz,* Function Spaces
173. *V. Ancona et al.,* Complex Analysis and Geometry
174. *E. Casas,* Control of Partial Differential Equations and Applications
175. *N. Kalton et al.,* Interaction Between Functional Analysis, Harmonic Analysis, and Probability
176. *Z. Deng et al.,* Differential Equations and Control Theory
177. *P. Marcellini et al.* Partial Differential Equations and Applications
178. *A. Kartsatos,* Theory and Applications of Nonlinear Operators of Accretive and Monotone Type
179. *M. Maruyama,* Moduli of Vector Bundles
180. *A. Ursini and P. Aglianò,* Logic and Algebra
181. *X. H. Cao et al.,* Rings, Groups, and Algebras
182. *D. Arnold and R. M. Rangaswamy,* Abelian Groups and Modules
183. *S. R. Chakravarthy and A. S. Alfa,* Matrix-Analytic Methods in Stochastic Models

184. *J. E. Andersen et al.*, Geometry and Physics
185. *P.-J. Cahen et al.*, Commutative Ring Theory
186. *J. A. Goldstein et al.*, Stochastic Processes and Functional Analysis
187. *A. Sorbi*, Complexity, Logic, and Recursion Theory
188. *G. Da Prato and J.-P. Zolésio*, Partial Differential Equation Methods in Control and Shape Analysis
189. *D. D. Anderson*, Factorization in Integral Domains
190. *N. L. Johnson*, Mostly Finite Geometries
191. *D. Hinton and P. W. Schaefer*, Spectral Theory and Computational Methods of Sturm–Liouville Problems
192. *W. H. Schikhof et al.*, *p*-adic Functional Analysis
193. *S. Sertöz*, Algebraic Geometry
194. *G. Caristi and E. Mitidieri*, Reaction Diffusion Systems
195. *A. V. Fiacco*, Mathematical Programming with Data Perturbations
196. *M. Křížek et al.*, Finite Element Methods: Superconvergence, Post-Processing, and A Posteriori Estimates
197. *S. Caenepeel and A. Verschoren*, Rings, Hopf Algebras, and Brauer Groups
198. *V. Drensky et al.*, Methods in Ring Theory
199. *W. B. Jones and A. Sri Ranga*, Orthogonal Functions, Moment Theory, and Continued Fractions
200. *P. E. Newstead*, Algebraic Geometry
201. *D. Dikranjan and L. Salce*, Abelian Groups, Module Theory, and Topology
202. *Z. Chen et al.*, Advances in Computational Mathematics
203. *X. Caicedo and C. H. Montenegro*, Models, Algebras, and Proofs
204. *C. Y. Yıldırım and S. A. Stepanov*, Number Theory and Its Applications
205. *D. E. Dobbs et al.*, Advances in Commutative Ring Theory
206. *F. Van Oystaeyen*, Commutative Algebra and Algebraic Geometry
207. *J. Kakol et al.*, *p*-adic Functional Analysis
208. *M. Boulagouaz and J.-P. Tignol*, Algebra and Number Theory
209. *S. Caenepeel and F. Van Oystaeyen*, Hopf Algebras and Quantum Groups
210. *F. Van Oystaeyen and M. Saorin*, Interactions Between Ring Theory and Representations of Algebras
211. *R. Costa et al.*, Nonassociative Algebra and Its Applications
212. *T.-X. He*, Wavelet Analysis and Multiresolution Methods
213. *H. Hudzik and L. Skrzypczak*, Function Spaces: The Fifth Conference
214. *J. Kajiwara et al.*, Finite or Infinite Dimensional Complex Analysis
215. *G. Lumer and L. Weis*, Evolution Equations and Their Applications in Physical and Life Sciences
216. *J. Cagnol et al.*, Shape Optimization and Optimal Design
217. *J. Herzog and G. Restuccia*, Geometric and Combinatorial Aspects of Commutative Algebra
218. *G. Chen et al.*, Control of Nonlinear Distributed Parameter Systems
219. *F. Ali Mehmeti et al.*, Partial Differential Equations on Multistructures

Additional Volumes in Preparation

partial differential equations on multistructures

proceedings of the conference held in Luminy, France

edited by

Felix Ali Mehmeti
Université de Valenciennes et du Hainaut Cambrésis
Valenciennes, France

Joachim von Below
Université du Littoral Côte d'Opale
Calais, France

Serge Nicaise
Université de Valenciennes et du Hainaut Cambrésis
Valenciennes, France

MARCEL DEKKER

MARCEL DEKKER, INC. NEW YORK • BASEL

ISBN: 0-8247-0565-3

This book is printed on acid-free paper.

Headquarters
Marcel Dekker, Inc.
270 Madison Avenue, New York, NY 10016
tel: 212-696-9000; fax: 212-685-4540

Eastern Hemisphere Distribution
Marcel Dekker AG
Hutgasse 4, Postfach 812, CH-4001 Basel, Switzerland
tel: 41-61-261-8482; fax: 41-61-261-8896

World Wide Web
http://www.dekker.com

The publisher offers discounts on this book when ordered in bulk quantities. For more information, write to Special Sales/Professional Marketing at the headquarters address above.

Copyright © 2001 by Marcel Dekker, Inc. All Rights Reserved.

Neither this book nor any part may be reproduced or transmitted in any form or by any means, electronic or mechanical, including photocopying, microfilming, and recording, or by any information storage and retrieval system, without permission in writing from the publisher.

Current printing (last digit):
10 9 8 7 6 5 4 3 2 1

PRINTED IN THE UNITED STATES OF AMERICA

Preface

The international conference on Partial Differential Equations on Multistructures was held at the Centre International de Rencontres Mathématiques (CIRM), Luminy, France. The conference brought together specialists from rather different research domains in partial differential equations who all investigate problems bearing a multistructure aspect, such as networks, ramified spaces, diffractions, or interfaces.

The analysis of multistructures arises in many applications, e.g. in neurobiology, electronics and continuum mechanics. The modelling of these problems is governed by partial differential equations in each of the components and by transition conditions between these. In the classical context, one considers a partial differential equation or a system of these in one domain subject to boundary conditions. The aim of the conference and of its proceedings was the investigation of partial differential equations on several domains connected by interaction zones. For some of the recent research already done we refer to F. Ali Mehmeti, J. von Below, J. Lagnese, G. Leugering, G. Lumer, S. Nazarov, St. Müller, S. Nicaise, J. P. Roth, E. Schmidt et al. Several groups are actually working in various aspects of this research field, which are developing in a very promising way not only in theory, but also in application. The contributions to the conference and its proceedings treat the general theme under the following aspects: specific modelling, spectral theory and operator theory, nonlinear analysis, variational calculus, integral equations, dynamical systems, control theory and numerical analysis.

The modelling of physical phenomena in heterogeneous media leads to the mathematical analysis on ramified spaces or, more generally, on multistructures that correspond to systems of partial differential equations under nonclassical boundary conditions. The latter ones play a key rôle since they describe the transition in the interaction zones and link the different domains to the whole multistructure. Many analogues to the classical domain case are now available for multistructure-like spaces. Especially the spectral theory reveals specific features of these spaces and, thereby, it helps to characterize and to describe the interaction models by means of their associated spectra. Correspondingly, the variational approach, the control theory, nonlinear dynamical systems and the numerical analysis, all reveal the specific character of the multistructure in question.

The articles of this volume are based on 14 of the 24 lectures of the conference and present new results in various topics of the analysis on multistructures. They include generalizations of classical theories to multistructures as well as specific characterizations and modellings of these. The volume gives a good overview of the actual research in that field and constitutes a valuable source for the interested researcher in partial differential equations, functional analysis and control theory who wants to participate in this research field of growing interest.

The organization of the conference was enabled and financed by

the Centre International de Rencontres Mathématiques (CIRM), Luminy, France,

the Université du Littoral Côte d'Opale, France,

the Université de Valenciennes et du Hainaut Cambrésis, France.

The organizers would like to express their gratitude to all of them, in particular to the staff of the conference center CIRM, notably to its Director J. P. Labesse, Mrs. A. Zeller–Meier and Mrs. S. Arnaud.

Moreover, the editors of this volume would like to express their gratitude to the contributors, to the referees and to Marcel Dekker, Inc., especially to Ms. M. Allegra and Ms. J. Paizzi, for their kind and generous cooperation in the production of the volume.

Felix Ali Mehmeti

Joachim von Below

Serge Nicaise

Contents

Contributors

FELIX ALI MEHMETI
Laboratoire de Mathématiques Appliquées et Calcul Scientifique, Université de Valenciennes et du Hainaut - Cambrésis, Le Mont Houy, F–59313 Valenciennes Cedex 9, France.

JOACHIM VON BELOW
Laboratoire de Mathématiques Pures et Appliquées Joseph Liouville, Université du Littoral Côte d'Opale, B.P. 699, F-62228 Calais Cedex, France.

MARIUS BOCHNIAK
Mathematische Institut A, Universität Stuttgart, Pfaffenwaldring 57, D–70511 Stuttgart, Germany.

MARYSE BOURLARD
Laboratoire de Mathématiques Appliquées et Calcul Scientifique, Université de Valenciennes et du Hainaut - Cambrésis, Le Mont Houy, F–59313 Valenciennes Cedex 9, France.

BERTRAND DEKONINCK
Laboratoire de Mathématiques Appliquées et Calcul Scientifique, Université de Valenciennes et du Hainaut - Cambrésis, Le Mont Houy, F–59313 Valenciennes Cedex 9, France.

TORSTEN FISCHER
Fakultät für Mathematik und Physik, Universität Bayreuth, D–95440 Bayreuth, Germany.

GILLES FREMIOT
Institut Elie Cartan, Université de Nancy I, B.P. 239, F–54506 Vandœuvre–les–Nancy Cedex, France.

DOMINIQUE LEGUILLON
Laboratoire de Modélisation en Mécanique, Université de Pierre et Marie Curie, F–75252 Paris Cedex 05, France.

MICHEL LENCZNER
Equipe de Mathématiques, Université de Franche Comté, Route de Gray, F–25030 Besançon Cedex, France.

GÜNTHER LEUGERING
Fachbereich Mathematik, Technische Universität Darmstadt, Schloßgartenstraße 7, D–64289 Darmstadt, Germany.

JOSÉ A. LUBARY
Departement de Matemàtica Aplicata II, Universitat Polytecnica de Catalunya, 5 Pau Gargallo, ES–08028 Barcelona, Spain.

GUNTER LUMER
Institut de Mathématiques et d'Informatique, Université de Mons–Hainaut, 6, Ave. du Champ de Mars, B–7000 Mons, Belgium.

ABDERRAHMAN MAGHNOUJI
Laboratoire de Mathématiques Appliquées et Calcul Scientifique, Université de Valenciennes et du Hainaut - Cambrésis, Le Mont Houy, F–59313 Valenciennes Cedex 9, France.

DENIS MERCIER
Laboratoire de Mathématiques Appliquées et Calcul Scientifique, Université de Valenciennes et du Hainaut - Cambrésis, Le Mont Houy, F–59313 Valenciennes Cedex 9, France.

FRANK NEUBRANDER
Departement of Mathematics, Lousiana State University, Baton Rouge, La. 70803-4918 U.S.A.

SERGE NICAISE
Laboratoire de Mathématiques Appliquées et Calcul Scientifique, Université de Valenciennes et du Hainaut - Cambrésis, Le Mont Houy, F–59313 Valenciennes Cedex 9, France.

LUC PAQUET
Laboratoire de Mathématiques Appliquées et Calcul Scientifique, Université de Valenciennes et du Hainaut - Cambrésis, Le Mont Houy, F–59313 Valenciennes Cedex 9, France.

OLEG M. PENKIN
Voronezh State University, 1 Universitatskaya Pl., 304693 Voronezh, Russia.

JEAN-MARC SAC-ÉPÉE
Département de Mathématiques, Université de Metz, Ile du Saulcy, F-57045 Metz Cedex 01, France.

ANNA-MARGARETE SÄNDIG
Mathematisches Institut A, Universität Stuttgart, Pfaffenwaldring 57, D-70511 Stuttgart, Germany.

JEANNINE SAINT JEAN-PAULIN
Département de Mathématiques, Université de Metz, Ile du Saulcy, F-57045 Metz Cedex 01, France.

JACQUELINE SANCHEZ-HUBERT
UFR Sciences, Mécanique, Université de Caen, Boulevard du Maréchal Juin, Campus 2, F-14032 Caen, France.

ÉVARISTE SANCHEZ PALENCIA
Laboratoire de Modélisation en Mécanique, Université de Pierre et Marie Curie, F-75252 Paris Cedex 05, France.

GHOUTI SENOUCI
Equipe de Mathématiques, Université de Franche Comté, Route de Gray, F-25030 Besançon Cedex, France.

JAN SOKOLOWSKI
Institut Elie Cartan, Université de Nancy I, B.P. 239, F-54506 Vandœuvre-les-Nancy Cedex, France.

Conference Program

Chairman: J. von Below

10:30 Opening session

11:00 J. Saint Jean Paulin, Metz (France)
Study of optimal control with rapidly oscillating coefficients

Chairman: F. Ali Mehmeti

14:00 A. Sändig, Stuttgart (Germany)
Sensitivity analysis for elastic structures in presence of stress singularities

15:00 J. Sokolowski, Nancy (France)
Shape sensitivity analysis for optimal control problems

16:30 P. Ferreira, Polytechnique, Palaiseau (France)
Diffraction at plane and curved gratings

17:30 T. Abboud, Polytechnique, Palaiseau (France)
Diffraction by gratings at low and high frequencies

Chairman: G. Lumer

8:30 S. Nicaise, Valenciennes (France)
Asymptotic expansion of the solution of a mixed Dirichlet-Ventcel problem with a small parameter

10:00 Y. Dermanjian, Marseille (France)
Courbes et relations de dispersion en élasticité

11:00 O. Poisson, Marseille (France)
Principe d'amplitude limite pour la propagation d'ondes acoustiques dans une bande multistratifiée, et prolongement méromorphe de la résolvante de l'opérateur associé

Chairman: S. Nicaise

15:00 M. Lenczner, Besançon (France)
Modélisation et stratégie de contrôle pour des systèmes avec actionneurs et capteurs distribués

16:30 D. Leguillon, Paris (France)
Three dimensional interface crack tip singularities with contact and friction

17:30 D. Mercier, Valenciennes (France)
Some systems of partial differential equations on two-dimensional networks

Chairman: G. Leugering

8:30 J. von Below, Littoral (France)
Can one hear the shape of a network?

10:00 B. Dekoninck, Valenciennes (France)
Spectrum and controllability of networks of beams

11:00 J. Lubary, Barcelona (Spain)
On the geometric and algebraic multiplicities for eigenvalue problems on graphs

Chairman: E. Schmidt

8:30 G. Leugering, Bayreuth (Germany)
On optimal control problems and domain decomposition for evolution problems on graphs

10:00 C. Cattaneo, Milano (Italy)
The spectrum of the continuous Laplacian on a graph

11:00 L. Fontana, Milano (Italy)
The Cauchy problem for the wave equation on finite, weighted one dimensional networks

Chairman: M. Costabel

14:00 D. Cioranescu, Paris (France)
On some homogenization problems

15:00 G. Lumer, Mons (Belgium)
Singular dynamics in general and for ramified systems

16:30 J. Sanchez Palencia, Paris (France)
Generalities on singular perturbations without limit in the space of finite energy

17:30 J. Sanchez-Hubert, Caen (France)
Model problem for boundary and internal layers in parabolic shells

Chairman: J. von Below

8:30 M. Costabel, Rennes (France)
Singularities of Maxwell interface problems

10:00 F. Ali Mehmeti, Valenciennes (France)
Various interaction problems

11:00 O. Penkin, Voronezh (Russia)
Some qualitative properties of the solutions of elliptic inequalities on stratified sets

Conference Participants

T. ABBOUD, Centre de Mathématiques Appliquées, Ecole Polytechnique, 91128 Palaiseau Cedex (France)
abboud@pyrenees.polytechnique.fr

F. ALI MEHMETI, Université de Valenciennes, MACS, Le Mont Houy, 59313 Valenciennes Cedex 9 (France)
alimehme@univ-valenciennes.fr

J. von BELOW, Université du Littoral Côte d'Opale, Département de Mathématiques, Laboratoire de Mathématiques Pures et Appliquées Joseph Liouville, 50 rue F. Buisson, B.P. 699, 62228 Calais Cedex (France)
Joachim.von.Below@lmpa.univ-littoral.fr

C. CATTANEO, Dipartimento di Matematica, Università di Milano, Via Saldini 50, 20133 Milano (Italy)
fontana@vmimat.mat.unimi.it

D. CIORANESCU, Analyse Numérique, Tour 55-65, 5ème étage, Université Pierre et Marie Curie, 4 place Jussieu, 75252 PARIS Cedex 05 (France)
cioran@ann.jussieu.fr

M. COSTABEL, Université de Rennes I, IRMAR, Campus de Beaulieu, 35042 Rennes Cedex (France)
costabel@lie.univ-rennes1.fr

E. CROC, Analyse Numérique, Tour 55-65, 5ème étage, Université Pierre et Marie Curie, 4 place Jussieu, 75252 PARIS Cedex 05 (France)
ecroc@ccr.jussieu.fr

C. DE COSTER, Université du Littoral Côte d'Opale, Département de Mathématiques, Laboratoire de Mathématiques Pures et Appliquées Joseph Liouville, 50 rue F. Buisson, B.P. 699, 62228 Calais Cedex (France)
decoster@lmpa.univ-littoral.fr

B. DEKONINCK, Université de Valenciennes, MACS, Le Mont Houy, 59313 Valenciennes Cedex 9 (France)
Bertrand.Dekoninck@univ-valenciennes.fr

Y. DERMENJIAN, Université de Provence, Centre de Mathématiques et d'Informatique, 39, rue Joliot-Curie, 13453 Marseille Cedex 13 (France)
yves.dermenjian@cmi.univ-mrs.fr

P. FERREIRA, Centre de Mathématiques Appliquées, Ecole Polytechnique, 91128 Palaiseau Cedex (France)
pedro@cmapx.polytechnique.fr

T. FISCHER, Universität Bayreuth, Fakultät für Mathematik und Physik, Lehrstuhl V Mathematik, 95440 Bayreuth (Germany)
Torsten.Fischer@uni-bayreuth.de

L. FONTANA, Dipartimento di Matematica, Università di Milano, Via Saldini 50, 20133 Milano (Italy)
fontana@vmimat.mat.unimi.it

D. LEGUILLON, Laboratoire de Modélisation en Mécanique, Université Pierre et Marie Curie, 4, place Jussieu, 75252 Paris Cedex 05 (France)
dol@ccr.jussieu.fr

M. LENCZNER, Université de Franche Comté, Equipe de Mathématiques, UMR 6623, Route de Gray, 25030 Besancon Cedex (France)
michel.lenczner@univ-fcomte.fr

G. LEUGERING, Fachbereich Mathematik, Technische Universität Darmstadt, Schloßgartenstraße 7, D–64289 Darmstadt (Germany)
Guenter.Leugering@uni-bayreuth.de

J. LUBARY, Departament de Matematica Aplicata II, Universitat Polytecnia de Catalunya, 5 Pau Gargallo, 08028 Barcelona (Spain)
lubary@ma2.upc.es

G. LUMER, Université de Mons-Hainaut, Institut de Mathématique et d'Informatique, Bâtiment Le Pentagone, 6 Avenue du Champ de Mars, 7000 Mons (Belgium)
Pierre.Dufour@umh.ac.be

D. MERCIER, Université de Valenciennes, MACS, Le Mont Houy, 59313 Valenciennes Cedex 9 (France)
Denis.Mercier@univ-valenciennes.fr

S. NICAISE, Université de Valenciennes, MACS, Le Mont Houy, 59313 Valenciennes Cedex 9 (France)
snicaise@univ-valenciennes.fr

O. PENKIN, Voronezh State University, 1 Universitatskaya Pl., 394693 Voronezh (Russia)
omp@wowmail.com

O. POISSON, Université de Provence, CMI, 39 rue F. Joliot-Curie, 13453 Marseille Cedex 13 (France)
poisson@protis.univ-mrs.fr

J. SAINT JEAN PAULIN, Université de Metz, Département de Mathématiques, Ile du Saulcy, 57045 Metz Cedex 01 (France)
sjpaulin@poncelet.sciences.univ-metz.fr

J. SANCHEZ-HUBERT, Université de Caen, UFR Sciences, Mécanique, Boulevard du Maréchal Juin, Campus 2, 14032 Caen (France)
sanchez@lmm.jussieu.fr

J. SANCHEZ PALENCIA, Laboratoire de Modélisation en Mécanique, Université Pierre et Marie Curie, 4, place Jussieu, 75252 Paris Cedex 05 (France)
sanchez@lmm.jussieu.fr

A. SÄNDIG, Universität Stuttgart, Math. Institut A, 57 Pfaffenwaldring, 70569 Stuttgart (Germany)
anna@mathematik.uni-stuttgart.de

E. SCHMIDT, Mac Gill University, Department of Math. and Statistics, 805 Sherbrooke St. West, Montréal, Quebec H3A2K6 (Canada)
gschmidt@mafalda.math.mcgill.ca

J. SOKOLOWSKI, Université de Nancy I, Institut Elie Cartan, B.P. 239, 54506 Vandoeuvre-les-Nancy Cedex (France)
sokolows@antares.iecn.u-nancy.fr

Transient vibrations of planar networks of beams: interaction of flexion, transversal and longitudinal waves.

F. ALI MEHMETI and B. DEKONINCK Université de Valenciennes et du Hainaut Cambrésis, I.S.T.V./M.A.C.S., B.P. 311, 59304 VALENCIENNES, Cedex - FRANCE. E-mail : alimehme@univ-valenciennes.fr

1 INTRODUCTION

In this paper, we study a mathematical model for in-plane vibrations of planar networks of beams, incorporating the interaction of longitudinal and transversal displacements at the nodes. We generalize the model proposed in [13] for the special case of a hexagonal network of identical beams, constructed for the description of vibrations of honeycombs. We construct a self-adjoint operator adapted to reformulate the concrete dynamic problem into an abstract wave equation, which is equivalent to an abstract principle of stationary action (see [3]). This allows us to prove the convergence of an expansion of the solution in terms of eigenfunctions of this operator. In view of the explicit study of these eigenfunctions in special geometric configurations, we formulate an associated characteristic equation for the general case.

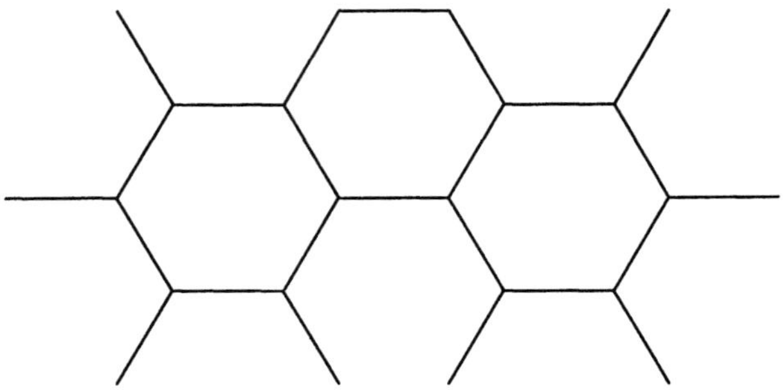

Fig. 1 The honeycombs as a network of beams

According to a hypothesis formulated by the biologists J. Tautz, M. Lindauer and D.C. Sandeman in [21], honeybees transmit with their waggle dance essential informations through the vibrations of the hexagonal surface of the honeycombs. This hypothesis leads to basic questions about this medium. As the wax is thicker and harder on the surface, it is reasonable to search a mathematical modelization of the honeycomb as a hexagonal network of beams. This point of view was initiated by N. Eichler in [13].

A lot of multiple-link structures, composed of a finite number of flexible elements, as strings, beams, plates and shells have been recently studied (see for instance [1, 3, 5, 15, 17, 18, 19]). For networks of beams, we refer to [14, 16, 7]. The case of Euler-Bernoulli beams was studied in [9, 10, 11].

In this paper, we consider not necessarily hexagonal planar networks of beams. All forces and deformations will be considered to be in that plane. We consider the spectral analysis of these structures and solutions of the dynamic problem.

In section 2, we first recall the modelization of a single beam as described in [13]. This includes the hypotheses of homogeneity, isotropy and linear elasticity of an isothermal beam. At the equilibrium, the beams are straight and have constant cross-sections. Under motion, the cross-sections stay plane and do not shear. The state of the beam is thus completely described by a vector function, whose components are the longitudinal and transversal displacements of the particles on its axis. The spatial character of each simple beam is therefore taken into account only by the position of the central axis. This description of the beam is quite unrealistic for large displacements. But this simplification can be admissible for the motions that we want to study. The deformation of the beam is governed by a diagonal matrix, namely the second derivative of the longitudinal displacement of each particle and the fourth derivative of the transverse one. Our model can be considered as a kind of Bresse/Euler-Bernoulli system, which is of fourth order in the space variable. It does not take into account any shearing and exhibits a rigid character of the beam. This rigidity allows the consideration of rigid nodes. Lagnese, Leugering and Schmidt use basically Bresse/Timoshenko systems in [15] (see §III.7 and §V.1.3), in the context of spectral theory and control. In these systems (which are of second order), rigid nodes have no meaning. Instead, one can consider the interaction of shearing and displacement at the nodes. The rigid character is thus replaced by interior deformation. Nevertheless in [15, §V.1.4] is given a model for plane network of Euler-Bernoulli beams without detailed study of its spectral theory. But there, the authors consider transversal motions vertical to the network planen and longitudinal displacements. In our model, the transversal motions are in the network plane. Beams are rigid, so we can impose boundary conditions such as the directions of the beams at its vertices. On the network, we get a system of equations coupled only by transmission conditions at the nodes. The beams are connected through rigid nodes in the sense of [15]. We get two sets of geometric transmission conditions (coming from rigidity) and two sets of dynamic transmission conditions (coming from the equilibrium of forces and moments). We prove that the spatial operator is a non-negative selfadjoint operator. This allows us to reformulate the concrete physical problem of the vibration of the structure into an abstract wave equation. At this point we observe, that the whole modelling procedure can be replaced by the application of a functionnal analytic principle of stationary action derived by F. Ali Mehmeti in [3, §4.2]. It allows us to establish the same model using only the geometric structure of the network and the expression for the potential energy of each beam. The equilibrium conditions for the forces and moments are not needed in this approach. The selfadjointness of the underlying operator permits to state a solution formula for the dynamic problem in terms of an eigenfunction expansion. The general form of the eigenfunctions, as described in (3.16, 3.17), suggests that the transient motion of the network consists of 3 types of displacements : longitudinal (acoustic) waves, transversal waves and flexion vibrations (with an infinite propagation speed). The different displacements are interacting at the nodes.

Section 3 of this article concerns the eigenvalue problem of the operator describing the network. We apply a method of matrix representation of the spectral problem developed by J. von Below in [5] for the Laplacian on graphs, and applied by S. Nicaise and B. Dekoninck to networks of Euler-Bernoulli beams. In [5, 11], the authors considered scalar quantities satisfying the equation of motion on the branches of networks coupled by transmission conditions. In mechanical applications, this describes planar networks of beams with displacements orthogonal to the plane of the network. In this configuration, we had interaction of transversal waves and flexion, whereas acoustic waves were not taken in account. In the present paper we consider in-plane-motions such that the displacement becomes a vectorial quantity including acoustic waves. We obtain a characteristic equation depending holomorphically on the spectral parameter, the coefficients of the equations and the full geometric structure of the physical network of beams. Recall that in [11], the angles between the beams were irrelevant. In special geometric situations (as studied in [13]), the resolution of the characteristic equation might be explicitly possible. This could lead to an explicit solution of the dynamic problem.

Acknowledgements : we thank Professor J. Tautz (Würzburg) for sending us the material and many useful comments on his hypothesis. Further we thank Dr. R. Strub-Röttgerding (Studienstiftung des Deutschen Volkes, Bonn) for having initiated the collaboration with Professor Tautz.

2 THE PHYSICAL MODEL

2.1 Notations

We consider $I\!R^3$ with the orthonormal basis $(\vec{\imath}, \vec{\jmath}, \vec{k})$, and denote by (u_0, y, w_0) the components of a vector in this basis. We consider here a planar network of beams. Each beam is supposed to be homogeneous, with a constant cross-section, and can be modelized by its axis Ω_j, $j = 1, \dots, N$, which is in the plane $(\vec{\imath}, \vec{k})$. Let l_j be the length of Ω_j. We denote by $(S_i)_{i=1,\dots,n}$ the n vertices of the network. We define :

$$N_i = \{j = 1, \dots, N : \Omega_j \text{ is adjacent to } S_i\},$$
$$M_j = \{i = 1, \dots, n : S_i \text{ is a vertex of } \Omega_j\}$$

We suppose that each beam is initially a straight segment of the plane. On each Ω_j, we choose a tangent unitary vector $\vec{e_{j1}}$ and define a normal unitary vector $\vec{e}_{j2}$ as $\vec{e}_{j1} \wedge \vec{\jmath}$. The choice of $\vec{e}_{j1}$ defines the orientation of the parameterization π_j of Ω_j by its arc length x_j. We use the following notations and abbreviations for any function $u_j \ : I\!R \times [0, l_j] \to I\!R$:

$u_j(., S_i) = u_j(., \pi_j^{-1}(S_i))$ for any vertex S_i adjacent to Ω_j,
$\ddot{u}_j(t, .)$ is the second time derivative,
$u_{j\,x_j}(., x_j)$ is the first spatial derivative,
$u_{j\,x_j^{(n)}}(., x_j)$ is the n-th spatial derivative.

We denote by $h_j(x_j)$ be the modulus of elasticity of a beam Ω_j at the point x_j (cf. the law of Hooke), by $a_j(x_j)$ the surface of a cross-section at this point, and by $b_j(x_j)$, the moment of this cross-section around the $\vec{\jmath}$-axis. We suppose that all these mechanical coefficients are constant along each beam, and so they will be denoted by h_j, a_j, b_j.
We define the $n \times N$ incidence matrix $D = (d_{ij})$:

$$d_{ij} = \begin{cases} 1 \text{ if } \pi_j(l_j) = S_i, \\ -1 \text{ if } \pi_j(0) = S_i, \\ 0 \text{ otherwise.} \end{cases}$$

Let $\mathcal{E} = (e_{ih})_{(i,h)=1,\cdots,n}$ be the adjacency matrix

$$e_{ih} = \begin{cases} 1 \text{ if there exists an edge } \Omega_{s(i,h)} \text{ between } S_i \text{ and } S_h, \\ 0 \text{ otherwise.} \end{cases}$$

Let $O = (o_{ih})_{(i,h)=1,\cdots,n}$ be the orientated adjacency matrix

$$o_{ih} = \begin{cases} 1 \text{ if } \Omega_{s(i,h)} \text{ is directed from } S_i \text{ to } S_h, \\ -1 \text{ if } \Omega_{s(i,h)} \text{ is directed from } S_h \text{ to } S_i, \\ 0 \text{ otherwise.} \end{cases}$$

We then define the deformation components $(u_j(x_j), w_j(x_j))$ of the beam Ω_j in the directions $\vec{e}_{j1}$ and $\vec{e}_{j2}$. We denote by φ_{ij} the angle between $\vec{e}_{j1}$ and the outer tangent vector to the deformed beam Ω_j at the vertex S_i, by $\vec{N}_{ij}, \vec{Q}_{ij}$ the forces created by a deformation at the node S_i in the direction of the exterior tangent vector and the direct normal vector. We denote by $\vec{M}_{ij}$ the couples in the $\vec{j}$-direction created by a deformation.

We denote by f_j and g_j the components in the basis $(\vec{e}_{j1}, \vec{e}_{j2})$ of an external distribution of force applied to the beam. As in [13], we suppose that the deformation of the beam Ω_j and these forces, moments and angles fulfill the following equations :

$$h_j\ a_j\ u_{jx_j^2}(x_j) = -f_j(x_j), \tag{2.1}$$

$$h_j\ b_j\ w_{jx_j^4}(x_j = g_j(x_j), \tag{2.2}$$

$$\vec{N}_{ij} = h_j\ a_j\ d_{ij}\ u_{jx_j}(S_i)\ \vec{e}_{j1}, \tag{2.3}$$

$$\vec{Q}_{ij} = -h_j\ b_j\ d_{ij}\ w_{x_j^{(3)}}(S_i)\ \vec{e}_{j2}, \tag{2.4}$$

$$\vec{M}_{ij} = -h_j\ b_j\ d_{ij}\ w_{jx_j^{(2)}}(S_i)\ \vec{j}, \tag{2.5}$$

$$\varphi_{ij} = -w_{jx_j}(S_i). \tag{2.6}$$

We will denote by α_j the oriented angle between the vectors $\vec{\imath}$ and $\vec{e}_{j1}$.
Let R_{α_j} be the matrix $\begin{pmatrix} \cos(\alpha_j) & -\sin(\alpha_j) \\ \sin(\alpha_j) & \cos(\alpha_j) \end{pmatrix}$.

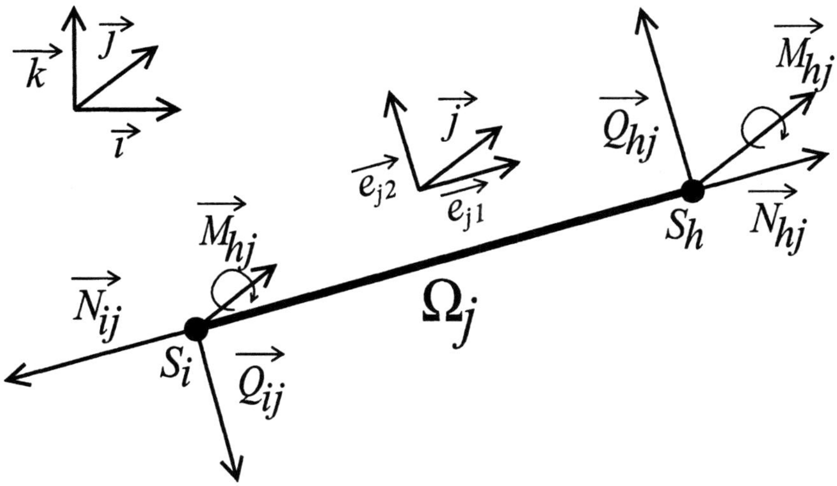

Fig. 2.1 The beam Ω_j, the bases, and the forces and moments at the nodes (the network plane is the slice of paper).

2.2 The equilibrium conditions

Let $\vec{F}_i^{ext}$ and $\vec{M}_i^{ext}$ be the exterior force and moment applied to S_i, in the canonical basis of $I\!R^3$.
We suppose that the nodes are rigid, i.e. that the displacement vectors and the rotations of each vertex are the same on each incident edge. We further suppose that the sum of the forces and the sum of the moments are vanishing at each node. This leads to the following transmission and boundary conditions for all vertices S_i, $i = 1, \ldots, n$:

$$\exists \left(u_0(S_i), w_0(S_i)\right) \; : \begin{pmatrix} u_0(S_i) \\ w_0(S_i) \end{pmatrix} = R_{\alpha_j} \begin{pmatrix} u_j(S_i) \\ w_j(S_I) \end{pmatrix}, \tag{2.7}$$

$$w_{jx_j}(S_i) = \text{const.}, \forall j \in N_i, \tag{2.8}$$

$$\sum_{j \in N_i} R_{\alpha_j} \begin{pmatrix} d_{ij}\; h_j\; a_j\; u_{jx_j}(S_i) \\ -d_{ij}\; h_j\; b_j\; w_{jx_j^{(3)}}(S_i) \end{pmatrix} = -\vec{F}_i^{ext} \tag{2.9}$$

$$\sum_{j \in N_i} h_j\; b_j\; d_{ij}\; w_{jx_j^{(2)}}(S_i)\; \vec{\jmath} = \vec{M}_i^{ext} \tag{2.10}$$

Note that, via the matrices R_{α_j}, the angles between the beams are taken into account.
We can now formulate the complete mathematical model for the vibrations of the network. We consider any functions (f_j, g_j), $i = 1, \cdots, N$ describing the forces applied to each beam Ω_j in the plane of the network. We denote by ρ_j the mass density of the beam k_j (which is supposed to be constant).
For all $j = 1, \cdots, N$, we search $(u_j, w_j) \; : I\!R \times \Omega_j \to I\!R^2$ satisfying (2.7), (2.8), (2.9), (2.10) and

$$\rho_j a_j \ddot{u}_j(t, x_j) - h_j a_j u_{j\, x_j^{(2)}}(t, x_j) = f_j(t, x_j), \tag{2.11}$$

$$\rho_j a_j \ddot{w}_j(t, x_j) + h_j b_j w_{j\, x_j^{(4)}}(t, x_j) = g_j(t, x_j). \tag{2.12}$$

for all $x_j \in l_j$ and all t.

2.3 The operator

We consider a network without applying any force or moment at the nodes. The equilibrium conditions are then (2.7), 2.8) and :

$$\sum_{j \in N_i} R_{\alpha_j} \begin{pmatrix} d_{ij}\; h_j\; a_j\; u_{jx_j}(S_i) \\ d_{ij}\; h_j\; b_j\; w_{jx_j^{(3)}}(S_i) \end{pmatrix} = 0 \tag{2.13}$$

$$\sum_{j \in N_i} h_j\; b_j\; d_{ij}\; w_{jx_j^{(2)}}(S_i) = 0 \tag{2.14}$$

We consider the operator $\mathcal{A}$ on the Hilbert space $H = \Pi_{j=1}^{N} L^2((0,l_j)) \times \Pi_{j=1}^{N} L^2((0,l_j))$, endowed with the usual product norm.

$$\left\{ \begin{array}{l} D(\mathcal{A}) = \{v = (u,w) \in H \ : (u_j, w_j) \in H^2((0,l_j)) \times H^4((0,l_j)) \\ \qquad\qquad \text{fulfilling (2.7), (2.8), (2.13), (2.14)} \ \forall j = 1, \cdots, N \ \}, \\ \forall v = (u,w) \in D(\mathcal{A}) \ : (\mathcal{A} \begin{pmatrix} u \\ w \end{pmatrix})_j = \begin{pmatrix} -h_j \ a_j \ u_{jx_j^{(2)}} \\ h_j \ b_j \ w_{jx_j^{(4)}} \end{pmatrix}. \end{array} \right.$$

THEOREM 2.1 *$\mathcal{A}$ is non-negative selfadjoint operator with a compact resolvent.*

Proof : We first prove that $\mathcal{A}$ is the Friedrichs' extension of the triple (H, V, a) defined by

$V := \{v = (u,w) \in H \ : (u_j, w_j) \in H^1((0,l_j)) \times H^2((0,l_j)) \text{ fulfilling (2.7), (2.8)}\}$,

$V \hookrightarrow H$, which is a Hilbert space for the inner product

$$(v_1, v_2)_V = \sum_{j=1}^{N} (u_{1j}, u_{2j})_{H^1(0,l_j)} + (w_{1j}, w_{2j})_{H^2(0,l_j)},$$

and the bilinear form $a : V \times V \to I\!R$ defined, for any $(v_1, v_2) \in V \times V$, by

$$a(v_1, v_2) = \sum_{j=1}^{N} h_j \int_0^{l_j} (a_j \, u_{1jx_j}(x_j) \, u_{jx_j}(x_j) + b_j \, w_{1jx_j^{(2)}}(x_j) \, w_{2jx_j^{(2)}}(x_j)) \, dx_j. \quad (2.15)$$

a is a continuous, symmetric, strongly V-coercive (i.e. $(v,v) \geq \alpha \, \|v\|_V^2$), bilinear form.
We further denote by $(\mathcal{A}_F, D(\mathcal{A}_F))$ the Friedrichs extension of the triple (H, V, a).
We first prove that $D(\mathcal{A}_F) \subset D(\mathcal{A})$. For some $v_1 = \begin{pmatrix} u \\ w \end{pmatrix} \in D(\mathcal{A}_F)$, we denote by $\tilde{v_1} = \mathcal{A}_F v_1 = \begin{pmatrix} \tilde{u}_1 \\ \tilde{w}_1 \end{pmatrix}$. Then we get for all $v_2 = \begin{pmatrix} u_2 \\ w_2 \end{pmatrix} \in V$:

$$(\tilde{v_1}, v_2)_{V \times V} = a(v_1, v_2)$$

And so, applying the Green's formula to the integrals in the expression of $a(v_1, v_2)$, we get the following identity :

$$\sum_{j=1}^{N} h_j \ a_j \int_0^{l_j} \tilde{u}_{1j} \ u_{2j} \ dx_j + \sum_{j=1}^{N} h_j \ b_j \int_0^{l_j} \tilde{w}_{1j} \ w_{2j} \ dx_j =$$

$$\sum_{j=1}^{N} h_j \ a_j \int_0^{l_j} u_{1jx_j^{(2)}} \ u_{2j} \ dx_j + \sum_{j=1}^{N} h_j \ b_j \int_0^{l_j} w_{1jx_j^{(4)}} \ w_{2j} \ dx_j$$

$$+ \sum_{j=1}^{N} h_j \ a_j \ [u_{1jx_j}(S_i) \ u_{2j}(S_i)]_0^{l_j} + \sum_{j=1}^{N} h_j \ b_j \ [w_{1jx_j^{(2)}}(S_i) \ w_{2jx_j}(S_i)]_0^{l_j}$$

$$+ \sum_{j=1}^{N} h_j \ b_j \ [w_{1jx_j^{(3)}}(S_i) \ w_{2j}(S_i)]_0^{l_j},$$

which can be written in the following way :

$$\sum_{j=1}^{N} h_j\, a_j \int_0^{l_j} \tilde{u}_{1j}\, u_{2j}\, dx_j + \sum_{j=1}^{N} h_j\, b_j \int_0^{l_j} \tilde{w}_{1j}\, w_{2j}\, dx_j = \tag{2.16}$$

$$\sum_{j=1}^{N} h_j\, a_j \int_0^{l_j} u_{1jx_j^{(2)}}\, u_{2j}\, dx_j + \sum_{j=1}^{N} h_j\, b_j \int_0^{l_j} w_{1jx_j^{(4)}}\, w_{2j}\, dx_j$$

$$+ \sum_{i=1}^{n} \sum_{j\in N_i} d_{ij}\, h_j\, a_j\, u_{1jx_j}(S_i)\, u_{2j}(S_i) + \sum_{i=1}^{n} \sum_{j\in N_i} d_{ij}\, h_j\, b_j\, w_{1jx_{j(2)}}(S_i)\, w_{2jx_j}(S_i)$$

$$+ \sum_{i=1}^{n} \sum_{j\in N_i} d_{ij}\, h_j\, b_j\, w_{1jx_j^{(3)}}(S_i)\, w_{2j}(S_i).$$

For all functions $v_2 = (u_2, w_2) \in \prod_{j=1}^{N} C_\circ^\infty(0, l_j) \times C_\circ^\infty(0, l_j)$, we get :

$$a(v_1, v_2) = \sum_{j=1}^{N} h_j\, a_j \int_0^{l_j} u_{1jx_j^{(2)}}\, u_{2j}\, dx_j + h_j\, b_j \int_0^{l_j} w_{1jx_j^{(4)}}\, w_{2j}\, dx_j.$$

Because $\prod_{j=1}^{N} C_\circ^\infty(0, l_j) \times C_\circ^\infty(0, l_j)$ is dense in H,

$$\tilde{u}_{1j} = u_{1jx_j^{(2)}} \quad \text{and} \quad \tilde{w}_{1j} = w_{1jx_j^{(4)}}, \forall j = 1, \cdots, N. \tag{2.17}$$

For each vertex S_i, we now choose a test function $v_2 = (u_2, w_2) \in V$ with the following properties :

- v_2 as a compact support included in $\bigcup_{j\in N_i} \left( ]0, l_j[\cup \pi_j^{-1}(S_i) \right)$, (2.18)
- $u_{2j}(S_i) = w_{2j}(S_i) = 0$, $\forall j \in N_i$,
- The common value of the derivatives $w_{2jx_j}(S_i)$ (see (2.8)) does not vanish.

From (2.16) and (2.17), we deduce that w_1 fulfils (2.9).
Last, for each vertex S_i, we further choose other test functions $v_2 \in V$ with the following properties :

- v_2 fulfils (2.18),
- $\exists (u_2^\circ, w_2^\circ) \neq (0,0)$, $\begin{pmatrix} u_{2j}(S_i) \\ w_{2j}(S_i) \end{pmatrix} = R_{\alpha_j} \begin{pmatrix} u_2^\circ \\ w_2^\circ \end{pmatrix}$,
- $w_{j\,x_j}(S_i) = 0$.

Then, we deduce from (2.16) that v_1 fulfils the identity :

$$\sum_{j\in N_i} \begin{pmatrix} d_{ij}\, h_j\, a_j\, u_{1j\,x_j}(S_i) \\ d_{ij}\, h_j\, b_j\, w_{1j\,x_j^{(3)}}(S_i) \end{pmatrix}^T \begin{pmatrix} u_{2j}(S_i) \\ w_{2j}(S_i) \end{pmatrix} = 0.$$

Because R_{α_j} is a unitary matrix, and applying (2.7), we get

$$\sum_{j\in N_i} \begin{pmatrix} d_{ij}\, h_j\, a_j\, u_{1j\,x_j}(S_i) \\ d_{ij}\, h_j\, b_j\, w_{1j\,x_j^{(3)}}(S_i) \end{pmatrix}^T R_{\alpha_j}^T \begin{pmatrix} u_{2j}^\circ(S_i) \\ w_{2j}^\circ(S_i) \end{pmatrix} = 0.$$

and so, v_1 has to fulfil (2.10). This proves $D(\mathcal{A}_F) \subset D(\mathcal{A})$.

We consider now the inclusion of $D(\mathcal{A})$ in $D(\mathcal{A}_F)$.
Let $v_1 \in D(\mathcal{A})$. For all $v_2 \in V$, we apply the Green's formula, the boundary conditions defining $D(\mathcal{A})$ and the identity (2.17), we get :

$$a(v_1, v_2) = (\mathcal{A}_F v_1, v_2)_{V\times V} = (\mathcal{A}v_1, v_2)_{V\times V}$$

and so

$$v_1 \in D(\mathcal{A}_F) \text{ and } \mathcal{A}v_1 = \mathcal{A}_F v_2.$$

The last assertion follows from the compact embedding of V into H (see [20, Th. III.7.C]). ■

We deduce from this last theorem that the spectrum of $\mathcal{A}$ is only made of non-negative eigenvalues λ_n, $n \in I\!N$ and that there exists an orthonormal basis of H denoted by $\{\varphi_n\}_{n\in I\!N}$ of eigenfunctions belonging to $D(A^\infty)$ associated with these eigenvalues.

2.4 The stationary action principle

We refer in this part to the proofs given in [3, §4.2]. We consider now the abstract wave equation on the network.

$$\ddot{v}(t) + \mathcal{A}v(t) = 0,\ v(t) \in D(\mathcal{A}),\ \forall t \in [0,T],\ T > 0, \qquad (2.19)$$
$$v(0) = v_0 \in D(\mathcal{A}),\ v(T) = v_1 \in D(\mathcal{A})$$

This is a reformulation of the concrete problem {(2.7), (2.8), (2.11), (2.12), (2.13), (2.14)}.
All the results given here are true for the Friedrichs' extensions of all triples (H, V, a) as described in section (2.3).

DEFINITION 2.2 *Let $T > 0$ be fixed.*
(i) Endow $X := C^1([0,T], H) \cap C^0([0,T), V)$ with the norm

$$\|v\|_X := \sup_{t\in[0,T]} \|v(t)\|_V + \sup \|\dot{u}(t)\|_H .$$

(ii) Let $v_1,\ v_2 \in D(\mathcal{A})$ be fixed. Write shortly $b := \{v_1,\ v_2\}$. Define the subset

$$M_b := \{v \in X : v(0) = v_1,\ v(T) = v_2\}$$

of X with fixed boundary values b.
(iii) Define $B : X \times X \to I\!R$ by

$$B(v_1, v_2) := \int_0^T [(\dot{v_1}(t), \dot{v_2}(t))_H - a(v_1(t), v_2(t))]dt.$$

Clearly B is a continuous, symmetric bilinear form.
(iv) Define $S : X \to I\!R$ by $S(v) := \frac{1}{2}B(v,v)$ for $v \in X$. S is called the action functional associated with a.
(v) Define $S_b := S|_{M_b}$ which is called the action functional associated with the bilinear form a and with the boundary values $b \subseteq D(\mathcal{A})$.

For the statement of our theorem, we need the notion of the Fréchet derivative in accumulation points of the domain.

DEFINITION 2.3 *Let X_1 be a normed space and X_2 a Banach space. Let $M \subseteq X_1$ be any subset and x_0 an accumulation point of M in X_1. Consider $F : M \to X_2$. We say that F is Fréchet differentiable in $x_0 \in M$ if and only if*

$$\exists T \in B(X_1, X_2) \ : F(x_0 + h_n) - F(x_0) - Th_n = o(\|h_n\|), \ n \to \infty$$

for all sequences $(h_n)_{n \in I\!N} \subseteq X$ such that $h_n \to 0$ for $n \to \infty$ and $x_0 + h_n \in M$ for all $n \in I\!N$.
We write $DF(x_0) := T$. For interior points, the above definition coincides with the usual Fréchet derivative, but the one-sided for functions $f : [a,b] \to I\!R$ in $x_0 = a, b$ falls under this definition too.

THEOREM 2.4 *For $u \in C^2([0,T], H) \cap M_b$ the equivalence*

$$DS_b(v) = 0 \Leftrightarrow \ddot{v}(t) + \mathcal{A}v(t) = 0, \forall t \ in[0,T] \tag{2.20}$$

holds. The right hand side of (2.20) is an equation in H.

This leads us to the following **Heuristic Principle** : if we accept in the framework of this part the Principle of Stationary Action $DS_b(u) = 0$ as a physical axiom, then we have the following statements :
(i) Assume that the evolution of a physical system can be described by functions $v : [0,T] \to H$, such that $\frac{1}{2}a(v(t), v(t))$ can be interpreted as the potential energy and $\frac{1}{2}(\dot{v}(t), \dot{v}(t))_H$ as the kinetic energy at the time t and such that the sum of the both is the total energy. Assume further that the subspace $V \hookrightarrow H$ encodes physical side-conditions, such that a, V and H completely define the data of the physical system (without stating any physical axioms as conservation laws, etc.), then equation (2.19) with the Friedrichs' extension $\mathcal{A} : D(\mathcal{A}) \to H$ of the triple (H, V, a) is a complete set of physical axioms for the system.
(ii) The conditions characterizing whether an element of V belongs to $D(\mathcal{A}) \subseteq V$ or not, are physical laws (whereas the conditions giving V and the coefficients of a are physical data).

This point of view gives another way to justify physically the dynamic transmission conditions (2.9, 2.10). On the other hand, the way theses conditions were established underlines the plausibility of the Principle of Stationary Action and the Heuristic Principle.

2.5 Solutions of the wave equation

The abstract wave equation (2.19) can easily be solved by the spectral theorem.

THEOREM 2.5 *Let $k \in I\!N$.*
(i) $\forall v_0 \in D(\mathcal{A}^{(k+1)/2}), v_1 \in D(\mathcal{A}^{k/2})$ there exists a unique solution

$$v(\cdot) \in C^{k+1-j}([0,\infty), D(\mathcal{A}^{j/2})),\ j = 0, \cdots, k+1 \tag{2.21}$$

of

$$\begin{cases} \ddot{v}(t) + \mathcal{A}v(t) = 0, \forall t \geq 0, \\ v(0) = v_0 \ ;\ \dot{v}(0) = v_1. \end{cases} \tag{2.22}$$

This solution is given by

$$v(t) = \cos(\sqrt{\mathcal{A}}t)v_0 + (\sqrt{\mathcal{A}})^{-1}\sin(\sqrt{\mathcal{A}}t)v_1. \tag{2.23}$$

(ii) The solution u in (i) satisfies further for $j = 0, \cdots, k$:

$$\left\|\mathcal{A}^{(j+1)/2}v(t)\right\|_H^2 + \left\|\mathcal{A}^{j/2}\dot{v}(t)\right\|_H^2 = const\ \forall t \geq 0. \tag{2.24}$$

The proof of the theorem 2.5 uses standard method (cf. for example [4, 12]). We can reformulate the solution using its expansion in the orthonormal basis of eigenfunctions $(\varphi_k)_{k\in I\!N}$ of $\mathcal{A}$ associated to the eigenvalues $(\lambda_k)_{k\in I\!N}$. We get the immediate

COROLLARY 2.6 *Let $k \in I\!N$, $v_0 \in D(\mathcal{A}^{(k+1)/2}), v_1 \in D(\mathcal{A}^{1/2})$. The unique solution $v \in C^1([0,\infty), H) \cap C^0([0,\infty), D(\mathcal{A}^{1/2}))$ is*

$$v(t) = \sum_{k\in I\!N}\left(\cos(\sqrt{\lambda_k}t)(v_0,\varphi_k)_H + \frac{1}{\sqrt{\lambda_k}}\sin(\sqrt{\lambda_k}t)(v_1,\varphi_k)_H\right)\varphi_k. \tag{2.25}$$

This proves the question of convergence of the formal solution of the dynamic problem unresolved in [13].
The problem is now to explicitly get the eigenvalues and eigenfunctions of $\mathcal{A}$ in order to study the solutions of the wave equation. This is the motivation of the second part of this paper.

3 THE EIGENVALUE PROBLEM

We use here a method developed by J. von Below in [5]. This requires the following notations.

3.1 Notations

We shall use the Hadamard product of two $n \times n$ matrices M and N defined by

$$M \cdot N = (m_{ih}\, n_{ih})_{n\times n}.$$

Furthermore, for any function $p : I\!R \to I\!R$, we define the matrix $p(M) = (p_{ih})_{n\times n}$ defined by

$$p_{ih} = \begin{cases} p(m_{ih}) \text{ if } e_{ih} = 1, \\ 0 \text{ otherwise.} \end{cases}$$

In particular, if $p(x) = x^r$, $r \in I\!R$, we write $M^{(r)}$ instead of $p(M)$. In the same spirit, we shall say that $M = N$ (resp. $M \neq N$) in the Hadamard sense if and only if $m_{ih} = n_{ih}$ for all $i, h = 1, \cdots, n$ such that $e_{ih} = 1$.
We set $e = (1)_{n\times 1}$ and for any vector u of $I\!R^n$, we define the diagonal matrix $Diag(u) = (\delta_{ih} v_i)_{n\times n}$. We introduce the matrices $A = (a_{ih})_{n\times n}$, $B = (b_{ih})_{n\times n}$, $H = (h_{ih})_{n\times n}$ and $L = (l_{ih})_{n\times n}$ of the mechanical constants and lengths of the beams defined by

$$\begin{cases} a_{ih} = a_{s(i,h)}\ ;\ b_{ih} = b_{s(i,h)}\ ;\ h_{ih} = h_{s(i,h)}\ ;\ l_{ih} = l_{s(i,h)}, \text{ if } e_{ih} = 1, \\ a_{ih} = b_{ih} = h_{ih} = l_{ih} = 0 \text{ otherwise.} \end{cases}$$

In the same way, we define the matrix $\mathcal{R}_\alpha$ of the angles α_j.
For all $v = \begin{pmatrix} u \\ w \end{pmatrix} \in H$ we define the $n \times n$ matrix functions on $(0,1)$ $U = (u_{ih})_{(i,h)=1,\cdots,n}$ and $W = (w_{ih})_{(i,h)=1,\cdots,n}$ by

$$u_{ih}(x) = e_{ih}\; u_{s(i,h)}\left(l_{s(i,h)}\left(\frac{1+d_{is(i,h)}}{2} - x\, d_{is(i,h)}\right)\right), \tag{3.1}$$

$$w_{ih}(x) = e_{ih}\; w_{s(i,h)}\left(l_{s(i,h)}\left(\frac{1+d_{is(i,h)}}{2} - x\, d_{is(i,h)}\right)\right). \tag{3.2}$$

The $x-$derivative of such matrix functions $U(x)$ are denoted by $U'(x)$, $U''(x)$, $U'''(x)$, $U''''(x)$.
We define the following vector space :

$$M(\Gamma) := \{U = (u_{ih})_{n\times n}, \forall i, h \in \{1, \cdots, n\} : (e_{ih} = 0 \Rightarrow u_{ih} = 0)\}.$$

Let λ^2 be an eigenvalue of $\mathcal{A}$ of associated eigenvector $v = \begin{pmatrix} u \\ w \end{pmatrix} \in D(\mathcal{A})$. Then v satisfies (2.7), (2.8), (2.13), (2.14) and

$$-h_j\; a_j\; u_{jx_j^{(2)}} = \lambda^2 u_j, \forall j \in 1, \ldots, N, \tag{3.3}$$

$$h_j\; b_j\; w_{jx_j^{(4)}} = \lambda^2 w_j, \forall j \in 1, \ldots, N, \tag{3.4}$$

$$v_j \in H^2((0, l_j)) \times H^4((0, l_j)), \forall j \in 1, \ldots, N. \tag{3.5}$$

3.2 The matrix representation

LEMMA 3.1 *$v = (u, w) \in D(\mathcal{A})$ is an eigenvector of $\mathcal{A}$ associated with the eigenvalue λ^2 ($\lambda \geq 0$), i.e. v satisfies (2.7), (2.8), (2.13), (2.14), and (3.3), (3.4), (3.5), if and only if U, W is a solution of the differential problem hereafter :*

$$U, W \in M(\Gamma) \text{ and } (u_{ih}, w_{ih}) \in H^2((0,1)) \times H^4((0,1)), \tag{3.6}$$
$$L^{(-2)} \cdot H \cdot A \cdot U''(x) = -\lambda^2 U(x), \ \forall x \in (0,1), \tag{3.7}$$
$$L^{(-4)} \cdot H \cdot B \cdot W''''(x) = \lambda^2 W(x), \ \forall x \in (0,1), \tag{3.8}$$
$$\exists \varphi_0^u, \varphi_0^w \in I\!R^n \ : \left\{ \begin{array}{l} U(0) = (\varphi_0^u e^T) \cdot \cos(\mathcal{R}_\alpha) + (\varphi_0^w e^T) \cdot \sin(\mathcal{R}_\alpha), \\ W(0) = -(\varphi_0^u e^T) \cdot \sin(\mathcal{R}_\alpha) + (\varphi_0^w e^T) \cdot \cos(\mathcal{R}_\alpha), \end{array} \right. \tag{3.9}$$
$$\exists \varphi_1^w \in I\!R^n \ : W'(0) = (\varphi_1^w e^T) \cdot L \cdot O, \tag{3.10}$$
$$[L^{(-2)} \cdot H \cdot B \cdot O \cdot W''(0)]e = 0, \tag{3.11}$$
$$[L \cdot H \cdot A \cdot \cos(\mathcal{R}_\alpha) \cdot U'(0)]e - [L^{(-3)} \cdot H \cdot B \cdot \sin(\mathcal{R}_\alpha) \cdot W'''(0)]e = 0 \tag{3.12}$$
$$[L \cdot H \cdot A \cdot \sin(\mathcal{R}_\alpha) \cdot U'(0)]e + [L^{(-3)} \cdot H \cdot B \cdot \cos(\mathcal{R}_\alpha) \cdot W'''(0)]e = 0 \tag{3.13}$$
$$U^T(1-x) = U(x), \ W^T(1-x) = W(x), \forall x \in [0,1]. \tag{3.14}$$

The proof of this lemma is inspired from the one given in [5, 12] and is made of immediate computations of the equations.

3.3 The characteristic equation

THEOREM 3.2 *If $v = \begin{pmatrix} u \\ w \end{pmatrix} \in \ker \mathcal{A}$, then on each edge Ω_j,*

$u_j(x_j) = u_{0j}$; $w_j(x_j) = w_{0j} + w_{1j}x_j$, for some real constants (u_{0j}, w_{0j}, w_{1j}).

Moreover :

$$\dim \ker \mathcal{A} \leq 2n. \tag{3.15}$$

Proof : $v = \begin{pmatrix} u \\ w \end{pmatrix} \in \ker \mathcal{A}$ iff

$$\forall v_1 \in V, \ \ a(v, v_1) = 0$$

and so

$$\|u_{jx_j}\|_{L^2((0,l_j))} = \|w_{jx_j^{(2)}}\|_{L^2((0,l_j))} = 0.$$

Each u_j and w_j is polynomial with $d^\circ(u_j) = 0$ and $d^\circ(w_j) \leq 1$.
Moreover, we get (3.15) because we have

$$U(0) = U(0)^T, \ W'(0) = W(0)^T - W(0),$$

because of (3.14). ∎

The case of a positive eigenvalue is treated with the methods developed in [5], [10], [11]. Let B_1, B_2, C_1, C_2 be the $n \times n$ matrices

$$\begin{aligned} B_1 &= L \cdot (H \cdot A)^{(-1/2)}, \\ B_2 &= L \cdot (H \cdot B)^{-1/4)}, \\ C_1 &= \sin(\lambda B_1), \\ C_2 &= \cosh(\sqrt{\lambda} B_2) \cdot \cos(\sqrt{\lambda} B_2) - \mathcal{E}. \end{aligned}$$

We make use of the general solutions

$$\begin{cases} U(x) = (\phi_0^u \cdot \cos(\lambda B_1) + \frac{1}{\lambda} B_1^{(-1)} \cdot \phi_1^u \cdot \sin(\lambda B_1), \\ W(x) = \sum_{i=0}^{3} \phi_i^w \cdot e_i^\lambda(x), \end{cases} \tag{3.16}$$

with

$$\begin{cases} e_0^\lambda(x) = \frac{1}{2}\Big(\cosh(\sqrt{\lambda} B_2 x) + \cos(\sqrt{\lambda} B_2 x)\Big), \\ e_1^\lambda(x) = \frac{1}{2\sqrt{\lambda}} B_2^{(-1)} \cdot \Big(\sinh(\sqrt{\lambda} B_2 x) + \sin(\sqrt{\lambda} B_2 x)\Big), \\ e_2^\lambda(x) = \frac{1}{2\lambda} B_2^{(-2)} \cdot \Big(\cosh(\sqrt{\lambda} B_2 x) - \cos(\sqrt{\lambda} B_2 x)\Big), \\ e_3^\lambda(x) = \frac{1}{2\lambda^{3/2}} B_2^{(-3)} \cdot \Big(\sinh(\sqrt{\lambda} B_2 x) - \sin(\sqrt{\lambda} B_2 x)\Big), \end{cases} \tag{3.17}$$

being functions of matrices defined in the Hadamard sense. We use the following notations :

$$\begin{aligned} a_1(\lambda) &= \sinh(\sqrt{\lambda} B_2) + \sin(\sqrt{\lambda} B_2) \\ b_1(\lambda) &= \cosh(\sqrt{\lambda} B_2) \cdot \sin(\sqrt{\lambda} B_2) + \sinh(\sqrt{\lambda} B_2) \cdot \cos(\sqrt{\lambda} B_2) \\ c_1(\lambda) &= \cosh(\sqrt{\lambda} B_2) - \cos(\sqrt{\lambda} B_2), \\ d_1(\lambda) &= \sinh(\sqrt{\lambda} B_2) \cdot \sin(\sqrt{\lambda} B_2), \\ c_2(\lambda) &= \sinh(\sqrt{\lambda} B_2) - \sin(\sqrt{\lambda} B_2), \\ d_2(\lambda) &= \cosh(\sqrt{\lambda} B_2) \cdot \sin(\sqrt{\lambda} B_2) - \sinh(\sqrt{\lambda} B_2) \cdot \cos(\sqrt{\lambda} B_2). \end{aligned}$$

REMARK 3.3 *The form of the general solutions (3.16, 3.17) explicits the interaction of three types of motion : while $U(x)$ and the trigonometric terms in (3.17) give longitudinal (acoustic) and transversal propagation waves for the solution of the dynamic problem, with a finite propagation speed on each beam, the* sinh *ans* cosh *terms in (3.17) give flexion vibrations with an infinite propagation speed.*

THEOREM 3.4 *Let $\lambda^2 > 0$ be an eigenvalue of $\mathcal{A}$ and $(\varphi_0^u, \varphi_0^w, \varphi_1^w)$ belong to an associated eigenvector (U, W) according to (3.9, 3.10). Then one of the following statement holds :*

a) $0 < \lambda = \lambda_k^{(j)} := \frac{\sqrt{h_j a_j}}{l_j} k\pi$ for some edge Ω_j joining the nodes S_i and S_h and for some $k \in I\!N$, with

$$\begin{aligned} &\varphi_{0h}^u \cos(\alpha_{ih}) + \varphi_{0h}^w \sin(\alpha_{ih}) = (-1)^k (\varphi_{0i}^u \cos(\alpha_{ih}) + \varphi_{0i}^w \sin(\alpha_{ih})), \quad (3.18) \\ &\phi_{1hi}^u = (-1)^{k+1} \phi_{1ih}. \end{aligned}$$

b) λ fulfils

$$\cos(\sqrt{\lambda}\frac{l_j}{(h_j b_j)^{1/4}}) = \frac{1}{\cosh(\sqrt{\lambda}\frac{l_j}{(h_j b_j)^{1/4}})}$$

for some edge Ω_j joining the nodes S_i and S_h and for some $k \in I\!N$. We set $\alpha = \cosh(\sqrt{\lambda}\frac{l_j}{(h_j b_j)^{1/4}})$ and $\varepsilon = sgn(\sin(\sqrt{\lambda}\frac{l_j}{(h_j b_j)^{1/4}})$. Then we get :

$$\begin{aligned}
&\sqrt{\lambda}B_{2ih}\sqrt{\alpha^2-1}\,[(-\varphi_{0h}^u\sin(\alpha_{ih})+\varphi_{0h}^w\cos(\alpha_{ih}))\\
&\quad -\varepsilon(-\varphi_{0i}^u\sin(\alpha_{ih})+\varphi_{0i}^w\cos(\alpha_{ih}))] =\\
&\quad (\alpha-\varepsilon)l_j O_{ih}\varphi_{1h}^w + \varepsilon(\alpha+\varepsilon)l_j O_{ih}\varphi_{1i}^w.
\end{aligned} \tag{3.19}$$

c) $C_1 \cdot C_2 \neq 0$ in the Hadamard sense (i.e. the case a) and b) are false for all edges), then we have the following characteristic equation :

$$\det \mathcal{D}(\lambda, H, A, B, L, O, \mathcal{R}_\alpha, \mathcal{E}) = 0 \tag{3.20}$$

$(\varphi_0^u, \varphi_0^w, \varphi_1^w)$ does not vanish lies in the kernel this matrix, which is defined by :

$$\mathcal{D}(\lambda, H, A, B, L, O, \mathcal{R}_\alpha, \mathcal{E}) = \begin{pmatrix} \mathcal{D}_{11} & \mathcal{D}_{12}, & \mathcal{D}_{13} \\ \mathcal{D}_{21} & \mathcal{D}_{22} & \mathcal{D}_{23} \\ \mathcal{D}_{31} & \mathcal{D}_{32} & \mathcal{D}_{33} \end{pmatrix} \tag{3.21}$$

with the $n \times n$ matrices $\mathcal{D}_{ij}$ given by

$$\begin{aligned}
\mathcal{D}_{11} = &\ \sqrt{\lambda}\Big[\Big(C_2^{(-1)} \cdot (H\cdot B)^{(1/2)} \cdot O \cdot \sin(\mathcal{R}_\alpha) \cdot c_1(\lambda)\Big)\\
&-Diag\Big[\Big(C_2^{(-1)} \cdot (H\cdot B)^{(1/2)} \cdot O \cdot \sin(\mathcal{R}_\alpha) \cdot d_1(\lambda)\Big)e\Big]\Big],\\
\mathcal{D}_{12} = &\ -\sqrt{\lambda}\Big[\Big(C_2^{(-1)} \cdot (H\cdot B)^{(1/2)} \cdot O \cdot \cos(\mathcal{R}_\alpha) \cdot c_1(\lambda)\Big)\\
&-Diag\Big[\Big(C_2^{(-1)} \cdot (H\cdot B)^{(1/2)} \cdot O \cdot \cos(\mathcal{R}_\alpha) \cdot d_1(\lambda)\Big)e\Big]\Big],\\
\mathcal{D}_{13} = &\ \Big(C_2^{(-1)} \cdot (H\cdot B)^{(3/4)} \cdot c_2(\lambda)\Big)\\
&+Diag\Big[\Big(C_2^{(-1)} \cdot (H\cdot B)^{(3/4)} \cdot d_2(\lambda)\Big)e\Big],
\end{aligned} \tag{3.22}$$

$$\begin{aligned}
\mathcal{D}_{21} = \quad & \Big(C_1^{(-1)} \cdot (H \cdot A)^{(1/2)} \cdot \cos(\mathcal{R}_\alpha)^{(2)}\Big) \\
& -Diag\Big[\Big(C_1^{(-1)} \cdot (H \cdot A)^{(1/2)} \cdot \cos(\mathcal{R}_\alpha)^{(2)} \cdot \cos(\lambda B_1)\Big)e\Big] \\
& +\sqrt{\lambda}\Big[\Big(C_2^{(-1)} \cdot (H \cdot B)^{(1/4)} \cdot \sin(\mathcal{R}_\alpha)^{(2)} \cdot a_1(\lambda)\Big) \\
& -Diag\Big[\Big(C_2^{(-1)} \cdot (H \cdot B)^{(1/4)} \cdot \sin(\mathcal{R}_\alpha)^{(2)} \cdot b_1(\lambda)\Big)e\Big]\Big], \\
\mathcal{D}_{22} = \quad & \Big(C_1^{(-1)} \cdot (H \cdot A)^{(1/2)} \cdot \cos(\mathcal{R}_\alpha) \cdot \sin(\mathcal{R}_\alpha)\Big) \\
& -Diag\Big[\Big(C_1^{(-1)} \cdot (H \cdot A)^{(1/2)} \cdot \cos(\mathcal{R}_\alpha) \cdot \sin(\mathcal{R}_\alpha) \cdot \cos(\lambda B_1)\Big)e\Big] \\
& -\sqrt{\lambda}\Big[\Big(C_2^{(-1)} \cdot (H \cdot B)^{(1/4)} \cdot \cos(\mathcal{R}_\alpha) \cdot \sin(\mathcal{R}_\alpha) \cdot a_1(\lambda)\Big) \\
& -Diag\Big[\Big(C_2^{(-1)} \cdot (H \cdot B)^{(1/4)} \cdot \cos(\mathcal{R}_\alpha) \cdot \sin(\mathcal{R}_\alpha) \cdot b_1(\lambda)\Big)e\Big]\Big], \\
\mathcal{D}_{23} = \quad & \Big(C_2^{(-1)} \cdot (H \cdot B)^{(1/2)} \cdot O \cdot \sin(\mathcal{R}_\alpha) \cdot c_1(\lambda)\Big) \\
& +Diag\Big[(C_2^{(-1)} \cdot (H \cdot B)^{(1/2)} \cdot O \cdot \sin(\mathcal{R}_\alpha) \cdot d_1(\lambda))e\Big], \\
\mathcal{D}_{31} = \quad & \Big(C_1^{(-1)} \cdot (H \cdot A)^{(1/2)} \cdot \sin(\mathcal{R}_\alpha) \cdot \cos(\mathcal{R}_\alpha)\Big) \\
& -Diag\Big[\Big(C_1^{(-1)} \cdot (H \cdot A)^{(1/2)} \cdot \sin(\mathcal{R}_\alpha) \cdot \cos(\mathcal{R}_\alpha) \cdot \cos(\lambda B_1)\Big)e\Big] \\
& -\sqrt{\lambda}\Big[\Big(C_2^{(-1)} \cdot (H \cdot B)^{(1/4)} \cdot \sin(\mathcal{R}_\alpha) \cdot \cos(\mathcal{R}_\alpha) \cdot a_1(\lambda)\Big) \\
& -Diag\Big[\Big(C_2^{(-1)} \cdot (H \cdot B)^{(1/4)} \cdot \sin(\mathcal{R}_\alpha) \cdot \cos(\mathcal{R}_\alpha) \cdot b_1(\lambda)\Big)e\Big]\Big], \\
\mathcal{D}_{32} = \quad & \Big(C_1^{(-1)} \cdot (H \cdot A)^{(1/2)} \cdot \sin(\mathcal{R}_\alpha)^{(2)}\Big) \\
& -Diag\Big[\Big(C_1^{(-1)} \cdot (H \cdot A)^{(1/2)} \cdot \sin(\mathcal{R}_\alpha)^{(2)} \cdot \cos(\lambda B_1)\Big)e\Big] \\
& +\sqrt{\lambda}\Big[\Big(C_2^{(-1)} \cdot (H \cdot B)^{(1/4)} \cdot \cos(\mathcal{R}_\alpha)^{(2)} \cdot a_1(\lambda) \\
& -Diag[(C_2^{(-1)} \cdot (H \cdot B)^{(1/4)} \cdot \cos(\mathcal{R}_\alpha)^{(2)} \cdot b_1(\lambda)\Big)e\Big]\Big], \\
\mathcal{D}_{33} = \quad & -\Big(C_2^{(-1)} \cdot (H \cdot B)^{(1/2)} \cdot O \cdot \cos(\mathcal{R}_\alpha) \cdot c_1(\lambda)\Big) \\
& -Diag\Big[\Big(C_2^{(-1)} \cdot (H \cdot B)^{(1/2)} \cdot O \cdot \cos(\mathcal{R}_\alpha) \cdot d_1(\lambda)\Big)e\Big].
\end{aligned}$$

The multiplicity of the eigenvalue λ^2 of $\mathcal{A}$ is

$$mult(\lambda^2) = \dim\ker\Big(\mathcal{D}(\lambda, H, A, B, L, O, \mathcal{R}_\alpha, \mathcal{E})\Big). \tag{3.23}$$

Proof : We take a fundamental solution of the form (3.16) and we isolate the components ϕ_1^u, ϕ_2^w and ϕ_3^w thanks to the system

$$\begin{cases} U(1)^T = U(0), \\ W(1)^T = W(0), \\ W'(1)^T = -W'(0). \end{cases}$$

This is equivalent to (3.14) in the case c) (see lemma 1.2.4 in [12]). We get :

$$\begin{cases} \frac{1}{\lambda}B_1^{(-1)} \cdot \sin(\lambda B_1) \cdot \phi_1^u & = \phi_0^{uT} - \phi_0^u \cdot \cos(\lambda B_1), \\ e_2^\lambda(1) \cdot \phi_2^w + e_3^\lambda(1) \cdot \phi_3^w & = \phi_0^{wT} - e_0^\lambda(1) \cdot \phi_0^w - e_1^\lambda(1) \cdot \phi_1^w, \\ e_1^\lambda(1) \cdot \phi_2^w + e_2^\lambda(1) \cdot \phi_3^w & = -\lambda^2 B_2^{(4)} \cdot e_3^\lambda(1) \cdot \phi_0^w - \phi_1^{wT} - e_0^\lambda(1) \cdot \phi_1^w, \end{cases}$$

which is equivalent to

$$\begin{cases} \frac{1}{\lambda}B_1^{(-1)} \cdot C_1 \cdot \phi_1^u & = \cos(\mathcal{R}_\alpha) \cdot \Big((e\varphi_0^{uT}) - (\varphi_0^u e^T) \cdot \cos(\lambda B_1)\Big) \\ & \quad + \sin(\mathcal{R}_\alpha) \cdot \Big((e\varphi_0^{wT}) - (\varphi_0^w e^T) \cdot \cos(\lambda B_1)\Big), \\ \frac{1}{2\lambda^2}B_2^{(-4)} \cdot C_2 \cdot \phi_2^w & = \frac{1}{2\lambda}B_2^{(-2)} \cdot \Big[-c_1(\lambda) \cdot \Big(-(e\varphi_0^{uT}) \cdot \sin(\mathcal{R}_\alpha) + (e\varphi_0^{wT}) \cdot \cos(\mathcal{R}_\alpha)\Big) \\ & \quad + d_1(\lambda) \cdot \Big(-(\varphi_0^u e^T) \cdot \sin(\mathcal{R}_\alpha) + (\varphi_0^w e^T) \cdot \cos(\mathcal{R}_\alpha)\Big)\Big] \\ & \quad + \frac{1}{2\lambda^{(3/2)}}B_2^{(-3)} \cdot \Big[-c_2(\lambda) \cdot (e\varphi_1^{wT}) \cdot L \cdot O^T + d_2(\lambda) \cdot (\varphi_1^w e^T) \cdot L \cdot O\Big], \\ \frac{1}{2\lambda^2}B_2^{(-4)} \cdot C_2 \cdot \phi_3^w & = \frac{1}{2\sqrt{\lambda}}B_2^{(-1)} \cdot \Big[a_1(\lambda) \cdot \Big(-(e\varphi_0^{uT}) \cdot \sin(\mathcal{R}_\alpha) + (e\varphi_0^{wT}) \cdot \cos(\mathcal{R}_\alpha)\Big) \\ & \quad - b_1(\lambda) \cdot \Big(-(\varphi_0^u e^T) \cdot \sin(\mathcal{R}_\alpha) + (\varphi_0^w e^T) \cdot \cos(\mathcal{R}_\alpha)\Big)\Big] \\ & \quad + \frac{1}{2\lambda}B_2^{(-2)} \cdot \Big[c_1(\lambda) \cdot (e\varphi_1^{wT}) \cdot L \cdot O^T - d_1(\lambda) \cdot (\varphi_1^w e^T) \cdot L \cdot O\Big], \end{cases}$$

thanks to (3.9), (3.10).
The cases a) and b) are direct computations of this system (for the corresponding component only) when it is degenerated on a beam $k_{s(i,h)}$. We get the result of the case c) thanks to (3.11), (3.12), (3.13) and the easily checked identities

$$\begin{cases} (M \cdot \mathcal{E} \cdot (e\varphi^T))e = M\varphi, \\ (M \cdot \mathcal{E} \cdot (\varphi e^T))e = [Diag(Me)]\varphi, \end{cases}$$

which hold for any $M \in I\!R^{n\times n}$ and any $\varphi \in I\!R^n$. ∎

REFERENCES

[1] F Ali Mehmeti. A characterisation of generalized C^∞ notion on nets, Integral Eq and Operator Theory 9 : 753-766, 1986.

[2] F Ali Mehmeti. Regular solutions of transmission and interaction problems for wave equations, Math Meth Appl Sc, 11 : 665-685, 1989.

[3] F Ali Mehmeti. Nonlinear waves in networks, Akademie Verlag, Berlin, Mathematical Research, 80, 1994.

[4] F Ali Mehmeti. Spectral Theory and L^∞-time decay estimates for Klein-Gordon equations on two half axes with transmission : the tunnel effect, Math Meth Appl Sci, 17 n° 9 : 697-752, 1994.

[5] J von Below. A characteristic equation associated to an eigenvalue problem on c^2-networks, Linear Algebra and Appl, 71 : 309-325, 1985.

[6] J von Below. Sturm-Liouville eigenvalue problems on networks, Math Meth Appl Sc, 10 : 383-395, 1988.

[7] A Borovskikh, R Mustafokulov, K Lazarev and Yu Pokornyi. A class of fourth-order differential equations on a spatial net, Doklady Math, 52 : 433-435, 1995.

[8] B. Dekoninck and S. Nicaise, The eigenvalue problem for networks of beams, Linear Algebra and its Applications, to appear.

[9] B Dekoninck and S Nicaise. Control of networks of Euler-Bernouilli beams, ESAIM-COCV,vol 4 (1999), pp 57-82.

[10] B Dekoninck and S Nicaise. The eigenvalue problem for networks of beams. In : I Antoniou and G Lumer eds, Generalized Functions, Operator Theory and Dynamical Systems, Chapman and Hall Research in Math, 1999, pp 335-344.

[11] B Dekoninck et S Nicaise. Spectre des réseaux de poutres, C R Acad Sci Paris, tome 326, Série 1 : 1249-1254, 1998.

[12] B Dekoninck. Spectre et contrôlabilité de réseaux de poutres, Thèse de doctorat, Université de Valenciennes, sept. 98.

[13] N Eichler. Mathematische Modellierung der mechanischen Eigenschaften einer Bienenwabe, Diplomarbeit, Referenten : J. Tautz, U. Helmke, Bayerische Julius-Maximilians-Universität Würzburg, August 1998.

[14] JE Lagnese, G Leugering and EJPG Schmidt. Control of planar networks of Timoshenko beams, SIAM J Control Optim, 31 : 780-811, 1993.

[15] JE Lagnese, G Leugering and EJPG Schmidt. Modeling, analysis and control of dynamic elastic multi-link structures, Birkhäuser, Boston, 1994.

[16] JE Lagnese, G Leugering and EJPG Schmidt. Modeling of dynamic networks of thin thermoelastic beams, Math Meth in the Appl Sci, 16 : 327-358, 1993.

[17] S Nicaise. Some results on spectral theory over networks applied to nerve impulse transmission, Lecture Notes in Math , 1171, Springer-Verlag, 1985, pp 532-541.

[18] S Nicaise. Diffusion sur les espaces ramifiés, Thesis, Université de Mons,1986.

[19] S Nicaise. Spectre des réseaux topologiques finis, Bull Sc Math , 2ème série, 111 : 401-413, 1987.

[20] RE Showalter. Hilbert space methods for partial differential equations, Monographs and Studies in Mathematics, 1, Pitman, Boston, 1977.

[21] J Tautz, M Lindauer and DC Sandeman. Transmission of vibration across honeycombs and its detection by bee leg receptors, J of Experimental Biology, 199 : 2585-2594, 1999.

Can One Hear the Shape of a Network?

JOACHIM VON BELOW LMPA Joseph Liouville, EA 2597, Université du Littoral Côte d'Opale, 50, rue F. Buisson, B.P. 699, F–62228 Calais Cédex (France)

1 INTRODUCTION

Already in 1892, in a book about spectrocospy, A. Shuster raised the question "... how to find a shape of a bell by means of the sounds which it is capable of sending out." [15] The inverse spectral problem hidden behind the physical context was first formulated in a mathematical setting by H. Weyl indirectly in 1911 and by S. Bochner in 1950, cf. [15]. Finally, in 1966, M. Kac published the famous paper [16] intitled "Can one hear the shape of a drum?" After a separation ansatz of the wave equation, the problem reads in mathematical terms as follows. Suppose that two bounded domains Ω_1 and Ω_2 in $\mathbb{R}^n$ are *isospectral* i.e. the spectra of their Laplacian under the homogeneous Dirichlet boundary condition (D) or under the Neumann boundary condition (N) coincide counting multiplicities. Does this imply that Ω_1 and Ω_2 are isometric, i.e. do they coincide up to rotations, reflections or translations?

$$\sigma\left(\Delta_{\Omega_1}^{D/N}\right) \overset{\text{m}}{=} \sigma\left(\Delta_{\Omega_2}^{D/N}\right) \quad \overset{?}{\Longrightarrow} \quad \Omega_1 \cong \Omega_2$$

Evidently, for $n = 1$ the answer is affirmative, as well as the inverse implication is true. In fact, a negative answer to the question has been given earlier on compact riemannian manifolds. In 1964 J. Milnor

[19] already gave examples of isospectral nonisometric 16–dimensional tori, while in 1971 M. Berger, P. Gauduchon and E. Mazet [9] showed that two–dimensional isospectral tori are isometric. Finally, for any $n \geq 2$, M.–F. Vignéras [26] (1980) constructed n–dimensional compact isospectral nonisometric manifolds. For euclidian domains, H. Urakawa [25] (1982) gave a counterexample in $\mathbb{R}^4$. But for plane domains the problem remained unsolved until 1992, when C. Gordon, D. Webb and S. Wolpert [13], [14] and P. Bérard [8] showed that one cannot determine the shape of a domain by the spectrum of its Laplacian, cf. also [11]. A distinctive feature of their counterexamples is that the domains involved are all polygonal. As it stands, the problem seems not yet to be solved for smooth domains.

For ramified spaces of dimension greater than 1, as treated e.g. in [2], [7], [18], [22], [23], the inverse spectral problem is settled by the negative answer for higher dimensional domains. But, it still remains to solve the problem on ramified networks with one–dimensional branches. As basic multistructures, these networks enjoyed an increasing interest during the last twenty years, see e.g. [1], [3]–[6], [17], [18], [21], [22] and [24]. In the present paper, we show that, in contrast to the one-dimensional domain case, one cannot recover the shape of a network from the spectrum of its Laplacian under the continuity condition at ramification nodes and the Kirchhoff condition at all vertices. Thus, in that regard, networks behave like higher dimensional objects. Moreover, we shall discuss the eigenvalue asymptotics as well as the distinction of network immanent eigenvalues from those stemming from single branches.

2 NETWORKS AND VERTEX TRANSITION

All networks in this paper are supposed to be finite $\mathcal{C}^2$-networks in the sense of [3] Chapter 2. By definition, a finite $\mathcal{C}^2$-network G is the union of the edges k_j of a topological graph Γ in $\mathbb{R}^m$ with finite sets of vertices $V = \{v_i | 1 \leq i \leq n\}$ and edges $K = \{k_j | 1 \leq j \leq N\}$. Moreover, the arc length parametrizations π_j are supposed to belong to $\mathcal{C}^2([0, l_j], \mathbb{R}^m)$. The arc length parameter of an edge k_j is denoted by x_j. The topological graph Γ belonging to G is assumed to be simple and connected, i.e. $\Gamma = (V, K)$ consists in a collection of the supports of N Jordan curves k_j with the following properties: Each k_j has its endpoints in the set V, any two vertices in V can be connected by a path with arcs in K, and any two edges $k_j \neq k_h$ satisfy $k_j \cap k_h \subset V$ and $|k_j \cap k_h| \leq 1$. Endowed with the induced topology G is a connected and compact space in $\mathbb{R}^m$. The valency of each vertex is denoted by $\gamma_i = \gamma(v_i) = |\{k \in K | v_i \in k\}|$. We distinguish the ramification nodes $V_r = \{v_i \in V | \gamma_i > 1\}$ from the boundary vertices $V_b = \{v_i \in V | \gamma_i = 1\}$. The orientation of Γ and G is given by the incidence matrix $D = (d_{ij})_{n\times N}$ with

$$d_{ij} = \begin{cases} 1 & \text{if } \pi_j(l_j) = v_i, \\ -1 & \text{if } \pi_j(0) = v_i, \\ 0 & \text{otherwise.} \end{cases}$$

The adjacency matrix $\mathfrak{A}(\Gamma) = (e_{ih})_{n\times n}$ of the graph Γ is defined by

$$e_{ih} = \begin{cases} 1 & \text{if } v_i \text{ and } v_h \text{ are adjacent in } \Gamma, \\ 0 & \text{else.} \end{cases}$$

Note that $\mathfrak{A}(\Gamma)$ is indecomposable since Γ is connected. Moreover, we set

$$s(i,h) = \begin{cases} s & \text{if } k_s \cap V = \{v_i, v_h\}, \\ 1 & \text{otherwise.} \end{cases}$$

For further graph theoretical terminology we refer to [10] and [27]. For a function $u : G \to \mathbb{R}$ we set $u_j := u \circ \pi_j : [0, l_j] \to \mathbb{R}$ and use the abbreviations

$$u_j(v_i) := u_j(\pi_j^{-1}(v_i)), \quad \partial_j u_j(v_i) := \frac{\partial}{\partial x_j} u_j(x_j)\Big|_{\pi_j^{-1}(v_i)} \quad \text{etc.}$$

As special subspaces of $\mathcal{C}(G)$ we introduce for $1 \leq k \in \mathbb{N}$ the Banach spaces $\mathcal{C}^k(G)$ endowed with the norm $|u|_{k,G} = \sum_{j=1}^N |u_j|_{\mathcal{C}^k([0,l_j])}$ by

$$\mathcal{C}^k(G) = \{u \in \mathcal{C}(G) | \forall j \in \{1, ..., N\} : u_j \in \mathcal{C}^k([0, l_j])\},$$

where the Banach space $\mathcal{C}^k([0, l_j])$ is endowed with the usual $\mathcal{C}^k-$ norm. For the Hilbert spaces $\mathcal{H}^k := \prod_{j=1}^N H^k[0, l_j]$ endowed with the usual H^k product norm, the Sobolev embedding theorem permits to evaluate the components at vertices for $k \geq 1$. Thus,

$$\mathcal{H}^k(G) := \left\{(w_j)_{N\times 1} \in \mathcal{H}^k \,|\, \exists u \in \mathcal{C}(G) \forall j \in \{1, ..., N\} : w_j = u \circ \pi_j\right\}$$

is a closed subspace of $\mathcal{H}^k$ for $k \geq 1$.

As the basic geometric transition condition at ramification nodes we impose the following continuity condition

$$\forall v_i \in V_r \; : \; k_j \cap k_s = \{v_i\} \implies u_j(v_i) = u_s(v_i), \tag{1}$$

that clearly is contained in the condition $u \in C(G)$. Moreover, at each vertex $v_i \in V$ we impose the classical Kirchhoff condition

$$\forall i \in \{1, \ldots, n\} \; : \; \sum_{j=1}^{N} d_{ij} \partial_j u_j(v_i) = 0. \tag{K}$$

In fact, the Kirchhoff condition (K) generalizes the Neumann boundary condition on an interval to a vertex transition condition in networks. For operators and function spaces on G, let the super- or subscript K indicate the validity of condition (K). Accordingly, the spaces $\mathcal{C}^1_K(G)$, $\mathcal{C}^2_K(G)$ and $\mathcal{H}^2_K(G)$ are well defined.

3 THE LAPLACIAN ON A NETWORK

The canonical Laplacian on a $\mathcal{C}^2$–network G is defined as the operator

$$\Delta_G^K = \left(u \mapsto \left(\partial_j^2 u_j\right)_{N \times 1} \right) \; : \; \mathcal{C}^2_K(G) \longrightarrow \prod_{j=1}^{N} \mathcal{C}([0, l_j])$$

or as an operator $\Delta_G^K \; : \; \mathcal{H}^2_K(G) \longrightarrow \mathcal{H}$. Owing to the results in [4], $\sigma(\Delta_G^K)$ is real and infinitely countable without finite accumulation point, and $\mathcal{H}$ possesses an orthonormal basis of eigensolutions of Δ_G in $\mathcal{C}^2_K(G)$. Moreover, we note that there are no positive eigenvalues:

$$\sigma(\Delta_G^K) \leq 0,$$

since due to (1) and (K):

$$\sum_{j=1}^{N} \int_0^{l_j} u_j \partial_j^2 u_j dx_j = -\sum_{j=1}^{N} \int_0^{l_j} (\partial_j u_j)^2 dx_j + \sum_{j=1}^{N} [u_j \partial_j u_j]_0^{l_j}$$

$$= -\sum_{j=1}^{N} \int_0^{l_j} (\partial_j u_j)^2 dx_j + \sum_{i=1}^{n} u(v_i) \underbrace{\sum_{j=1}^{N} d_{ij} \partial_j u_j(v_i)}_{=0} \leq 0.$$

Of course, one can also consider the Laplacian Δ_G under other transition conditions, especially under replacing the Kirchhoff condition by the homogeneous Dirichlet condition at some vertices. But, as it stands, in view of the results in [3]–[5], the present case is the essential model case that can also help in treating other ones.

Let us precise the inverse spectral problem in question. By definition, two networks G_1 and G_2 are *isometric* ($G_1 \cong G_2$) if there is an homeomorphism $H : G_1 \longrightarrow G_2$ such that for each edge $k \subset G_1$, $H\big|_k$ is an isometric diffeomorphism onto some edge of G_2. In particular, H is length preserving. Moreover, the underlying abstract graphs Γ_1 and Γ_2 are called *isomorphic* as graphs ($\Gamma_1 \simeq \Gamma_2$) if there is a bijection $V(\Gamma_1) \longrightarrow V(\Gamma_2)$ that preserves the adjacency relation between vertices. If G_1 and G_2 are isometric networks, then the underlying abstract graphs Γ_1 and Γ_2 are isomorphic as graphs.

PROBLEM *Suppose that for two networks G_1 and G_2 in $\mathbb{R}^m$ the spectra of $\Delta^K_{G_1}$ and $\Delta^K_{G_2}$ coincide counting multiplicities. Does this imply that G_1 and G_2 are isometric?*

$$\sigma\left(\Delta^K_{G_1}\right) \stackrel{\mathrm{m}}{=} \sigma\left(\Delta^K_{G_2}\right) \quad \stackrel{?}{\Longrightarrow} \quad G_1 \cong G_2 \tag{2}$$

For the underlying abstract graphs Γ_1 and Γ_2, we are led to the reduced problem

$$\sigma\left(\Delta^K_{\Delta_1}\right) \stackrel{\mathrm{m}}{=} \sigma\left(\Delta^K_{\Delta_2}\right) \quad \stackrel{?}{\Longrightarrow} \quad \Gamma_1 \simeq \Gamma_2 \tag{3}$$

In Section 4, we shall show that the implication (3) is wrong, which in turn shows the same for (2). We can restrict ourselves to the case where all edge lengths are equal:

$$\forall j \in \{1, ..., N\} : \; l_j = 1 \tag{4}$$

In fact, under condition (4), the definition of a C^2-network implies that $G_1 \cong G_2 \iff \Gamma_1 \simeq \Gamma_2$. The eigenvalue problem for Δ^K_G reads

$$0 \neq u \in C^2_K(G) \quad \text{and} \quad \partial^2_j u_j = -\lambda u_j \quad \text{for} \quad 1 \leq j \leq N. \tag{5}$$

Following the transformations in [3], we formulate Problem (5) as an equivalent matrix differential boundary eigenvalue problem incorporating the adjacency structure of the network. For that purpose we recall that the Hadamard matrix product is defined as

$$(a_{ik})_{n\times n} \star (b_{ik})_{n\times n} = (a_{ik}b_{ik})_{n\times n}\,.$$

For a function $u : G \to \mathbb{R}$ we set for $x \in [0,1]$

$$u_{ih}(x) = e_{ih} u_{s(i,h)} \left(\frac{1 + d_{is(i,h)}}{2} - x d_{is(i,h)} \right)$$

and

$$\varphi = (u(v_i))_{n \times 1}, \qquad \mathbf{U} = (u_{ih})_{n \times n}, \qquad \mathbf{e} = \mathbf{e}_n = (1)_{n \times 1}.$$

Thus,

$$\mathbf{U}(0) = \underbrace{\begin{pmatrix} \varphi_1 & \varphi_1 & \cdots & \varphi_1 & \varphi_1 \\ \varphi_2 & \varphi_2 & \cdots & \varphi_2 & \varphi_2 \\ \vdots & \vdots & \ddots & \vdots & \vdots \\ \varphi_n & \varphi_n & \cdots & \varphi_n & \varphi_n \end{pmatrix}}_{\varphi \mathbf{e}^*} \star \mathfrak{A},$$

and (5) becomes equivalent to the following differential boundary eigenvalue problem (6) for the matrix $\mathbf{U}$:

$$(5) \iff (6) \begin{cases} u_{ih} \in C^2([0,1]) & \text{for } 1 \le i,h \le n & (6.1) \\ e_{ih} = 0 \Rightarrow u_{ih} = 0 & \text{for } 1 \le i,h \le n & (6.2) \\ \mathbf{U}'' = -\lambda \mathbf{U} & \text{in } [0,1] & (6.3) \\ \mathbf{U}(0) = \varphi \mathbf{e}^* \star \mathfrak{A} & \text{(continuity in } V_r) & (6.4) \\ \mathbf{U}^*(x) = \mathbf{U}(1-x) & \text{for } x \in [0,1] & (6.5) \\ \mathbf{U}'(0)\mathbf{e} = 0 & \text{(K)} & (6.6) \end{cases}$$

We set

$$\Phi := \mathbf{U}(0) = \varphi \mathbf{e}^* \star \mathfrak{A}, \qquad \Psi := \mathbf{U}'(0),$$

and recall the following elementary rules for a $n \times n$–matrix M:

$$(M \star \mathbf{e}\varphi^*)\,\mathbf{e} = M\varphi \qquad\qquad (M \star \varphi \mathbf{e}^*)\,\mathbf{e} = \mathrm{Diag}(M\mathbf{e})\varphi \qquad (7)$$

For $\lambda = 0$, any solution $\mathbf{U}$ of (6) satisfies $\mathbf{U}(x) = \Phi + x\,(\Phi^* - \Phi)$ and $(\Phi^* - \Phi)\,\mathbf{e} = 0$, which implies $\mathrm{Diag}_i\left(\gamma_i^{-1}\right) \mathfrak{A}(\Gamma)\,\varphi = \varphi$. Since $\mathfrak{A}$ is indecomposable, the Perron–Frobenius Theorem [20] yields that 0 is a simple eigenvalue of Δ_G^K with eigenfunctions $\mathbf{U} = \mathrm{const.}\mathfrak{A}$

For $\lambda > 0$, a fundamental solution of (6.3) is given by

$$\mathbf{U}(x) = \cos(x\sqrt{\lambda})\Phi + \frac{\sin(x\sqrt{\lambda})}{\sqrt{\lambda}}\Psi \tag{8}$$

In the case $\sin\sqrt{\lambda} \neq 0$, (6.5) and (6.6) yield

$$\mathbf{U}(1) = \Phi^* = \Phi\cos\sqrt{\lambda} + \frac{\sin\sqrt{\lambda}}{\sqrt{\lambda}}\Psi,$$

$$\Psi = \frac{\sqrt{\lambda}}{\sin\sqrt{\lambda}}\left(\mathbf{e}\varphi^* - \cos\sqrt{\lambda}\,\varphi\mathbf{e}^*\right) \star \mathfrak{A},$$

and using (7)

$$(\mathfrak{A} \star \mathbf{e}\varphi^*)\,\mathbf{e} - \cos\sqrt{\lambda}\,(\mathfrak{A} \star \varphi\mathbf{e}^*)\,\mathbf{e} = 0.$$

Thus we are led to the following characteristic equation

$$\mathfrak{A}(\Gamma)\varphi = \cos\sqrt{\lambda}\,\mathrm{Diag}_i\,(\gamma_i)\,\varphi. \tag{9}$$

This is part of the following theorem that we recall from Section 5 of [3] in the special case (4) here.

THEOREM 1 *Under condition (4) and using the above definitions, $\lambda \in \sigma(-\Delta_G^K)$ iff either $\varphi = 0$, $\Psi \neq 0$ and $\sin\sqrt{\lambda} = 0$ or φ is an eigenvector belonging to the eigenvalue $\cos\sqrt{\lambda}$ of the row–stochastic matrix*

$$\mathbf{Z} = \mathrm{Diag}_i\,(\gamma_i^{-1})\,\mathfrak{A}(\Gamma).$$

Moreover, the multiplicities $m(\lambda)$ are

$$m(0) = 1,$$

$$m(\lambda) = \dim\ker\left(\mathbf{Z} - \cos\sqrt{\lambda}\,\mathbf{I}_n\right), \quad \textit{if } \sin\sqrt{\lambda} \neq 0,$$

$$m\left(\pi^2 4k^2\right) = N - n + 2,$$

$$m\left(\pi^2(2k+1)^2\right) = N - n + 2, \quad \textit{if } \Gamma \textit{ is bipartite},$$

$$m\left(\pi^2(2k+1)^2\right) = N - n, \quad \textit{if } \Gamma \textit{ is not bipartite}.$$

Note that the multiplicities depend only on $n, N, \gamma_1, \ldots, \gamma_n$ and the adjacency matrix $\mathfrak{A}$. Thus, the spectrum of the Laplacian Δ_G^K is entirely determined by combinatorial quantities of the underlying graph Γ. As an immediate consequence of Theorem 1, we note the following result contained in the results [3] by von Below 1985 and independently shown by Ali Mehmeti [1] 1986 and by Nicaise [21] 1987. Let $b(z)$ denote the number of eigenvalues of $-\Delta_G^K$ in $[0, z]$. Using the above multiplicity formulae, we find $b((2\pi k)^2) = 2Nk + 1$ and

$$\lim_{x \to \infty} \frac{b(x)}{\sqrt{x}} = \frac{N}{\pi}.$$

This yields the

COROLLARY 1 *Let $\{\lambda_k \,|\, k \in \mathbb{N}\}$ denote the monotonically increasing sequence of eigenvalues of $-\Delta_G^K$ under condition (4). Then*

$$\lim_{k \to \infty} \frac{\lambda_k}{k^2} = \frac{\pi^2}{N^2}.$$

In view of the results of the next section it will be of interest to determine common invariants of isospectral networks. Here we mention only the following

COROLLARY 2 *Under condition (4), isospectral networks have the same number of vertices and the same number of edges. Moreover, networks that are isospectral with a bipartite network are also bipartite.*

Proof: By the Lemma in Section 5 of [3], the eigenvalues of $\mathbf{Z}$ are real, $\mathbb{R}^n$ possesses a basis formed by eigenvectors of $\mathbf{Z}$, and the underlying graph Γ is bipartite iff $-1 \in \sigma(\mathbf{Z})$. By the Perron–Frobenius Theorem [20], all eigenvalues of $\mathbf{Z}$ of modulus 1 are simple. Let G_1 and G_2 be two $\mathcal{C}^2$–networks in $\mathbb{R}^m$ and indicate all corresponding entities by 1 and 2 respectively. Now suppose that $\sigma\left(\Delta_{G_1}^K\right)$ and $\sigma\left(\Delta_{G_2}^K\right)$ coincide counting multiplicities. Then apply Theorem 1 in order to conclude $\sigma(\mathbf{Z_1}) \overset{\mathrm{m}}{=} \sigma(\mathbf{Z_2})$. This shows the last assertion. Furthermore, the multiplicities formulae in Theorem 1 imply

$$n_1 = \sum_{\mu \in \sigma(\mathbf{Z_1})} \dim\ker\left(\mathbf{Z_1} - \mu\,\mathbf{I}_n\right) = \sum_{\mu \in \sigma(\mathbf{Z_2})} \dim\ker\left(\mathbf{Z_2} - \mu\,\mathbf{I}_n\right) = n_2$$

and, finally, $N_1 = N_2$. □

Thus, under condition (4), for subclasses of networks that are characterized only by the number of vertices n or only by the number of edges N, the implications (2) and (3) are true within the subclass. For instance, isospectral paths are isometric, isospectral circuits are isometric, or isospectral star shaped networks, i.e. $|V_r| = 1$, are isometric.

4 COUNTEREXAMPLES

Recall that a graph Γ is called γ–regular if all vertex valencies γ_i are equal to γ. In this case the characteristic equation (9) reads

$$\mathfrak{A}(\Gamma)\varphi = \gamma \cos \sqrt{\lambda}\, \varphi. \tag{10}$$

Combining (10) with the Lemma in Section 5 of [3], we obtain the

COROLLARY 3 *Suppose that Γ is a regular. Then the following conditions are equivalent:*

(a) Γ *is bipartite.*

(b) $-1 \in \sigma(\mathbf{Z})$

(c) $\sigma(\mathbf{Z}) \overset{\mathrm{m}}{=} -\sigma(\mathbf{Z})$

(d) $\cos \sqrt{\sigma(-\Delta_G^K)} \overset{\mathrm{m}}{=} -\cos \sqrt{\sigma(-\Delta_G^K)}$

COROLLARY 4 *Suppose that Γ is a regular. If Γ is bipartite or a circuit of odd length, then*

$$\gamma \cos \sqrt{\sigma(-\Delta_G^K)} = \sigma\left(\mathfrak{A}(\Gamma)\right).$$

If Γ is neither bipartite nor a circuit of odd length, then

$$\gamma \cos \sqrt{\sigma(-\Delta_G^K)} = \sigma\left(\mathfrak{A}(\Gamma)\right) \cup \{-\gamma\}.$$

Proof: The regular nonbipartite connected graphs fulfilling $N = n$ are exactly the circuits of odd length. Now apply Theorem 1, (10) and Corollary 3. □

In the regular case, the multiplicities are $m(0) = 1$ and for $\lambda > 0$

$$m(\lambda) = \dim\ker\left(\mathfrak{A}(\Gamma) - \gamma\cos\sqrt{\lambda}\,\mathbf{I}_n\right), \text{ if } \sin\sqrt{\lambda} \neq 0,$$

$$m(\lambda) = \begin{cases} N - n + 2, \\ N - n, \end{cases}, \text{ if } \sin\sqrt{\lambda} = 0,$$

and do only depend on n, N and the adjacency matrix $\mathfrak{A}$. This can be applied in order to find two isospectral networks G_1 and G_2 with underlying topological graphs Γ_1 and Γ_2, respectively, such that the corresponding problems (5) have the same eigenvalues counted according to their multiplicities, while Γ_1 and Γ_2 are not isomorphic as abstract graphs. We only have to find graphs Γ_1 and Γ_2 that are regular, nonisomorphic, and isospectral as graphs, i.e. $\sigma(\mathfrak{A}(\Gamma_1)) = \sigma(\mathfrak{A}(\Gamma_2))$ with identical multiplicities. A large variety of such examples can be obtained with the aid of a result of A. J. Hoffman, see e.g. Theorem 6.1 in [12], which states that for any natural number M there is an integer n_0 such that for any $n \geq n_0$ there exist M nonisomorphic isospectral regular connected graphs with n vertices. Obviously, all these must have the same valency. As a concrete example, we display the following two 4–regular graphs Γ_1 and Γ_2 with $n = 12$ and $N = 24$

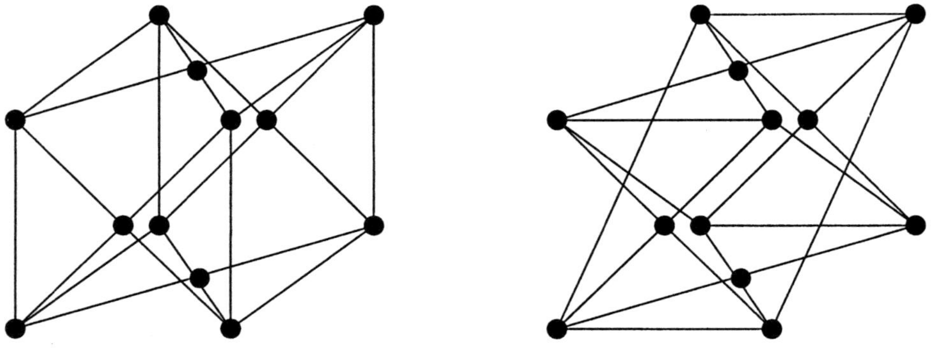

Fig.1 Two isospectral networks that are not isometric.

as depicted in Fig.1. The corresponding networks G_1 and G_2 can be easily realized under condition (4). We readily compute

$$\sigma(\mathfrak{A}(\Gamma_1)) = \sigma(\mathfrak{A}(\Gamma_2)) = (4, 2, 2, 2, 0, 0, 0, -2, -2, -2, -2, -2).$$

But, both graphs are not isomorphic, since the left one Γ_1 is planar while the right one Γ_2 is not realizable in $\mathbb{R}^2$, see Fig.2, where in each graph the two half-moon like vertices have to be identified. On the

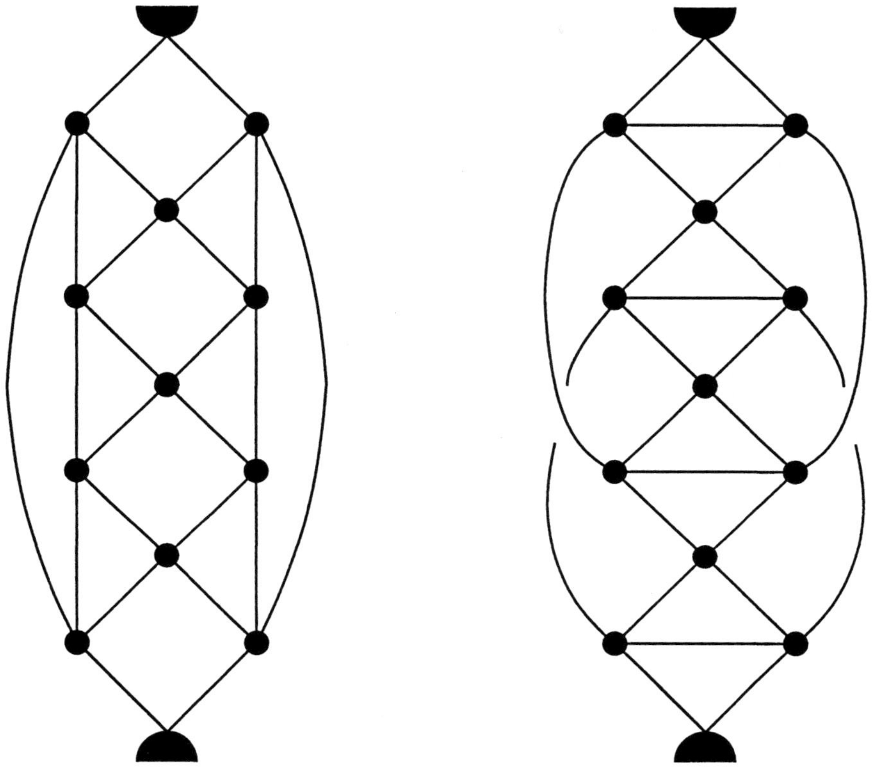

Fig.2 The nonisomorphic underlying graphs.

other hand, by Theorem 1, (10) and (11) we conclude that the spectra $\sigma\left(\Delta_{G_1}^K\right)$ and $\sigma\left(\Delta_{G_2}^K\right)$ coincide counting multiplicities. Thus, we can resume:

THEOREM 2 *In general, the shape of a C^2-network G cannot be reconstructed from the spectrum of its Laplacian Δ_G^K. There exist pairs of regular isospectral networks that are not isometric.*

The same phenomenon can occur when the underlying abstract graphs are isomorphic, but different edge lengths are admitted. Roth [24] has already given an example of a nonisometric pair of networks with the same multiple underlying graph and with the same eigenvalues of the Laplacian. For the simple graphs considered here, we consider the case of a weighted Laplacian on a network where the diffusion rates on all edges are given by the squares of the edge lengths. The corresponding eigenvalue problem including a consistent Kirchhoff condition reads

$$
(12)\quad \begin{cases} u \in C^2(G), & \\ l_j^2 \partial_j^2 u_j = -\lambda u_j & \text{in } [0,1] \text{ for } 1 \le j \le N, \\ \sum_{j=1}^{N} d_{ij} l_j^2 \partial_j u_j(v_i) = 0 & \text{for all } v_i \in V. \end{cases}
$$

Note that the consistency of the Kirchhoff condition is indispensable in order to ensure the symmetry of the Laplacian, see [3] Section 7 and [4], except for a very special class of networks containing one inconsistent ramification node, see [5]. Using the transformations of Section 3 with the modifications

$$
u_{ih}(x) = e_{ih} u_{s(i,h)} \left(l_{s(i,h)} \left[\frac{1 + d_{is(i,h)}}{2} - x d_{is(i,h)} \right] \right)
$$

and

$$
\mathbf{L} = \left(e_{ih} l_{s(i,h)} \right)_{n \times n},
$$

the equivalent matrix differential boundary eigenvalue problem incorporating the adjacency structure of the network becomes:

$$
(13)\quad \begin{cases} u_{ih} \in C^2([0,1]) & \text{for } 1 \le i,h \le n \\ e_{ih} = 0 \Rightarrow u_{ih} = 0 & \text{for } 1 \le i,h \le n \\ \mathbf{U}'' = -\lambda \mathbf{U} & \text{in } [0,1] \\ \mathbf{U}(0) = \varphi \mathbf{e}^* \star \mathfrak{A} & \\ \mathbf{U}^*(x) = \mathbf{U}(1-x) & \text{for } x \in [0,1] \\ [\mathbf{L} \star \mathbf{U}'(0)]\, \mathbf{e} = 0 & \end{cases}
$$

By the results of Section 5 of [3], all the assertions of Theorem 1 remain valid for the eigenvalues of (12) and (13) with the row–stochastic matrix

$$\mathbf{Z} = \mathrm{Diag}\,(\mathbf{Le})^{-1}\,\mathbf{L}.$$

Thus, in order to find the desired example, we only have to ensure that for some nonisometric networks the corresponding matrices $\mathbf{Z}$ all have the same eigenvalues with multiplicities counted and the same zero pattern. Take for instance $\Gamma = \mathcal{K}_3$ to be the circuit of length 3 and G as indicated in Fig.3 with $(l_1+l_2)(l_1+l_3)(l_2+l_3) = 8l_1l_2l_3$. Then the set of eigenvalues of (12) with multiplicities counted coincides with the one in the special case $l_1 = l_2 = l_3 = 1$ given by (14), (15) and (17) in Section 5 for $n = 3$, since the common characteristic polynomial for all matrices $\mathbf{Z}$ reads $-\lambda^3 + \frac{3}{4}\lambda + \frac{1}{4}$.

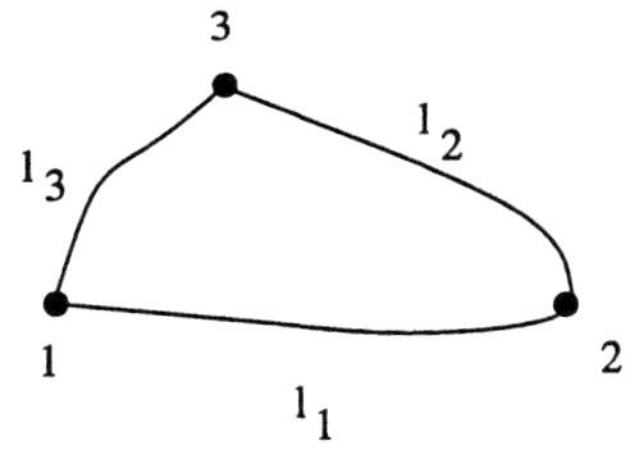

$$(l_1 + l_2)(l_1 + l_3)(l_2 + l_3) = 8l_1l_2l_3$$

Fig.3 A family of nonisometric isospectral networks with isomorphic underlying graph.

5 NETWORK IMMANENT EIGENVALUES

To what extent can one hear a network, i.e. which frequencies are immanent to the network system and cannot be derived from the spectrum of the Laplacian on a single edge under 0–Dirichlet or Neumann boundary conditions? Let us discuss this problem for the complete graph on n vertices $\mathcal{K}_n$. Its adjacency matrix $\mathfrak{A} = \mathbf{e}_n\mathbf{e}_n^* - \mathbf{I}_n$ has the simple eigenvalue $n-1$ and the $(n-1)$-fold eigenvalue -1 leading to the following eigenvalue sequences of (5) by Theorem 1:

$$\lambda = 0,\ m(0) = 1,\ \varphi = \mathbf{e}_n \tag{14}$$

$$\lambda = 4\pi^2k^2,\ k \neq 0,\ m(\lambda) = 2 + \frac{1}{2}n(n-3),\ \varphi = \mathbf{e}_n \tag{15}$$

$$\lambda = \pi^2(2k+1)^2,\ k \neq 0,\ m(\lambda) = \frac{1}{2}n(n-3),\ \varphi = 0, \tag{16}$$

$$\cos\sqrt{\lambda} = -(n-1)^{-1}, \quad m(\lambda) = n-1,\ \varphi \in \ker \mathbf{ee}^* \tag{17}$$

The results apply for instance to the wave equation on a tetrahedron network $\mathcal{K}_4$. Using the relation $\mathbf{U}'(1) = -\Psi^\star$, all eigenvalues (frequencies) and corresponding eigensolutions are given in terms of the transformations of Section 3 by (8) and Theorem 1 as follows:

$$\lambda = 0,\ m(0) = 1,\ \mathbf{U} = r\left(\mathbf{e}_4\mathbf{e}_4^* - \mathbf{I}_4\right),\ r \in \mathbb{R} \tag{18}$$

$$\begin{aligned} &\lambda = 4\pi^2k^2,\ 1 \leq k \in \mathbb{N},\ m(\lambda) = 4, \\ &\quad \mathbf{U}(x) = r\cos(x2\pi k)\left(\mathbf{e}_4\mathbf{e}_4^* - \mathbf{I}_4\right) + \frac{\sin(x2\pi k)}{2\pi k}\Psi \star \mathfrak{A}, \\ &\quad r \in \mathbb{R}, \Psi^* = -\Psi, \Psi\mathbf{e}_4 = 0 \end{aligned} \tag{19}$$

$$\begin{aligned} &\lambda = \pi^2(2k+1)^2,\ k \in \mathbb{N},\ m(\lambda) = 2,\ \varphi = 0, \\ &\quad \mathbf{U}(x) = \frac{\sin(x\pi(2k+1))}{\pi(2k+1)}\Psi \star \mathfrak{A},\ \Psi^* = \Psi, \Psi\mathbf{e}_4 = 0 \end{aligned} \tag{20}$$

$$\begin{aligned} &\lambda = \left((-1)^k \arccos\left(\frac{-1}{3}\right) + \frac{2k+1-(-1)^k}{2}\pi\right)^2,\ k \in \mathbb{N}, \\ &\quad \text{i.e. } \cos\sqrt{\lambda} = -\frac{1}{3},\ m(\lambda) = 3,\ \mathbf{e}_4^*\varphi = 0, \\ &\quad \mathbf{U}(x) = \cos(x\sqrt{\lambda})\varphi\mathbf{e}_4^* \star \mathfrak{A} + \frac{\sin(x\sqrt{\lambda})}{\sin\sqrt{\lambda}}\left(\mathbf{e}_4\varphi^* - \cos(\sqrt{\lambda})\varphi\mathbf{e}_4^*\right) \star \mathfrak{A} \end{aligned} \tag{21}$$

On the one hand, solutions $\mathbf{U}$ with $\Psi = 0$ in (19) correspond just to single edge solutions with Neumann boundary conditions glued together such that they form a continuous function on G. On the other hand, any $\mathbf{U} \neq 0$ with $\varphi = 0$ leads to an eigensolution on some edge with 0–Dirichlet conditions. Conversely, $u_0(x) := \sin(\pi kx)$ on $[0,1]$ can be extended to an eigensolution of $\mathcal{K}_4$ of the type (19) for $0 < k \equiv 0 \bmod 2$ and of the type (20) for $k \equiv 1 \bmod 2$ as displayed in Fig.4 and Fig.5. The bold edges indicate the circuits along which $u_{s(i,h)} = u_0$ on the edges in the circuit orientation, while on the remaining ones $u_{s(i,h)} \equiv 0$. In this way, a basis of the eigenspace in the case (19) is given by the matrix $\mathfrak{A} = \mathbf{e}_4\mathbf{e}_4^* - \mathbf{I}_4$, corresponding to the Neumann boundary condition on each edge $\Psi = 0$ and $\Phi = \Phi^*$, and the following matrices that correspond to the 0–Dirichlet condition on each edge $\Phi = 0$ and $\Psi^* = -\Psi$, see Fig.4.

$$\begin{pmatrix} 0 & +1 & -1 & 0 \\ -1 & 0 & +1 & 0 \\ +1 & -1 & 0 & 0 \\ 0 & 0 & 0 & 0 \end{pmatrix}, \begin{pmatrix} 0 & +1 & 0 & -1 \\ -1 & 0 & 0 & +1 \\ 0 & 0 & 0 & 0 \\ +1 & -1 & 0 & 0 \end{pmatrix}, \begin{pmatrix} 0 & 0 & +1 & -1 \\ 0 & 0 & 0 & 0 \\ -1 & 0 & 0 & +1 \\ +1 & 0 & -1 & 0 \end{pmatrix}$$

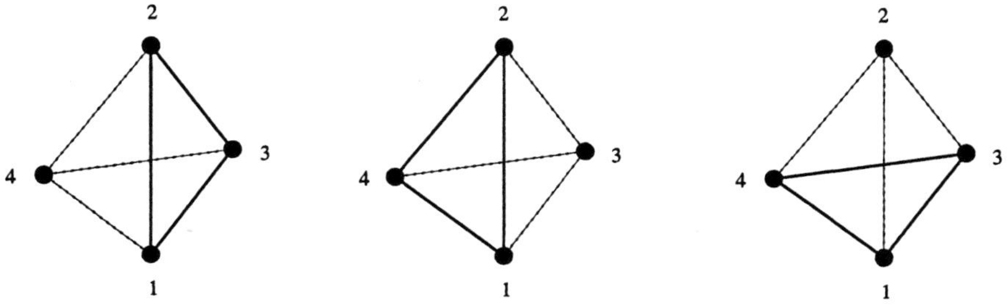

Fig. 4

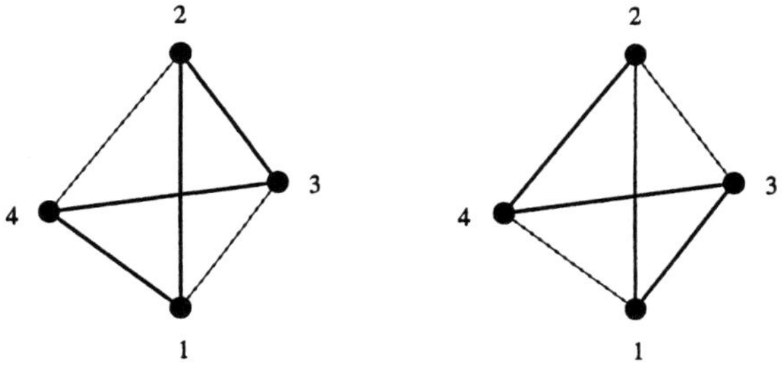

Fig. 5

A basis of the eigenspace in the case (20) is given by the matrices

$$\begin{pmatrix} 0 & +1 & 0 & -1 \\ +1 & 0 & -1 & 0 \\ 0 & -1 & 0 & +1 \\ -1 & 0 & +1 & 0 \end{pmatrix}, \begin{pmatrix} 0 & +1 & -1 & 0 \\ +1 & 0 & 0 & -1 \\ -1 & 0 & 0 & +1 \\ 0 & -1 & +1 & 0 \end{pmatrix}$$

that correspond again to the 0–Dirichlet condition on each edge $\Phi = 0$ and $\Psi^* = \Psi$, see Fig.5. Thus, the eigenvalues (21) are the only ones emanating from the network system and neither from a single clamped branch, nor a single branch with Neumann boundary condition. As for the acoustic interpretation, we observe that $\left(\arccos\left(\frac{-1}{3}\right)\right)^2$ is not compatible with the basic tone of a single edge, while for the triangle network $\mathcal{K}_3$ the corresponding network eigenvalue satisfying $\cos\sqrt{\lambda} = -\frac{1}{2}$ is compatible to the basic tone of a single edge: $\lambda = \frac{4}{9}\pi^2$. That is why tetrahedra sound so bad while triangles sound much better ...

Mutatis mutandis, the same distinction between single edge eigenvalues and network immanent ones holds for general networks. The eigenvalues of (5) belonging to the eigenvalues μ of the matrix $\mathbf{Z} = \mathrm{Diag}_i\left(\gamma_i^{-1}\right)\mathfrak{A}(\Gamma)$ with $|\mu| < 1$ are always immanent to the whole network system and cannot emanate from a single branch. In the bipartite case, there is an additional non zero node vector that belongs to the simple matrix eigenvalue $\mu = \cos\sqrt{\lambda} = -1$ and that can only occur in the whole system. Note that $0 \neq \Phi = -\Phi^*$ is impossible in nonbipartite networks. Note further that a single edge spectrum with zero boundary conditions can only be embedded if and only if Γ contains circuits. On a tree these embeddings are impossible. This also becomes clear by the multiplicity formulae in Theorem 1 that read for a tree $m\left(\pi^2k^2\right) = 1$ with $\varphi \in \mathbb{R}\mathbf{e}_n$ and $m(\lambda) = \dim\ker\left(\mathfrak{A}(\Gamma) - \cos\sqrt{\lambda}\,\mathbf{I}_n\right)$ for $\sin\sqrt{\lambda} \neq 0$ with $\varphi \neq 0$ as in (9).

REFERENCES

1. F. Ali Mehmeti. A characterisation of generalized C^∞–notion on nets. Integral Equ. Operator Theory 9:753–766, 1986.

2. F. Ali Mehmeti, S. Nicaise. Nonlinear interaction problems. Nonlinear Analysis, Theory, Meth. & Appl. 20:27–61, 1993.

3. J. von Below. A characteristic equation associated to an eigenvalue problem on c^2-networks. Lin. Alg. Appl. 71:309 - 325, 1985.

4. J. von Below. Sturm - Liouville eigenvalue problems on networks. Math. Meth. Applied Sciences 10:383 - 395, 1988.

5. J. von Below. Kirchhoff laws and diffusion on networks. Lin. Alg. Appl. 121: 692 - 697, 1989.

6. J. von Below. Parabolic network equations. 1st ed. Tübingen 1993, 3rd ed. to appear.

7. J. von Below, S. Nicaise. Dynamical interface transition in ramified media with diffusion. Commun. Partial Differential Equ. 21:255 - 279, 1996.

8. P. Bérard. Transplantation et isospectralité. Math. Annalen. 292:547–559, 1992.

9. M. Berger, P. Gauduchon, E. Mazet. Le spectre d'une variété riemannienne. Lecture Notes Math. 194, Springer Verlag Berlin, 1971.

10. N. L. Biggs. Algebraic graph theory. Cambridge Tracts Math.67, Cambridge University Press, 1967.

11. S. J. Chapman. Drums that sound the same. Amer. Math. Monthly 102:124–138, 1995.

12. D. M. Cvetcović, M. Doob, H. Sachs. Spectra of graphs. Academic Press New York, 1980.

13. C. Gordon, D. L. Webb, S. Wolpert. Isospectral plane domains and surfaces via Riemannian orbifolds. Inventiones Math. 110:1–22, 1992.

14. C. Gordon, D. L. Webb, S. Wolpert. One cannot hear the shape of a drum. Bull. AMS 27:134–138, 1992.

15. V. Isakov. Inverse problems for partial differential equations. Springer–Verlag, New York, 1998.

16. M. Kac. Can one hear the shape of a drum? Amer. Math. Monthly 73 II:1–23, 1966.

17. J. E. Lagnese, G. Leugering, E. J. P. G. Schmidt. Modeling, analysis and control of dynamic elastic multilink structures. Birkhäuser Verlag Bosten, 1994.

18. G. Lumer. Espaces ramifiés et diffusions sur des réseaux topologiques. C. R. Acad. Science Paris, Sér. A 201:627 - 630, 1980.

19. J. Milnor. Eigenvalues of the Laplace operator of certain manifolds. Pr. Nat. Ac. Sci. USA 51:542, 1964.

20. H. Minc. Nonnegative matrices. J. Wiley & Sons Inc. New York, 1988.

21. S. Nicaise. Spectre des réseaux topologiques finis. Bull. Sc. Math. 2e Série 111:401 - 413, 1987.

22. S. Nicaise. Le Laplacien sur les réseaux deux - dimensionnels polygonaux topologiques. J. Math. Pures Appl. 67:93 - 113, 1988.

23. S. Nicaise. Polygonal interface problems. Methoden und Verfahren der Mathematischen Physik, vol. 39, Peter Lang Verlag Frankfurt am Main, 1993.

24. J.-P. Roth. Le spectre du Laplacien sur un graphe, in: Proceedings of the Colloque J. Deny Orsay 1983, Lect. Not. Math. 1096:521 - 529, 1984.

25. H. Urakawa. Bounded domains which are isospectral but not congruent. Ann. Sci. Ecole Norm. Sup. 15:441–456, 1982.

26. M.-F. Vignéras. Variétés riemanniennes isospectrales et non isométriques. Ann. Math. 112:21–32, 1980.

27. R. J. Wilson. Introduction to graph theory. Oliver & Boyd Edinburgh, 1972.

Sensitivity analysis of 2D interface cracks

M. BOCHNIAK Mathematisches Institut A, Universität Stuttgart, Germany

A.–M. SÄNDIG Mathematisches Institut A, Universität Stuttgart, Germany

1 INTRODUCTION

The influence of the shape of the domain on the deformation behaviour of linear elastic bodies is of great importance in form optimization and fracture mechanics:

- The numerical solution of shape optimization problems by Newton type or steepest descent algorithms requires the knowledge of the Gâteaux derivative of corresponding cost and objective functionals with respect to a variation of the boundary.
- The Griffith criterion for the initiation of the crack growth states that the crack begins to grow when the energy release rate exceeds a critical value. Here the energy release rate is defined as the variation of the total potential energy of the cracked body under a small virtual perturbation of the crack tips.

The mathematical sensitivity analysis is usually performed under severe restrictions on the smoothness of the boundary which are not always satisfied in practice. The existence and the regularity of the material and the local derivative of the perturbed displacement and stress fields are thoroughfully investigated for the Dirichlet and the Neumann problem in smooth domains (see the monographs [9, 19] and the references therein). Corresponding results for the case of mixed boundary value problems in piecewise smooth domains were proved only recently in [3, 4].
In case of smooth domains it is known (see e.g. [19]), that the Gâteaux derivative of integral functionals depends only on the perturbation of the boundary in the direction of the normal vector. This holds also for piecewise smooth domains provided the stress singularities are not too strong [3]. For domains with cracks this is no longer true (see e.g. [15, 7, 12] for the variation of the potential energy). This result was generalized to a wide class of integral functionals by the authors in [3] who showed that the variation of functionals depends on the perturbation of the

boundary in the normal direction and also on the perturbation of the crack tips in the tangential direction. This result was proved recently also in an abstract form by Fremiot/Sokolowski [8]. In the present paper we show, that the result, proved in [3] for 2D cracks in homogeneous bodies, can be extended to 2D straight interface cracks. We rely on the existence and regularity results f or the material and the shape derivative proved by the authors in [4]. The method applied here is similar to the one used in [3] and is based on a combination of the material derivative approach [18, 19] and ideas from [6].

2 FORMULATION OF THE PROBLEM

2.1 Two-dimensional boundary transmission problems

Let $\Omega_1, \Omega_2 \subset \mathbb{R}^2$ be two bounded polygonal domains such that $\partial\Omega_1 \cap \partial\Omega_2 \neq \emptyset$ and let S be the finite set of isolated boundary points where stress singularities can occur, i.e. corner points and points where the boundary conditions change. We denote by $\Gamma \subset \partial\Omega_1 \cap \partial\Omega_2$ an interface between Ω_1 and Ω_2 and divide the outer boundary pieces into $\partial\Omega_j \setminus (\overline{\Gamma} \cup S) = \Gamma_j^D \cup \Gamma_j^N$. We assume that $\Gamma_1^D \cup \Gamma_2^D \neq \emptyset$ and that the domain Ω is locally diffeomorph in a neighbourhood of each point from S to a wedge. Furthermore we suppose that an interface crack occurs and denote by T_1, T_2 the crack tips (see Fig. 1, where $\overline{\Gamma}_j = \overline{\Gamma}_j^D \cup \overline{\Gamma}_j^N, j = 1, 2$).

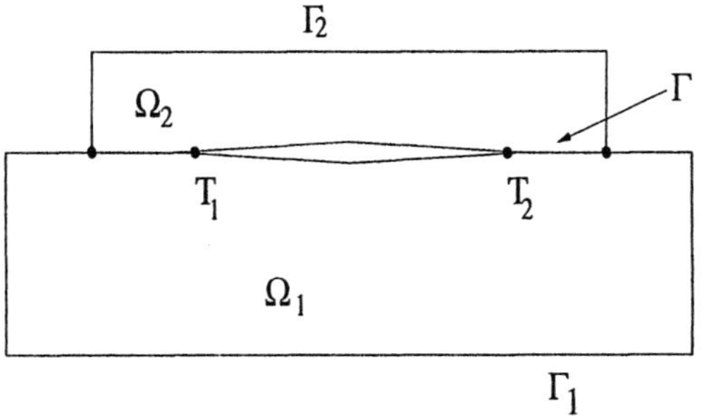

Fig. 1 *2D coupled structure*

The domains Ω_j are occupied by different isotropic linear elastic materials. The displacement vectors $\mathbf{u}_j$, $j = 1, 2$, in Ω_j satisfy the following transmission problem

$$\begin{aligned} L_j \mathbf{u}_j := -\mathrm{div}\sigma_j(\mathbf{u}_j) &= \mathbf{f}_j \quad \text{in } \Omega_j, \\ \mathbf{u}_j &= 0 \quad \text{on } \Gamma_j^D, \\ \sigma_j(\mathbf{u}_j)\mathbf{n}_j &= \mathbf{h}_j \quad \text{on } \Gamma_j^N, \\ \mathbf{u}_1 - \mathbf{u}_2 &= 0 \quad \text{on } \Gamma, \\ \sigma_1(\mathbf{u}_1)\mathbf{n}_1 + \sigma_2(\mathbf{u}_2)\mathbf{n}_2 &= 0 \quad \text{on } \Gamma. \end{aligned} \tag{2.1}$$

Here, $\sigma_j(\mathbf{u}_j)$ is the linearized stress tensor with the components

$$\sigma_{j,kl}(\mathbf{u}_j) = \lambda_j \delta_{kl} \mathrm{div}\mathbf{u}_j + \mu_j \left(\frac{\partial \mathbf{u}_{j,x_k}}{\partial x_l} + \frac{\partial \mathbf{u}_{j,x_l}}{\partial x_k} \right),$$

and Lamé constants λ_j, μ_j. Furthermore, $(\mathbf{u}_{j,x_1}, \mathbf{u}_{j,x_2})$ are the Cartesian components of $\mathbf{u}_j$ and $\mathbf{n}_j$ is the exterior normal vector on $\partial\Omega_j$. We suppose that the crack is stress free, i.e. homogeneous Neumann data are prescribed on the crack. In the following we assume for simplicity that $\lambda_j + \mu_j > 0$ and $\mu_j > 0$, $j = 1, 2$. These conditions guarantee that the transmission problem (2.1) is elliptic [17].

The behaviour of the two–dimensional displacement fields $(\mathbf{u}_1, \mathbf{u}_2)$ in a neighbourhood of a corner point $P \in S \cap \overline{\Gamma}$ can be described by an asymptotic expansion, provided the right hand sides of (2.1) are sufficiently smooth [13]

$$\begin{pmatrix} \mathbf{u}_1 \\ \mathbf{u}_2 \end{pmatrix} (r,\omega) = \sum_{s=1}^{N} \sum_{k=1}^{m_g(\alpha_s)} \sum_{l=0}^{\kappa_{k,s}-1} c_{s,k,l} \begin{pmatrix} \mathbf{v}_{1,s}^{l,k} \\ \mathbf{v}_{2,s}^{l,k} \end{pmatrix} (r,\omega) + O(r), \tag{2.2}$$

with singular functions

$$\begin{pmatrix} \mathbf{v}_{1,s}^{l,k} \\ \mathbf{v}_{2,s}^{l,k} \end{pmatrix} (r,\omega) = r^{\alpha_s} \sum_{q=0}^{l} \frac{1}{q!} (\log r)^q \begin{pmatrix} \varphi_{1,s}^{l-q,k} \\ \varphi_{2,s}^{l-q,k} \end{pmatrix} (\omega), \tag{2.3}$$

where (r,ω) are polar coordinates with origin in P. The functions $(\mathbf{v}_{1,s}^{l,k}, \mathbf{v}_{2,s}^{l,k})$ are solutions of the problem (2.1) with vanishing right–hand sides in the double–wedge $C_P = \{(r,\omega) : 0 < r < \infty, \omega \in (\omega_0, \omega_1) \cup (\omega_1, \omega_2)\}$, which is locally diffeomorph to $\Omega = \Omega_1 \cup \Omega_2$ in a neighbourhood of P. Writing the boundary transmission problem in polar coordinates (r,ω) and applying the Mellin transform with respect to r we obtain a generalized eigenvalue problem $\mathcal{A}_P(\alpha)u = 0$, whose eigensolutions coincide with the functions $(\varphi_{1,s}^{0,k}, \varphi_{2,s}^{0,k})$. The complex exponents α_s are eigenvalues of the operator pencil $\mathcal{A}_P$ and the functions $\{(\varphi_{1,s}^{l,k}, \varphi_{2,s}^{l,k}) : 1 \le k \le m_g(\alpha_s), 0 \le l \le \kappa_{k,s} - 1\}$ form a system of Jordan–chains of $\mathcal{A}_P$. Furthermore, N is the number of eigenvalues in the strip $0 < \mathrm{Re}\alpha < 1$, $m_g(\alpha_s)$ denotes the number of Jordan–chains to the eigenvalue α_s and $\kappa_{k,s}$ is the length of the k–th Jordan–chain. In the neighbourhood of a corner point $P \notin S \cap \overline{\Gamma}$ with $P \in \partial\Omega_j$ the field $\mathbf{u}_j$ has an asymptotics analogously to (2.2).

In case of a crack tip $T_p, p = 1, 2$, the eigenvalues of $\mathcal{A}_{T_p}$ with the smallest positive real part are $1/2 \pm i\epsilon$ with

$$\epsilon = \frac{1}{2\pi} \log \frac{\mu_1 + \kappa_1 \mu_2}{\mu_2 + \kappa_2 \mu_1},$$

where $\kappa = 3 - 4\nu$ for the plane–strain problem, $\kappa = (3 - \nu)/(1 + \nu)$ for the plane–stress problem and $\nu = \lambda/(2\lambda + 2\mu)$ is the Poisson's ratio. The corresponding singular functions (2.3) and coeffcients $c_{s,k,l}$ are complex valued but conjugate to each other. Therefore it is possible to replace the singular functions by two real valued functions with real valued coeffcients. They can be given explicitly in terms of Kolosov/Muskhelishvili potentials [16, 5]. It is known that the Cartesian components u_{x_1}, u_{x_2} of a displacement field can be expressed by two complex potentials $\Phi(z)$ and $\Psi(z)$

$$2\mu(u_{x_1} + iu_{x_2}) \quad = \quad \kappa\Phi(z) - (z - \overline{z})\overline{\Phi'(z)} - \overline{\Psi(z)}.$$

The displacement fields $\mathbf{u}_1, \mathbf{u}_2$ have in the neighbourhood of an interface crack tip $T_p, p = 1, 2$ the asymptotics

$$\begin{pmatrix} \mathbf{u}_1 \\ \mathbf{u}_2 \end{pmatrix} = \sum_{k=1}^{2} c_{k,p} \begin{pmatrix} \mathbf{v}_{1,k}^p \\ \mathbf{v}_{2,k}^p \end{pmatrix} + O(|\mathbf{x} - T_p|), \tag{2.4}$$

where $\mathbf{x} = (x_1, x_2)$ and $\mathbf{v}_{1,1}^p, \mathbf{v}_{1,2}^p$ are generated by the potentials

$$\Phi_{1k}^p(z) = d_k e^{-\pi\epsilon}(z - T_p)^{1/2 - i\epsilon}, \quad \Psi_{1k}^p(z) = \overline{d_k} e^{\pi\epsilon}(z - T_p)^{1/2 + i\epsilon},$$

and the real functions $\mathbf{v}_{2,1}^p, \mathbf{v}_{2,2}^p$ are generated by the potentials

$$\Phi_{2k}^p(z) = d_k e^{\pi\epsilon}(z - T_p)^{1/2 - i\epsilon}, \quad \Psi_{2k}^p(z) = \overline{d_k} e^{-\pi\epsilon}(z - T_p)^{1/2 + i\epsilon}.$$

Here is

$$d_k = \delta_k \frac{1}{\sqrt{8\pi}\cosh(\pi\epsilon)(1/2 - i\epsilon)}, \quad \delta_1 = 1, \delta_2 = i.$$

The coefficients $c_{k,p}$ in (2.4) are given according to [5, 2] by

$$c_{k,p} = \sum_{j=1}^{2} \left\{ \int_{\Omega_j} \mathbf{f}_j \cdot \mathbf{w}_{j,k}^p d\mathbf{x} + \int_{\partial\Omega_j} \left(\sigma_j(\mathbf{u}_j)\mathbf{n}_j \cdot \mathbf{w}_{j,k}^p - \mathbf{u}_j \cdot \sigma_j(\mathbf{w}_{j,k}^p)\mathbf{n}_j \right) ds_{\mathbf{x}} \right\},$$

where $\mathbf{w}_{1,k}^p, \mathbf{w}_{2,k}^p$ are generated by the potentials

$$\tilde{\Phi}_{1k}^p(z) = b_k e^{-\pi\epsilon}(z - T_p)^{-1/2 - i\epsilon}, \quad \tilde{\Psi}_{1k}^p(z) = \overline{b_k} e^{\pi\epsilon}(z - T_p)^{-1/2 + i\epsilon},$$

and

$$\tilde{\Phi}_{2k}^p(z) = b_k e^{\pi\epsilon}(z - T_p)^{-1/2 - i\epsilon}, \quad \tilde{\Psi}_{2k}^p(z) = \overline{b_k} e^{-\pi\epsilon}(z - T_p)^{-1/2 + i\epsilon},$$

respectively. Here is

$$b_k = \beta_k \frac{\cosh(\pi\epsilon)\sqrt{8}\mu_1\mu_2}{(\mu_1 + \mu_2 + \kappa_1\mu_2 + \kappa_2\mu_1)\sqrt{\pi}}, \quad \beta_1 = 1, \ \beta_2 = -i.$$

The normalization factors b_k are required, because the singular functions, which satisfy the homogeneous problem (2.1) in an infinite sector, are defined up to a constant factor.

The coefficients $c_{k,p}$ can also be expressed as (see e.g. [2])

$$c_{k,p} = \sum_{j=1}^{2} \left\{ \int_{\Omega_j} \mathbf{f}_j \cdot \mathbf{W}_{j,k}^p d\mathbf{x} + \int_{\Gamma_j^N} \mathbf{h}_j \cdot \mathbf{W}_{j,k}^p ds_{\mathbf{x}} \right\}, \tag{2.5}$$

where $\mathbf{W}^p_{1,k}, \mathbf{W}^p_{2,k}$ satisfy the homogeneous problem (2.1) and admit the decomposition

$$\mathbf{W}^p_{j,k} = \mathbf{w}^p_{j,k} + \mathbf{w}^{p,reg}_{j,k} \tag{2.6}$$

with $\mathbf{w}^{p,reg}_{j,k} \in \mathbf{H}^1(\Omega_j)$.

2.2 Shape sensitivity

In order to describe the shape sensitivity of the displacement fields $(\mathbf{u}_1, \mathbf{u}_2)$ under a small perturbation of the crack tips we consider the boundary transmission problem

$$\begin{aligned} L_j \mathbf{u}_{j,\varepsilon} &= \mathbf{f}_j && \text{in } \Omega_j \\ \mathbf{u}_{j,\varepsilon} &= 0 && \text{on } \Gamma^D_j, \\ \sigma_j(\mathbf{u}_{j,\varepsilon})\mathbf{n}_{j,\varepsilon} &= \mathbf{h}_{j,\varepsilon} && \text{on } \Gamma^N_{j,\varepsilon}, \\ \mathbf{u}_{1,\varepsilon} - \mathbf{u}_{2,\varepsilon} &= 0 && \text{on } \Gamma_\varepsilon, \\ \sigma_1(\mathbf{u}_{1,\varepsilon})\mathbf{n}_{1,\varepsilon} + \sigma_2(\mathbf{u}_{2,\varepsilon})\mathbf{n}_{2,\varepsilon} &= 0 && \text{on } \Gamma_\varepsilon. \end{aligned} \tag{2.7}$$

Here the perturbed configuration $(\Omega_j, \Gamma^D_j, \Gamma^N_{j,\varepsilon}, \Gamma_\varepsilon)$ is the image of a fixed reference configuration $(\Omega_j, \Gamma^D_j, \Gamma^N_j, \Gamma)$ under the C^∞–diffeomorphism $\Theta_\varepsilon = I + \varepsilon\Theta$ with Θ given by

$$\Theta = \sum_{p=1}^{2} \eta_p \begin{pmatrix} h_p \\ 0 \end{pmatrix}$$

with $\varepsilon \in [0, \varepsilon_0]$. Here η_p is a cut–off function with support in the neighbourhood of T_p and h_p, $p = 1, 2$, are two given real numbers which describe the directions of the perturbation of both crack tips. If only one crack tip is moved then we can simply choose $h_1 = 1, h_2 = 0$.
Thus we have

$$\Gamma_\varepsilon = \Theta_\varepsilon(\Gamma), \quad \Gamma^N_{j,\varepsilon} = \Theta_\varepsilon(\Gamma^N_j).$$

Note that this perturbation moves small neighbourhoods of the crack tips in the direction of the x_1–axis and lets the remaining parts of the domains fixed.

The sensitivity of the displacement fields $(\mathbf{u}_{1,\varepsilon}, \mathbf{u}_{2,\varepsilon})$ can be described with the help of its material derivative

$$\dot{\mathbf{u}}_j = \frac{d}{d\varepsilon}(\mathbf{u}_{j,\varepsilon} \circ \Theta_\varepsilon)\bigg|_{\varepsilon=0} = \lim_{\varepsilon \to 0} \frac{\mathbf{u}_{j,\varepsilon} \circ \Theta_\varepsilon - \mathbf{u}_{j,0}}{\varepsilon} \tag{2.8}$$

or its shape derivative

$$\mathbf{u}'_j = \dot{\mathbf{u}}_j - \nabla \mathbf{u}_{j,0} \cdot \Theta. \tag{2.9}$$

Our goal is to investigate the sensitivity of functionals associated with the elastic fields $\mathbf{u}_{j,\varepsilon}$ and $\sigma_j(\mathbf{u}_{j,\varepsilon})$ of the form

$$J(\mathbf{u}_{1,\varepsilon}, \mathbf{u}_{2,\varepsilon}) = \sum_{j=1}^{2} \int_{\Omega_j} F_j(\mathbf{u}_{j,\varepsilon}, \sigma_j(\mathbf{u}_{j,\varepsilon}))d\mathbf{x}_\varepsilon. \tag{2.10}$$

The functions F_j satisfy for a positive constant c the growth conditions

$$F_j(p,q) \leq a(p)(c+|q|^2),\ \partial_p F_j(p,q) \leq a(p)(c+|q|^2),\ \partial_q F_j(p,q) \leq a(p)(c+|q|) \tag{2.11}$$

for some $a \in C(\mathbb{R}^2)$ and all $p \in \mathbb{R}^2$, $q \in \mathbb{R}^4$. This guarantees that the functional (2.10) is well defined for all displacements $\mathbf{u}_{j,\varepsilon} \in [W^1_{2+\delta}(\Omega_j)]^2$ and all stresses $\sigma_j(\mathbf{u}_{j,\varepsilon}) \in [L_{2+\delta}(\Omega_j)]^4$ with a small $\delta > 0$ [1, Lemma 9.5]. Note that the growth conditions (2.11) allow at most quadratic dependence on the stress tensor σ_j, an assumption which is satisfied by the energy integrals corresponding to all differential operators of second order.

Our goal is to derive formulae for the sensitivity of the functional J with respect to the perturbation mapping Θ_ε, i.e. we want to calculate the shape derivative

$$dJ(\mathbf{u}_{1,0}, \mathbf{u}_{2,0}, \Theta) = \lim_{\varepsilon \to 0} \frac{J(\mathbf{u}_{1,\varepsilon}, \mathbf{u}_{2,\varepsilon}) - J(\mathbf{u}_{1,0}, \mathbf{u}_{2,0})}{\varepsilon}$$

and to express $dJ(\mathbf{u}_{1,0}, \mathbf{u}_{2,0}, \Theta)$ as an integral over $\partial\Omega_1 \cup \partial\Omega_2$.

Some common functionals, which satisfy (2.11), are the potential energy

$$J(\mathbf{u}_{1,\varepsilon}, \mathbf{u}_{2,\varepsilon}) = \sum_{j=1}^{2} \frac{1}{2} \int_{\Omega_j} \sigma_j(\mathbf{u}_{j,\varepsilon}) : e(\mathbf{u}_{j,\varepsilon}) d\mathbf{x}_\varepsilon,$$

where $e(\mathbf{u}) = \frac{1}{2}(\nabla\mathbf{u} + \nabla\mathbf{u}^\top)$ is the strain tensor, and the functional

$$J(\mathbf{u}_{1,\varepsilon}, \mathbf{u}_{2,\varepsilon}) = \frac{1}{2} \sum_{j=1}^{2} \int_{\Omega_j} (\mathbf{u}_{j,\varepsilon} - \tilde{\mathbf{u}}_j)^2 d\mathbf{x},$$

with given $\tilde{\mathbf{u}}_j \in \mathbf{L}_2(\Omega_j)$, which appears often as an constraint in shape optimization.

3 EXISTENCE AND REGULARITY RESULTS FOR THE MATERIAL AND THE SHAPE DERIVATIVE

The existence and the regularity of the material derivative $\dot{\mathbf{u}}_j$ defined by (2.8) was investigated in [4] in weighted Sobolev spaces which describe not only the regularity of functions in the interior of the domain but also their behaviour near singular points of the boundary.

For $P \in S$ let C_P be the infinite cone which coincides with Ω in a neighbourhood of P.

DEFINITION 1 For $d \in \mathbb{N}_0$ and $\beta \in \mathbb{R}$ we define the weighted Sobolev space $V^d_\beta(C_P)$ as the closure of $C_0^\infty(\overline{C}_P \setminus P)$ with respect to the norm

$$||u||_{V^d_\beta(C_P)} = \sum_{|\alpha| \leq d} ||r_P^{\beta-d+|\alpha|} D^\alpha u||_{L_2(C_P)}.$$

The space $V_\beta^{d-1/2}(\partial C_P)$ consists for $d \in \mathbb{N}$ of traces on ∂C_P of functions in $V_\beta^d(C_P)$ equipped with the norm

$$\|u\|_{V_\beta^{d-1/2}(\partial C_P)} = \inf \|v\|_{V_{\beta 2}^d(C_P)},$$

where the infimum is taken over the set of all functions $v \in V_\beta^d(C_P)$ such that $v = u$ on ∂C_P.

DEFINITION 2 Let $S = \{P_1, \ldots, P_Q\}$ be the set of singular boundary points. For $d \in \mathbb{N}_0, \vec{\beta} = (\beta_1, \ldots, \beta_Q) \in \mathbb{R}^Q$ we define the space $V_{\vec{\beta}}^d(\Omega)$ as the closure of $C_0^\infty(\overline{\Omega} \setminus S)$ with respect to the norm which is assembled by means of a partition of unity from the local norms $V_{\beta_q}^d(C_{P_q})$. Let $\zeta_q \in C_0^\infty(\mathbb{R}^2)$ with $0 \leq \zeta_q \leq 1$ be such that $\zeta_q = 1$ near the conical point $P_q, q = 1, \ldots, Q$, and $\zeta_q = 0$ near $P_l \in S$ with $l \neq q$. We set $\zeta_0 := 1 - \zeta_1 - \cdots - \zeta_Q$. Then the norm in $V_{\vec{\beta}}^d(\Omega)$ is defined by

$$\|u\|_{V_{\vec{\beta}}^d(\Omega)} = \left(\|\zeta_0 u\|_{W_2^d(\Omega)}^p + \sum_{q=1}^{Q} \|\zeta_q u\|_{V_{\beta_q}^d(C_{P_q})}^2 \right)^{1/2}.$$

For $d \in \mathbb{N}$ we denote by $V_{\vec{\beta}}^{d-1/2}(\partial\Omega)$ the space of traces on $\partial\Omega \setminus S$ of functions in $V_{\vec{\beta}}^d(\Omega)$.

For the sake of shortness we write $\mathbf{V}_{\vec{\beta}}^d(\Omega)$ and $\mathbf{V}_{\vec{\beta}}^{d-1/2}(\partial\Omega)$ instead of $[V_{\vec{\beta}}^d(\Omega)]^k$ and $[V_{\vec{\beta}}^{d-1/2}(\partial\Omega)]^k$.

We denote by $\Sigma(\mathcal{A}_P)$ the spectrum of the operator pencil $\mathcal{A}_P$ corresponding to the singular point $P \in S$ and define

$$a_0(P) = \min\{\Re\alpha : \alpha \in \Sigma(\mathcal{A}_P), \Re\alpha > 0\}.$$

Let us suppose that the transformed boundary forces $\mathbf{h}_{j,\varepsilon} \circ \Theta_\varepsilon$ of (2.7) admit a Taylor expansion with respect to ε

$$\mathbf{h}_{j,\varepsilon} \circ \Theta_\varepsilon = \mathbf{h}_j + \varepsilon \dot{\mathbf{h}}_j + \varepsilon^2 \mathbf{h}_{j,R}(\varepsilon). \tag{3.12}$$

THEOREM 1 [4] Let $\vec{\beta}, \vec{\gamma} \in \mathbb{R}^Q$ satisfy

$$\begin{aligned} \beta_q &:= \tfrac{3}{2} + \delta,\ \gamma_q := \tfrac{3}{2} - \delta \quad \text{for} \quad P_q \in \{T_1, T_2\}, \\ \beta_q &:= \gamma_q = 2 - a_0(P_q) + \delta \quad \text{for} \quad P_q \in S \setminus \{T_1, T_2\} \end{aligned}$$

with an arbitrarily small $\delta > 0$. We suppose that $\mathbf{f}_j \in V_{\vec{\gamma}}^1(\Omega_j)$, $\mathbf{h}_j, \dot{\mathbf{h}}_j, \mathbf{h}_{j,R} \in V_{\vec{\gamma}}^{3/2}(\Gamma_j^N)$. Then we have $\mathbf{u}_{j,0} \in \mathbf{V}_{\vec{\beta}}^3(\Omega_j)$, $\dot{\mathbf{u}}_j \in \mathbf{V}_{\vec{\beta}}^3(\Omega_j)$, $\mathbf{u}_j' \in \mathbf{V}_{\vec{\beta}}^2(\Omega_j)$ and the asymptotics (2.4) is valid for $\mathbf{u}_{j,0}$. Furthermore the following estimate, which guarantees the existence of the material derivative $\dot{\mathbf{u}}_j$, holds with a positive real constant c

$$\sum_{j=1}^{2} \|\mathbf{u}_{j,\varepsilon} \circ \Theta_\varepsilon - \mathbf{u}_{j,0} - \varepsilon \dot{\mathbf{u}}_j\|_{\mathbf{V}_{\vec{\beta}}^3(\Omega_j)} \leq c\varepsilon^2.$$

REMARK Theorem 1 was proved in [4] for simplicity under the additional assumption, that the weights β_q, γ_q are the same for all singular points $P_q \in S$. This restriction is not necessary and the proof from [4] is valid also in our case. Furthermore, the first condition for $\beta_q, P_q \in \{T_1, T_2\}$, implies that $\dot{\mathbf{u}}_j$ has an asymptotics with singular terms of the order $O(r^{1/2})$, whereas the second condition for $\beta_q, P_q \in S \setminus \{T_1, T_2\}$, says that there is no asymptotic expansion in general.

4 SENSITIVITY OF FUNCTIONALS

In this and the following sections we write shortly $\sigma_{j,0}$ for $\sigma_j(\mathbf{u}_{j,0})$ and denote by $\dot{\sigma}_j, \sigma_j'$ the material and the shape derivative of $\sigma_j(\mathbf{u}_{j,\varepsilon})$, respectively.

Let us assume that the scalar functions F_j in (2.10) are twice continuously differentiable with respect to all its arguments and satisfy the growth conditions (2.11). Standard formulae for the sensitivity of integrals over varying domains (see e.g. [18, 19]) yield taking into account that $\langle \Theta, \mathbf{n}_{j,0} \rangle = 0$ on $\partial\Omega_j$

$$\begin{aligned} dJ(\mathbf{u}_{1,0}, \mathbf{u}_{2,0}, \Theta) &= \sum_{j=1}^{2} \int_{\Omega_j} \{ \langle \partial_{\mathbf{u}} F_j(\mathbf{u}_{j,0}, \sigma_{j,0}), \dot{\mathbf{u}}_j \rangle + \partial_\sigma F_j(\mathbf{u}_{j,0}, \sigma_{j,0}) : \dot{\sigma}_j \\ &+ F_j(\mathbf{u}_{j,0}, \sigma_{j,0}) \mathrm{div}\Theta \} \, d\mathbf{x}. \end{aligned} \tag{4.13}$$

and

$$dJ(\mathbf{u}_{1,0}, \mathbf{u}_{2,0}, \Theta) = \sum_{j=1}^{2} \int_{\Omega_j} \{ \langle \partial_{\mathbf{u}} F_j(\mathbf{u}_{j,0}, \sigma_0), \mathbf{u}_j' \rangle + \partial_\sigma F_j(\mathbf{u}_{j,0}, \sigma_0) : \sigma_j' \} \, d\mathbf{x}. \tag{4.14}$$

Theorem 1 guarantees, that all integrals in (4.13) and (4.14) exist. Following some ideas from [6] we replace $\partial_\sigma F_j(\mathbf{u}_{j,0}, \sigma_{j,0}) : \sigma_j'$ in (4.14) by $\partial_\sigma F_j(\mathbf{u}_{j,0}, \sigma_{j,0}) : (C_j : e(\mathbf{u}_j'))$ and use Green's first formula applied to a field $\mathbf{z}_j$ with $\sigma_j(\mathbf{z}_j) = C_j : \partial_\sigma F_j(\mathbf{u}_{j,0}, \sigma_{j,0})$ and to the field $\mathbf{u}_j'$

$$\begin{aligned} \int_{\Omega_j} (C_j : \partial_\sigma F_j(\mathbf{u}_{j,0}, \sigma_{j,0})) : e(\mathbf{u}_j') d\mathbf{x} &= \int_{\partial\Omega_j} \langle \mathbf{u}_j', (C_j : \partial_\sigma F_j(\mathbf{u}_{j,0}, \sigma_{j,0})) \mathbf{n}_{j,0} \rangle ds_{\mathbf{x}} \\ &- \int_{\Omega_j} \langle \mathbf{u}_j', \mathrm{div}(C_j : \partial_\sigma F_j(\mathbf{u}_{j,0}, \sigma_{j,0})) \rangle d\mathbf{x}. \end{aligned}$$

Here, C_j is Hooke's tensor with the components $c_{j,stkl} = \lambda_j \delta_{st} \delta_{kl} + \mu_j (\delta_{sk} \delta_{tl} + \delta_{sl} \delta_{tk})$ and $e(\mathbf{u}_j')$ is the linearized strain tensor.
In this way we obtain

$$dJ(\mathbf{u}_{1,0}, \mathbf{u}_{2,0}, \Theta) = \sum_{j=1}^{2} \Bigg\{ \int_{\Omega_j} \langle \partial_{\mathbf{u}} F_j(\mathbf{u}_{j,0}, \sigma_{j,0}) - \mathrm{div}(C_j : \partial_\sigma F_j(\mathbf{u}_{j,0}, \sigma_{j,0})), \mathbf{u}_j' \rangle d\mathbf{x}$$

$$+ \int_{\partial\Omega_j} \langle (C_j : \partial_\sigma F_j(\mathbf{u}_{j,0}, \sigma_{j,0}))\mathbf{n}_{j,0}, \mathbf{u}_j' \rangle \Bigg\} ds_{\mathbf{x}}. \tag{4.15}$$

We introduce adjoint displacement fields $\mathbf{z}_1, \mathbf{z}_2$ in Ω_1, Ω_2 as solutions of the boundary transmission problem

$$\begin{array}{rcll} -\mathrm{div}\sigma_j(\mathbf{z}_j) & = & \partial_{\mathbf{u}} F_j(\mathbf{u}_{j,0}, \sigma_{j,0}) & \\ & & -\mathrm{div}(C_j : \partial_\sigma F_j(\mathbf{u}_{j,0}, \sigma_{j,0})) & \text{in} \quad \Omega_j, \\ \mathbf{z}_j & = & 0 & \text{on} \quad \Gamma_j^D, \\ \sigma_j(\mathbf{z}_j)\mathbf{n}_{j,0} & = & (C_j : \partial_\sigma F_j(\mathbf{u}_{j,0}, \sigma_{j,0}))\mathbf{n}_{j,0} & \text{on} \quad \Gamma_j^N, \\ \mathbf{z}_1 - \mathbf{z}_2 & = & 0 & \text{on} \quad \Gamma, \\ \sigma_1(\mathbf{z}_1)\mathbf{n}_1 + \sigma_2(\mathbf{z}_2)\mathbf{n}_2 & = & 0 & \text{on} \quad \Gamma. \end{array} \tag{4.16}$$

Using the adjoint fields $\mathbf{z}_1, \mathbf{z}_2$ we can write (4.15) simply as

$$dJ(\mathbf{u}_{1,0}, \mathbf{u}_{2,0}, \Theta) = \sum_{j=1}^{2} \left\{ -\int_{\Omega_j} \langle \mathrm{div}\sigma_j(\mathbf{z}_j), \mathbf{u}_j' \rangle d\mathbf{x} + \int_{\Gamma_j^N} \langle \sigma_j(\mathbf{z}_j)\mathbf{n}_{j,0}, \mathbf{u}_j' \rangle \right\} ds_{\mathbf{x}}, \tag{4.17}$$

since the shape derivative $(\mathbf{u}_1', \mathbf{u}_2')$ satisfies the homogeneous problem (2.1) [3, 4].

LEMMA 1 Let the assumptions of Theorem 1 be satisfied. Then the boundary transmission problem (4.16) has a unique weak solution $(\mathbf{z}_1, \mathbf{z}_2)$ which belongs to $\mathbf{V}_{\vec{\beta}}^3(\Omega_1) \times \mathbf{V}_{\vec{\beta}}^3(\Omega_2)$ and admits in the neighbourhood of the crack tip T_p, $p = 1, 2$, an asymptotics analogous to (2.4).

Proof: Under the assumptions of Theorem 1 we have $\mathbf{u}_{j,0} \in \mathbf{V}_{\vec{\beta}}^3(\Omega_j), \sigma_{j,0} \in [V_{\vec{\beta}}^2(\Omega_j)]^4$. Therefore $\partial_{\mathbf{u}} F_j(\mathbf{u}_{j,0}, \sigma_{j,0}), \mathrm{div}(C_j : \partial_\sigma F_j(\mathbf{u}_{j,0}, \sigma_{j,0})) \in \mathbf{V}_{\vec{\beta}}^1(\Omega_j)$ and $(C_j : \partial_\sigma F_j(\mathbf{u}_{j,0}, \sigma_{j,0}))\mathbf{n}_{j,0} \in \mathbf{V}_{\vec{\beta}}^{3/2}(\Gamma_j^N)$ because of the growth conditions (2.11). Thus we can apply the regularity results for boundary transmission problems from [13] and obtain the assertion.

LEMMA 2 The shape derivative can be decomposed as

$$\mathbf{u}_j' = \gamma \sum_{p=1}^{2} h_p \sum_{k=1}^{2} c_{k,p}(\mathbf{u}_{1,0}, \mathbf{u}_{2,0}) \mathbf{W}_{j,k}^p$$

with

$$\gamma = \left(\frac{\kappa_1 + 1}{\mu_1} + \frac{\kappa_2 + 1}{\mu_2} \right) \frac{1}{8\cosh^2(\pi\epsilon)}. \tag{4.18}$$

Proof: Similiar as in [19, 3] for the case of boundary value problems, it can be shown that the shape derivative $(\mathbf{u}_1', \mathbf{u}_2')$ satisfies the transmission problem

$$
\begin{array}{rcll}
-\mathrm{div}\sigma_i(\mathbf{u}_j') & = & 0 & \text{in } \Omega_j, \\
\mathbf{u}_j' & = & -\langle\Theta, \mathbf{n}_{j,0}\rangle\partial_n \mathbf{u}_{j,0} & \text{on } \Gamma_j^D, \\
\sigma_j(\mathbf{u}_j')\mathbf{n}_{j,0} & = & \langle\Theta, \mathbf{n}_{j,0}\rangle(\mathbf{f}_{j,0} + \kappa_j \mathbf{h}_{j,0}) & \\
 & & -\mathrm{div}_\Gamma(\langle\Theta, \mathbf{n}_{j,0}\rangle\sigma_{j,0}^T) & \text{on } \Gamma_j^N, \\
\mathbf{u}_1' - \mathbf{u}_2' & = & -\langle\Theta, \mathbf{n}_{1,0}\rangle(\partial_n \mathbf{u}_{1,0} - \partial_n \mathbf{u}_{2,0}) & \text{on } \Gamma, \\
\sigma_1(\mathbf{u}_1')\mathbf{n}_{1,0} + \sigma_2(\mathbf{u}_2')\mathbf{n}_{2,0} & = & \langle\Theta, \mathbf{n}_{1,0}\rangle(\mathbf{f}_{1,0} - \mathbf{f}_{2,0}) & \\
 & & -\mathrm{div}_\Gamma(\langle\Theta, \mathbf{n}_{1,0}\rangle(\sigma_{1,0}^T - \sigma_{2,0}^T)) & \text{on } \Gamma.
\end{array} \tag{4.19}
$$

Here κ_j denotes the curvature of $\partial\Omega_j$,

$$\sigma_{j,0}^T := \sigma_j(\mathbf{u}_{j,0})\mathbf{n}_{j,0} - \langle\sigma_j(\mathbf{u}_{j,0})\mathbf{n}_{j,0}, \mathbf{n}_{j,0}\rangle\mathbf{n}_{j,0}$$

is the tangential component of the stress tensor and div_Γ is the tangential divergence operator. In our case we have $\langle\Theta, \mathbf{n}_{j,0}\rangle = 0$ on $\partial\Omega_j$, and so the shape derivative $(\mathbf{u}_1', \mathbf{u}_2')$ satisfies the homogeneous problem (4.19). Since $\mathbf{u}_j'$ belongs to $V_\beta^2(\Omega_j)$ and coincides with $\dot{\mathbf{u}}_j \in H^1(\Omega_j)$ outside of a small neighbourhood of the crack tips, it follows from results about the kernel of general elliptic boundary value problems in $V_\beta^2(\Omega_j)$ [11, 12] that $\mathbf{u}_j'$ is a linear combination of the weight functions $\mathbf{W}_{j,k}^p$ (see (2.6))

$$\mathbf{u}_j' = \sum_{p,k=1}^{2} \gamma_{k,p}\mathbf{W}_{j,k}^p$$

with coefficients $\gamma_{k,p}$ to be determined. They can be calculated with the help of the identity $\mathbf{u}_j' = \dot{\mathbf{u}}_j - \nabla\mathbf{u}_{j,0}\cdot\Theta$. Since $\dot{\mathbf{u}}_j \in \mathbf{H}^1(\Omega_j)$ and $\mathbf{u}_j', \nabla\mathbf{u}_{j,0} \in L_2(\Omega_j)$, so the main parts of the asymptotics of $\mathbf{u}_j'$ and $-\nabla\mathbf{u}_{j,0}\cdot\Theta$ must coincide. Noting that $\nabla\mathbf{u}_{j,0}\cdot\Theta = h_p\partial_{x_1}\mathbf{u}_{j,0}$ near the crack tip T_p and that [5]

$$\partial_{x_1}\mathbf{v}_{j,k}^p = -\gamma\mathbf{w}_{j,k}^p$$

with γ defined by (4.18) we conclude that

$$\nabla\mathbf{u}_{j,0}\cdot\Theta = -\gamma h_p \sum_{k=1}^{2} c_{k,p}(\mathbf{u}_{1,0}, \mathbf{u}_{2,0})\mathbf{w}_{j,k}^p + O(|\mathbf{x} - T_p|)$$

near the crack tip T_p. Thus the assertion follows.

THEOREM 2 Let the assumptions of Theorem 1 be satisfied. Then we have

$$dJ(\mathbf{u}_{1,0}, \mathbf{u}_{2,0}, \Theta) = \gamma\sum_{p=1}^{2} h_p \sum_{k=1}^{2} c_{k,p}(\mathbf{u}_{1,0}, \mathbf{u}_{2,0})c_{k,p}(\mathbf{z}_{1,k}, \mathbf{z}_{2,k}).$$

Proof: Since $\mathbf{u}_j'$ is a linear combination of the weight functions $\mathbf{W}_{j,1}^p, \mathbf{W}_{j,2}^p, p = 1, 2$, we can interpret the representation (4.17) as a special case of formula (2.5) and obtain the assertion.

EXAMPLE: In case of the energy functional

$$J(\mathbf{u}_{1,\varepsilon}, \mathbf{u}_{2,\varepsilon}) = \sum_{j=1}^{2} \frac{1}{2} \int_{\Omega_j} \sigma_j(\mathbf{u}_{j,\varepsilon}) : e(\mathbf{u}_{j,\varepsilon}) d\mathbf{x}_{\varepsilon},$$

the adjoint fields $\mathbf{z}_1, \mathbf{z}_2$ coincide with the displacement fields $\mathbf{u}_{1,0}, \mathbf{u}_{2,0}$. In this way we obtain a generalization of the well known Irwin formula [15, 7, 12] to the case of interface cracks [5]

$$dJ(\mathbf{u}_{1,0}, \mathbf{u}_{2,0}, \Theta) = \gamma \sum_{p=1}^{2} h_p \sum_{k=1}^{2} c_{k,p}(\mathbf{u}_{1,0}, \mathbf{u}_{2,0})^2.$$

This formula can be used to realize for interface cracks the Griffith criterion, which states that the crack begins to grow, when the energy release rate $\mathcal{G}$ exceeds a critical value $\mathcal{G}_{critical}$

$$\mathcal{G} := -dJ(\mathbf{u}_{1,0}, \mathbf{u}_{2,0}, \Theta) \geq \mathcal{G}_{critical}.$$

References

[1] J Appell, PP Zabrejko. Nonlinear Superposition Operators. Cambridge University Press, Cambridge, 1990.

[2] M Bochniak, A–M Sändig. Computation of generalized stress intensity factors for bonded elastic materials. Math. Model. Numer. Anal. 33:853–878, 1999.

[3] M Bochniak, A–M Sändig. Sensitivity analysis of elastic structures in presence of stress singularities. Arch. Mech. 51:155-171, 1999.

[4] M Bochniak, A–M Sändig. Sensitivity analysis of stress intensity factors in 2d coupled elastic structures. Preprint 99/06, SFB 404, Universität Stuttgart, 1999.

[5] Y–Z Chen, N Hasebe. Eigenfunction expansion and higher order weight functions of interface cracks. J. Appl. Mech. 61:843–849, 1994.

[6] K Dems, Z Mróz. Shape sensitivity in mixed Dirichlet–Neumann boundary–value problems and associated class of path–independent integrals. Eur. J. Mech. A/Solids 14:169–203, 1995.

[7] P Destuynder, M Djaoua. Sur une interprétation mathématique de l'integrale de Rice en théorie de la rupture fragile. Math. Meth. Appl. Sci. 3:70–87, 1981.

[8] G Fremiot, J Sokolowski. The structure theorem for the Eulerian derivative of shape functionals defined in domains with cracks. Preprint, Université Henri Poincaré Nancy, 1999.

[9] EJ Haug, KK Choi, V Komkov. Design Sensitivity Analysis of Structural Systems. Academic Press Inc., Orlando, 1986.

[10] VA Kondrat'ev. Boundary problems for elliptic equations in domains with conical or angular points. Trans. Moscow Math. Soc. 16:209–292, 1967.

[11] VG Maz'ya, BA Plamenevsky. On the coefficients in the asymptotics of solutions of elliptic boundary value problems in domains with conical points. Math. Nachr. 76:29–60, 1977 = Amer. Math. Soc. Transl. (2) 123:57–88, 1984.

[12] SA Nazarov, BA Plamenevsky. Elliptic Problems in Domains with Piecewise Smooth Boundaries. Walter de Gruyter, Berlin, 1994.

[13] S Nicaise, A–M Sändig. General interface problems I, II. Math. Meth. Appl. Sci. 17:395–429, 431–450, 1994.

[14] S Nicaise, A–M Sändig. Transmission problems for the Laplace and Elasticity operators: Regularity and boundary integral formulation. Math. Models and Methods in Appl. Sci. 9:855–898, 1999.

[15] JR Rice. Mathematical analysis in the mechanics of fracture. In: H. Liebowitz, ed. Fracture Vol. II. Academic Press, 1968, pp. 191–311.

[16] JR Rice. Elastic fracture mechanics concepts for interfacial cracks. J. Appl. Mech. 55:98–103, 1988.

[17] A–M Sändig. The Shapiro–Lopatinskij condition for boundary value and transmission problems for the Lamé system. Preprint 98/14, SFB 404, Univ. Stuttgart, 1998.

[18] J Simon. Differentiation with respect to the domain in boundary value problems. Numer. Funct. Anal. Optimiz. 2:649–687, 1980.

[19] J Sokolowski, JP Zolesio. Introduction to Shape Optimization. Springer Verlag, Berlin, 1992.

On the asymptotic expansion of the solution of a Dirichlet-Ventcel problem with a small parameter

M. BOURLARD, A. MAGHNOUJI, S. NICAISE and L. PAQUET, Université de Valenciennes et du Hainaut Cambrésis, ISTV, MACS, Le Mont Houy, F-59313 Valenciennes Cedex 9, France

1 Introduction

Let Ω denote a bounded open set of $\mathbf{R}^2$ with a polygonal boundary $\Gamma = \bigcup_{j=1}^{N} \overline{\Gamma_j}$, where Γ_j is an open straight line segment, and $\omega_j (\neq 0, 2\pi)$ is the interior angle at $S_j = \overline{\Gamma_j} \cap \overline{\Gamma_{j+1}}$, $\forall j \in \mathbf{Z}/N\mathbf{Z}$. For the sake of simplicity, we impose the Ventcel condition on one straight line segment Γ_2, denoted by Γ_V and we impose Dirichlet boundary condition on $\Gamma_D = \Gamma \setminus \overline{\Gamma_V}$ (our results clearly extend to any partition of Γ into Γ_V and Γ_D). The Dirichlet-Ventcel problem is

$$(P_\epsilon) \begin{cases} -\Delta u_\epsilon = f & \text{in } \Omega, \\ \gamma u_\epsilon = 0 & \text{on } \Gamma_D, \\ B_\epsilon u_\epsilon = g & \text{on } \Gamma_V, \end{cases}$$

where $B_\epsilon = \frac{\partial}{\partial \nu} - \epsilon \frac{\partial^2}{\partial \tau^2}$, $\epsilon > 0$ is a small parameter, $\frac{\partial}{\partial \nu}$ is the outer normal derivative, and $\frac{\partial}{\partial \tau}$ is the tangential derivative.

The boundary condition on Γ_V (called Ventcel's boundary condition [9]) corresponds to the heat exchange between the body Ω and a thin layer of thickness δ with a thermal conductivity ε/δ. When ε tends to zero, the layer becomes more and more isolating.

The specificity of problems (P_ε) is that they change of type as $\varepsilon \to 0$, degenerating to a mixed Dirichlet-Neumann problem

$$(P_0)\begin{cases} -\Delta u^0 = f & \text{in } \Omega, \\ \gamma u^0 = 0 & \text{on } \Gamma_D, \\ \frac{\partial u^0}{\partial \nu} = g & \text{on } \Gamma_V. \end{cases}$$

This is a singular perturbed problem [10]. Here the main difficulty relies on the singular behaviour at the corners where the boundary conditions change. Indeed if $j = 1$ or 2, the first singularity near S_j of problem (P_ε) behaves like $r_j^{\frac{\pi}{\omega_j}}$ [9] (r_j being the distance to S_j), while for the limit problem (P_0) it behaves like $r_j^{\frac{\pi}{2\omega_j}}$ [7, 4, 3]. The abrupt change of the singular behaviour at the limit is an old question and was recently analysed in [15, 13, 14, 6, 2] for some singular perturbed problems using multiscale asymptotics techniques. Our aim is then to extend the above techniques to our problem following [2].

As in the above works the lack of regularity of the solution of problem (P_0) does not allow an asymptotic expansion of u_ε in the form

$$u_\varepsilon \simeq u^0 + \varepsilon u^1 + \varepsilon^2 u^2 + \cdots,$$

each u^n, with $n \geq 1$, being solution of (P_0) with $f = 0$ and $g = \frac{\partial^2 u^{n-1}}{\partial \tau^2}$. We then need to correct this expansion in an appropriate way. As in [2], we do not use the general strategy of matching asymptotic expansions (see for instance [5]) but we construct alternatively the outer and inner terms. Namely with the help of the singular exponents of problem (P_0) associated with S_j, for $j = 1$ and 2, we define a strictly increasing sequence $(\delta_n)_{n\in\mathbf{N}}$ of nonnegative real numbers containing the nonnegative integers. For a fixed $n \in \mathbf{N}$ and for sufficiently smooth data f and g, we show that the full expansion of u_ε is

$$u_\varepsilon = \sum_{n=0}^{N} \left\{ \varepsilon^{\delta_n} u^{\delta_n} + w^n(\varepsilon) \right\} + \tilde{r}_\varepsilon^N,$$

where $\tilde{r}_\varepsilon^N$ is a remainder satisfying

$$\|\tilde{r}_\varepsilon^N\|_{H^1(\Omega)} = o(\varepsilon^{\delta_N}),$$

the outer term u^{δ_n} is solution of problem (P_0) with data which are obtained iteratively and $w^n(\varepsilon)$ are boundary layer terms which decay (as $r^{-\eta}$ with $\eta > 0$) at infinity.

2 Preliminary results

We first recall some Sobolev spaces on the edge Γ_V of Ω: let $\overset{\circ}{H}{}^1(\Gamma_V)$ denote the closure of $\mathcal{D}(\Gamma_V)$ in $H^1(\Gamma_V)$, and let $\tilde{H}^s(\Gamma_V)$, $0 \leq s \leq 1$, be the real interpolation space between $\overset{\circ}{H}{}^1(\Gamma_V)$ and $L^2(\Gamma_V)$ of order $1-s$. Let $H^{-s}(\Gamma_V)$, $0 < s \leq 1$, denote the dual space of $\tilde{H}^s(\Gamma_V)$.

The variational formulation of problem (P_ϵ) is:

$$(1) \quad \int_\Omega \nabla u_\epsilon \nabla v \, dx + \epsilon \int_{\Gamma_V} \frac{\partial u_\epsilon}{\partial \tau} \frac{\partial v}{\partial \tau} \, d\sigma = \int_\Omega f \, v \, dx + \langle g; \gamma v \rangle, \forall v \in V,$$

where γ is the trace operator with respect to Γ and by definition,

$$V = \{u \in H^1(\Omega), \gamma u = 0 \text{ on } \Gamma_D \text{ and } \gamma u \in \overset{\circ}{H}{}^1(\Gamma_V)\}.$$

Note that the existence and uniqueness of the solution $u_\epsilon \in V$ of (1) directly follows from the Lax-Milgram lemma if $f \in L^2(\Omega)$ and $g \in H^{-1}(\Gamma_V)$. Similarly, if $\epsilon = 0$, problem (1) has a unique solution $u^0 \in V_0$ (called the weak solution of (P_0)) if moreover $g \in H^{-\frac{1}{2}}(\Gamma_V)$, where

$$V_0 = \{u \in H^1(\Omega), \gamma u = 0 \text{ on } \Gamma_D\}.$$

It is well known [7, 4, 3] that the weak solution u^0 of (P_0) has a singular behaviour near the corners S_j that we shall now describe. Since later on we will only need this behaviour near S_j with $j = 1$ and 2, we restrict ourselve to this case (see [7, 4, 3] for the behaviour near the other vertices). For $j \in \{1, 2\}$, the set Λ_j of singular exponents at S_j is defined by

$$\Lambda_j = \{\frac{\pi}{2\omega_j} + \frac{k\pi}{\omega_j}, k \in \mathbf{Z}\}.$$

REMARK 2.1 For simplicity, we assume from now on that the domain fulfils a supplementary hypothesis, namely $\Lambda_j \cap \mathbf{Z} = \emptyset$, $\forall j \in \{1, 2\}$.

PROPOSITION 2.2 *Let $M \in \mathbf{N} \setminus \{0\}$ and let the data satisfy $f \in H^{M-1}(\Omega)$ and $g \in H^{M-\frac{1}{2}}(\Gamma_V)$. Then the weak solution $u^0 \in V_0$ of the mixed Dirichlet-Neumann problem (P_0) admits the decomposition*

$$(2) \qquad u^0 = u^0_{reg(M)} + \sum_{j=1,2} \sum_{\alpha \in \Lambda_j \cap]0,M[} c_{j,\alpha}(u^0) \chi_j \, S_{j,\alpha},$$

where the regular part $u^0_{reg(M)}$ belongs to $H^{1+M}(\Omega\backslash\mathcal{V})$, $\mathcal{V} = \cup_{k=3}^N B(S_k, \eta)$, for any $\eta > 0$ small enough such that $S_j \notin \mathcal{V}$, for $j = 1$ and 2; the coefficients $c_{j,\alpha}$ are real constants; the singular functions are, for $\alpha \in \Lambda_j$,

$$S_{j,\alpha}(r_j, \theta_j) = r_j^\alpha \psi_{j,\alpha}(\theta_j)\,, \tag{3}$$

(r_j, θ_j) being the polar coordinates with center S_j, the functions $\psi_{j,\alpha}$ are smooth functions on $[0, \omega_j]$, and finally χ_j denotes, here and below, a radial cut-off function equal to 1 in a neighbourhood of S_j and equal to 0 near the other vertices.

We now state a useful estimate whose proof is based on a usual trace theorem and the Poincaré-Friedrichs inequality (see [1] for the details).

PROPOSITION 2.3 *Assume that $f \in L^2(\Omega)$ and $g \in H^{-\frac{1}{2}}(\Gamma_V)$. Then the weak solution $u_\epsilon \in V$ of problem (P_ϵ) fulfils the estimate:*

$$||u_\epsilon||_{H^1(\Omega)} \leq C\left\{||f||_{L^2(\Omega)} + ||g||_{H^{-\frac{1}{2}}(\Gamma_V)}\right\}, \tag{4}$$

where C denotes, here and in the following, a positive constant independent of ε.

We finish this section by giving the singular behaviour of a sequence of solutions u^n of mixed Dirichlet-Neumann problems defined iteratively as follows: Starting from u^0 solution of (P_0), we define $u^n \in V_0$, for $n \geq 1$, as the unique solution of

$$\begin{cases} -\Delta u^n = f^n & \text{in } \Omega, \\ u^n = 0 & \text{on } \Gamma_D, \\ \frac{\partial}{\partial \nu} u^n = \frac{\partial^2}{\partial \tau^2} u^{n-1}_{reg(1)} + g^n & \text{on } \Gamma_V, \end{cases} \tag{5}$$

where $f^n \in L^2(\Omega)$ and $g^n \in H^{\frac{1}{2}}(\Gamma_V)$.

An induction argument based on Proposition 2.2 leads to the

PROPOSITION 2.4 *Let $n \in \mathbf{N} \setminus \{0\}$ and let the data in (5) satisfy $f^k \in H^n(\Omega)$ and $g^k \in H^{n+\frac{1}{2}}(\Gamma_V)$, for all $k = 0, \cdots, n$ (by definition we take $f^0 = f$ and $g^0 = g$). Then the unique solution $u^n \in V_0$ of problem (5) admits the following decomposition:*

$$u^n = u^n_{reg(1)} + \sum_{j=1,2} \sum_{\alpha \in \Lambda_j \cap]0, n+1[} c_{j,\alpha} \chi_j\, S^{(E(\alpha))}_{j,\alpha}, \tag{6}$$

where the coefficients $c_{j,\alpha}$ are real constants and $S^{(E(\alpha))}_{j,\alpha}$ are defined by (9) hereafter ($E(\alpha)$ being the integer part of α).

3 Singularities at infinity

To obtain the asymptotic development of u_ϵ, one of the main steps is to analyse the so-called exterior problem associated with problem (P_ϵ). Namely if $j = 1$ or 2, performing the change of variables

$$(r_j, \theta_j) \to (\frac{r_j}{\epsilon}, \theta_j),$$

the problem (P_ϵ) is transformed into

$$(7) \qquad \begin{cases} -\Delta w = f_\varepsilon & \text{in } C, \\ \gamma w = 0 & \text{on } \Gamma_-, \\ Bw = g_\epsilon & \text{on } \Gamma_+, \end{cases}$$

where $B = B_1$, C is the cone with vertex S_j that coincides locally with Ω, Γ_+ (respectively Γ_-) denotes the half-line that coincides with Γ_V (respectively Γ_D) in a neighbourhood of S_j. Here and below, we omit the index j for simplicity.

As usual [15, 2], the corner layer terms will be obtained with the help of some solutions K of the homogeneous problem (7):

$$(8) \qquad \begin{cases} \Delta K = 0 & \text{in } C, \\ \gamma K = 0 & \text{on } \Gamma_-, \\ BK = 0 & \text{on } \Gamma_+. \end{cases}$$

More precisely, we will associate to each positive $\alpha \in \Lambda$ a solution $K(\alpha)$ of (8) which behaves like S_α at infinity. The complete asymptotic behaviour at infinity of such a $K(\alpha)$, namely its decomposition into explicit terms and a remainder which decays (as $r^{-\eta_\alpha}$, for some $\eta_\alpha > 0$) at infinity is the main goal of this section.

For all $\alpha \in \Lambda$, let $S_\alpha^{(0)}$ denote the singular function S_α and $S_\alpha^{(l)}$, $l \in \mathbf{N}\backslash\{0\}$, be a particular solution of

$$(9) \qquad \begin{cases} \Delta S_\alpha^{(l)} = 0 & \text{in } C, \\ \gamma S_\alpha^{(l)} = 0 & \text{on } \Gamma_-, \\ \frac{\partial S_\alpha^{(l)}}{\partial \nu} = \frac{\partial^2}{\partial \tau^2} S_\alpha^{(l-1)} & \text{on } \Gamma_+. \end{cases}$$

By Theorem 4.22 of [12] and by iteration, we easily deduce that for all $\alpha \in \Lambda$ and $l \in \mathbf{N}$, a solution $S_\alpha^{(l)}$ of (9) exists in the form

$$(10) \qquad S_\alpha^{(l)} = r^{\alpha - l} \sum_{q=0}^{l} (\log r)^q \psi_{\alpha,q}^{(l)}(\theta),$$

for some smooth functions $\psi_{\alpha,q}^{(l)}$ such that

$$r^{\alpha-l}\psi_{\alpha,l}^{(l)}(\theta) = \begin{cases} c_\alpha^{(l)} S_{\alpha-l}^{(0)}(r,\theta) & \text{if } \alpha - l \in \Lambda, \\ 0 & \text{if } \alpha - l \notin \Lambda, \end{cases}$$

for some constant $c_\alpha^{(l)}$. From now on, we take these functions as particular solutions of (9).

Let us recall that the weighted Sobolev space of Kondratiev's type $H_\mu^s(C)$, for $s \in \mathbf{N}$ and $\mu \in \mathbf{R}$, is defined by

$$H_\mu^s(C) = \{u | r^{\mu+|\beta|-s} D^\beta u \in L^2(C), \forall |\beta| \leq s\}.$$

For $\alpha \in \Lambda$, $\alpha > 0$, we define the space

$$(11)\quad \mathcal{K}_\alpha = \left\{ \begin{array}{l} K \in H^1_{loc}(C) \text{ such that } \gamma K \in H^1_{loc}(\Gamma_+), \\ K \text{ is solution of problem (8)}, \\ \exists s \geq \alpha + 5/2, \exists \xi \in]0,1[,]\alpha, \alpha+\xi[\cap \Lambda = \emptyset, \\ \forall \mu \in \mathbf{R} : s - \mu - 1 \in]\alpha, \alpha+\xi[, (1-\chi)K \in H_\mu^s(C) \end{array} \right\}.$$

By the Sobolev embedding theorem and Theorem AA.3 of [3] (see Corollary 3.7), for any $K \in \mathcal{K}_\alpha$ and $\mu \in \mathbf{R} : s - \mu - 1 \in]\alpha, \alpha + \xi[$, there exist $C > 0$ and $R > 0$ such that (compare with [2])

$$|K(r,\theta)| \leq C r^{s-\mu-1}, \forall r \geq R.$$

PROPOSITION 3.1 *Let $K \in \mathcal{K}_\alpha$ and $P \in \mathbf{N}$. There exists $R_P > 0$ such that K admits the following expansion*

$$(12)\qquad K(r,\theta) = (K^{(0)} + ... + K^{(P)} + v_P)(r,\theta), \forall r \geq R_P,$$

where for all $l = 1, ..., P$, $K^{(l)}$ is solution of

$$\begin{cases} \Delta K^{(l)} = 0 & \text{in } C, \\ \gamma K^{(l)} = 0 & \text{on } \Gamma_-, \\ \frac{\partial K^{(l)}}{\partial \nu} = \frac{\partial^2 K^{(l-1)}}{\partial \tau^2} & \text{on } \Gamma_+, \end{cases}$$

$K^{(0)}$ is solution of the same problem with a zero right-hand side, v_P belongs to $H_\mu^{s-P-1}(C)$, with s and μ as in (11) such that $s - l - \mu \notin \Lambda$, $\forall l = 1, ..., P+1$, and $s - \mu - 1$ close enough to α so that $]s - \mu - 2 - l, s - \mu - 1 - l[\cap \Lambda =]\alpha - l - 1, \alpha - l] \cap \Lambda$, $\forall l = 0, ..., P$. Moreover

$$(13)\qquad K^{(l)} = \sum_{k=0}^{l} \sum_{\beta \in \Lambda \cap]\alpha-k-1, \alpha-k]} c_\beta(K) S_\beta^{(l-k)}.$$

For the proof, we refer to [1], where we apply the results of [7] in an induction argument on P.

The analysis of the space $\mathcal{K}_\alpha$ requires the variational formulation of the nonhomogeneous problem:

$$\text{(14)} \qquad \begin{cases} \Delta \tilde{K} = g_1 & \text{in } C, \\ \gamma \tilde{K} = 0 & \text{on } \Gamma_-, \\ B\tilde{K} = g_2 & \text{on } \Gamma_+. \end{cases}$$

To this end we introduce the variational space

$$\begin{aligned} \mathcal{V}(C) &= \Big\{ u | (1+r^2)^{-1/2} (\log(2+r))^{-1} u \in L^2(C), \nabla u \in (L^2(C))^2, \\ &\qquad \gamma u = 0 \text{ on } \Gamma_-, (1+r^2)^{-1/2} \gamma u \in L^2(\Gamma_+) \text{ and } \frac{\partial(\gamma u)}{\partial \tau} \in L^2(\Gamma_+) \Big\}. \end{aligned}$$

By the Lax-Milgram lemma, we prove the

PROPOSITION 3.2 *If g_1 belongs to $\mathcal{C}^\infty(C)$, g_1 is equal to zero for r sufficiently small and for r sufficiently large, and if there exists $\epsilon > 0$ such that $(1+r)^{1/2+\epsilon} g_2 \in L^2(\Gamma_+)$, then problem (14) admits a unique weak solution $\tilde{K} \in \mathcal{V}(C)$ whose variational formulation is*

$$\int_C \nabla \tilde{K} \, \nabla v \, dx + \int_{\Gamma_+} \frac{\partial \tilde{K}}{\partial \tau} \frac{\partial v}{\partial \tau} \, d\sigma = \int_{\Gamma_+} g_2 \gamma v \, d\sigma - \int_C g_1 v \, dx, \forall v \in \mathcal{V}(C).$$

For every $\beta \in \Lambda$, $\beta > 0$, we define

$$\text{(15)} \qquad \tilde{K}^\beta = \sum_{m=0}^{E(\beta)} S_\beta^{(m)}.$$

LEMMA 3.3 *Let $\alpha \in \Lambda$, $\alpha > 0$. Every $K \in \mathcal{K}_\alpha$ admits the expansion*

$$\text{(16)} \qquad K = \tilde{K}_\chi + (1-\chi) \sum_{\beta \in \Lambda \cap]0,\alpha]} c_\beta(K) \tilde{K}^\beta,$$

where $\tilde{K}_\chi \in \mathcal{V}(C)$ is the unique solution of (14) with

$$\begin{aligned} g_1 &= -\Delta \left[(1-\chi) \sum_{\beta \in \Lambda \cap]0,\alpha]} c_\beta(K) \tilde{K}^\beta \right], \\ g_2 &= -B \left[(1-\chi) \sum_{\beta \in \Lambda \cap]0,\alpha]} c_\beta(K) \tilde{K}^\beta \right]. \end{aligned}$$

Proof: Let $P = E(\alpha)$. Owing to (13), $K^{(P)} = K_1^{(P)} + K_2^{(P)}$, where

$$K_1^{(P)} = \sum_{k=0}^{P} \sum_{\beta \in \Lambda \cap]P-k, \alpha-k]} c_\beta(K) S_\beta^{(P-k)}. \tag{17}$$

We deduce from Proposition 3.1 that

$$K - (1-\chi)\left(K^{(0)} + \dots + K^{(P-1)} + K_1^{(P)}\right) = \chi K + (1-\chi) v_P + (1-\chi) K_2^{(P)}.$$

Let $\tilde{K}_\chi$ denote the left-hand side of the previous identity. We show that $\tilde{K}_\chi \in \mathcal{V}(C)$ by proving that each term has this regularity. First the fact that $K \in \mathcal{K}_\alpha$ directly implies that $\chi K \in \mathcal{V}(C)$. Since $s \geq P + 5/2$, the regularity $v_P \in H_\mu^{s-P-1}(C)$ implies that $(1-\chi) v_P \in \mathcal{V}(C)$. By (10), $K_2^{(P)}$ can be written as

$$K_2^{(P)} = \sum_{k=0}^{P} \sum_{\beta \in \Lambda, \alpha - P - 1 < \beta - P + k \leq 0} d_{\beta,k}(K) r^{\beta - P + k} \sum_{q=0}^{Q(\beta,k)} (\log r)^q \psi_{q,\beta,k}(\theta).$$

Since $\Lambda \cap \mathbf{Z} = \emptyset$ (see Remark 2.1), we always have $\beta - P + k < 0$ in the sum of the previous identity. This implies $(1-\chi) K_2^{(P)} \in \mathcal{V}(C)$.

Thanks to (13) and (17), we have the identity

$$K^{(0)} + \dots + K^{(P-1)} + K_1^{(P)} = \sum_{\beta \in \Lambda \cap]0, \alpha]} c_\beta(K) \tilde{K}^\beta.$$

As K is solution of the homogeneous problem (8), $\tilde{K}_\chi$ is solution of (14) with g_1 anf g_2 as defined in the Lemma. Its uniqueness follows from Proposition 3.2 since g_1 anf g_2 have the requested regularity. ∎

THEOREM 3.4 *Let $\alpha \in \Lambda$, $\alpha > 0$. Then the dimension of the space $\mathcal{K}_\alpha$ is given by*

$$dim\ \mathcal{K}_\alpha = \ card\ \{\beta \in \Lambda, 0 < \beta \leq \alpha\}. \tag{18}$$

Proof: By Lemma 3.3 every $K \in \mathcal{K}_\alpha$ admits the expansion (16), where the number of the coefficients $c_\beta(K)$ (and thus the degree of freedom) is exactly card $\{\beta \in \Lambda, 0 < \beta \leq \alpha\}$. In other words, we have dim $\mathcal{K}_\alpha \leq$ card $\{\beta \in \Lambda, 0 < \beta \leq \alpha\}$.

Let us establish the inverse inequality. Let c_β, for $\beta \in \Lambda \cap]0, \alpha]$, denote some fixed coefficients. We are looking for $\tilde{K} \in \mathcal{V}(C)$ solution of

$$\begin{cases} \Delta \tilde{K} = -\Delta \left[(1-\chi) \sum_{\beta \in \Lambda \cap]0,\alpha]} c_\beta \tilde{K}^\beta\right] & \text{in } C, \\ \gamma \tilde{K} = 0 & \text{on } \Gamma_-, \\ B\tilde{K} = -B\left[(1-\chi) \sum_{\beta \in \Lambda \cap]0,\alpha]} c_\beta \tilde{K}^\beta\right] & \text{on } \Gamma_+. \end{cases}$$

Thanks to Proposition 3.2, this problem admits a unique solution $\tilde{K} \in \mathcal{V}(C)$. We conclude by checking that

$$T = \tilde{K} + (1-\chi) \sum_{\beta \in \Lambda \cap]0,\alpha]} c_\beta \tilde{K}^\beta \tag{19}$$

belongs to $\mathcal{K}_\alpha$. ∎

Remark 3.5 *As the functions $(1-\chi)\tilde{K}^\beta$, $\beta \in \Lambda \cap]0,\alpha]$, are linearly independent modulo $\mathcal{V}(C)$, the expansion (19) implies that the set $\{(1-\chi)\tilde{K}^\beta, \beta \in \Lambda \cap]0,\alpha]\}$ constitutes a basis of $\mathcal{K}_\alpha$ modulo $\mathcal{V}(C)$.*

Corollary 3.6 *Let $\alpha \in \Lambda$, $\alpha > 0$. There exists a unique $K(\alpha) \in \mathcal{K}_\alpha$ such that*

$$K(\alpha) = S_\alpha^{(0)} + S_\alpha^{(1)} + \ldots + S_\alpha^{(E(\alpha)-1)} + S_\alpha^{(E(\alpha))} + Y_\alpha, \tag{20}$$

where $Y_\alpha + \chi\tilde{K}^\alpha$ belongs to $\mathcal{V}(C)$.

Moreover, for all $P \in \mathbf{N}$, $P > E(\alpha)$, Y_α admits the expansion

$$Y_\alpha = \sum_{l=E(\alpha)+1}^{P} S_\alpha^{(l)} + \sum_{k=0}^{P-E(\alpha)} \sum_{\beta \in \Lambda \cap]\alpha-P+k-1,0[} c_\beta(K(\alpha)) S_\beta^{(k)} + v_P, \tag{21}$$

where $(1-\chi)v_P \in H_\mu^{s-P-1}(C)$.

Proof: For the first assertion, we argue as in the proof of Theorem 3.4, but we choose some particular coefficients c_β: namely $c_\beta = 0$, for all $\beta \in \Lambda \cap]0,\alpha[$, and $c_\alpha = 1$. Then (19) becomes $T = \tilde{K} + (1-\chi)\tilde{K}^\alpha$. The claim follows then from (15) by defining $Y_\alpha = \tilde{K} - \chi\tilde{K}^\alpha$.

Now let us prove (21). Owing to Remark 3.5, for any $K \in \mathcal{K}_\alpha$ and $\beta \in \Lambda \cap]0,\alpha]$, the coefficient $c_\beta(K)$ is unique. Considering the choice that we made hereover, it is clear that for $\beta \in \Lambda \cap]0,\alpha]$, $c_\beta(K(\alpha)) = \delta_\beta^\alpha$. The identity (13) implies that $K^{(l)} = S_\alpha^{(l)}$ for $l = 0, \ldots, E(\alpha)-1$, and for $l \geq E(\alpha)$ we have

$$K^{(l)} = S_\alpha^{(l)} + \sum_{k=E(\alpha)}^{l} \sum_{\beta \in \Lambda \cap]\alpha-k-1,\alpha-k],\beta<0} c_\beta(K(\alpha)) S_\beta^{(l-k)}. \tag{22}$$

Replacing $K^{(l)}$ by this expression in (12), we obtain (21). ∎

COROLLARY 3.7 *Let $\alpha \in \Lambda$, $\alpha > 0$, and let $K(\alpha) \in \mathcal{K}_\alpha$ and Y_α be defined as in Corollary 3.6. Then there exist positive constants C_α, R_α and η_α such that*

$$|Y_\alpha(r,\theta)| \leq C_\alpha r^{-\eta_\alpha}, \forall r \geq R_\alpha. \tag{23}$$

Proof: Similarly to (21), we prove that there exists $R_\alpha > 0$ such that

$$Y_\alpha(r,\theta) = \left(\sum_{\beta \in \Lambda \cap]\alpha - E(\alpha) - 1, 0[} c_\beta(K(\alpha)) S_\beta + v_{E(\alpha)} \right)(r,\theta), \forall r \geq R_\alpha, \tag{24}$$

where $v_{E(\alpha)} \in H^{s-E(\alpha)-1}_\mu(C)$. This regularity of $v_{E(\alpha)}$ and Theorem AA.3 of [3] imply, setting $r = e^t$,

$$e^{\eta t}(1-\chi)(e^t) v_{E(\alpha)}(e^t, \theta) \in H^{s-E(\alpha)-1}(\mathbf{R} \times]0,\omega[)$$

with $\eta = \mu - s + E(\alpha) + 2$. Since $s - E(\alpha) - 2 > 0$, by the Sobolev embedding theorem, there exists $C_\alpha > 0$ such that

$$\left| e^{\eta t}(1-\chi)(e^t) v_{E(\alpha)}(e^t,\theta) \right| \leq C_\alpha, \forall t \in \mathbf{R}, \forall \theta \in]0,\omega[. \tag{25}$$

The estimate (23) now follows from (24), (25) and the condition $\mu > -\alpha - \xi + s - 1$ in the definition (11) of $\mathcal{K}_\alpha$. ■

4 The corner layer terms

The aim of this section is to give the asymptotic expansion of the problem solved by the corner layer terms.

Let us first start with some notation used very often hereafter. By $p[\ln \varepsilon]$, we mean a polynomial in $\ln \varepsilon$ with coefficients in $C^\infty(\bar{\Omega})$ or in $C^\infty(\bar{\Gamma}_V)$ according to the context (excluding any other dependence on ε); moreover if some indices and/or exponents are added to p, they will specify the dependance of p with respect to these quantities and not its degree. For any $\varepsilon > 0$ and all function v defined in polar coordinates (r,θ), we denote by v_ε the function

$$v_\varepsilon(r,\theta) = v(\frac{r}{\varepsilon},\theta).$$

Let us further introduce the (discrete) set

$$\Delta = \{l_1 \frac{\pi}{\omega_1} + l_2 \frac{\pi}{\omega_2} + l_3 \geq 0, l_1, l_2 \in \mathbf{N}, l_3 \in \mathbf{Z}\},$$

and define iteratively the sequence $(\delta_n)_{n\in\mathbf{N}}$ of elements of Δ as follows: $\delta_0 = 0$ and

$$\Delta_n = \{\delta = l_1\frac{\pi}{\omega_1} + l_2\frac{\pi}{\omega_2} + l_3 \in \Delta, l_3 \geq -2E(\delta_n) \text{ and } \delta > \delta_n\},$$
$$\delta_{n+1} = \min\{\delta, \delta \in \Delta_n\}, \forall n \in \mathbf{N}.$$

Note that the sequence $(\delta_n)_{n\in\mathbf{N}}$ satisfies $\delta_n \to +\infty$, as $n \to +\infty$, and that the sets Δ_n are almost increasing with respect to n, since we have

$$\Delta_n \subset \Delta_{n+1} \cup \{\delta_{n+1}\}, \forall n \in \mathbf{N}.$$

These two properties imply that the sequence contains any element of Δ_m, in other words for any $m \in \mathbf{N}$ and any $\delta \in \Delta_m$ there exists $n \in \mathbf{N}$ such that $\delta = \delta_n$.

For all $n \in \mathbf{N}$, the corner layer term $w^n(\varepsilon)$ associated with the vertex S is defined by

$$w^n(\varepsilon) = \sum_{\alpha\in\Lambda\cap]0,E(\delta_n)+1[} d_\alpha^n(\varepsilon)\chi\varepsilon^\alpha Y_{\alpha,\varepsilon}, \tag{26}$$

where the coefficients $d_\alpha^n(\varepsilon)$ will be precised in the next section but have the general form

$$d_\alpha^n(\varepsilon) = \varepsilon^{\delta_n - E(\alpha)} p_\alpha^n(\ln\varepsilon), \tag{27}$$

where p_α^n are polynomials.

In view of Corollary 3.6, we first start with the asymptotic behaviour of $S_{\alpha,\varepsilon}^{(l)}$ with respect to ε.

Lemma 4.1 *For all $\alpha \in \Lambda$ and all $l \in \mathbf{N}$, the unique solution $S_\alpha^{(l)}$ of (9) satisfies*

$$S_{\alpha,\varepsilon}^{(l)} = \varepsilon^{l-\alpha} \sum_{\substack{j\in\{0,\cdots,l\}\\ \alpha-l+j\in\Lambda}} p_{l-j}(\ln\varepsilon) S_{\alpha-l+j}^{(j)}, \tag{28}$$

for all $\varepsilon > 0$, where p_j is a polynomial of degree j satisfying $p_0(0) = 1$.

Proof: The assertion is proved by induction on l using Theorem 4.22 of [12]. ∎

We now look at the problem solved by $\chi\varepsilon^\alpha Y_{\alpha,\varepsilon}$.

LEMMA 4.2 *Fix $\alpha \in \Lambda, \alpha > 0$ and $\varepsilon > 0$ and let us set $w = \chi\varepsilon^\alpha Y_{\alpha,\varepsilon}$. Then for all $P \in \mathbf{N}$, w is solution of the following problem*

$$(29)\begin{cases} -\Delta w = \sum\limits_{l=E(\alpha)+1}^{P} \varepsilon^l F_\alpha^l[\ln\varepsilon] \\ + \sum\limits_{l=0}^{P-E(\alpha)} \sum\limits_{\beta\in\Lambda\cap]\alpha-P+l-1,0[} \varepsilon^{l+\alpha-\beta} F_{\alpha,\beta}^l[\ln\varepsilon] + F_\alpha^{(P)}(\varepsilon) & \text{in } \Omega, \\ \gamma w = 0 & \text{on } \Gamma_D, \\ B_\varepsilon w = \varepsilon^{1+\alpha}\chi\dfrac{\partial^2}{\partial\tau^2}S_{\alpha,\varepsilon}^{(E(\alpha))} + \sum\limits_{l=E(\alpha)+1}^{P} \varepsilon^{1+l} g_\alpha^l[\ln\varepsilon] \\ + \sum\limits_{l=0}^{P-E(\alpha)} \sum\limits_{\beta\in\Lambda\cap]\alpha-P+l-1,0[} \varepsilon^{1+l+\alpha-\beta} g_{\alpha,\beta}^l[\ln\varepsilon] + g_\alpha^{(P)}(\varepsilon) & \text{on } \Gamma_V, \end{cases}$$

where the remainders $F_\alpha^{(P)}(\varepsilon)$ and $g_\alpha^{(P)}(\varepsilon)$ satisfy

$$(30)\qquad \varepsilon||F_\alpha^{(P)}(\varepsilon)||_{L^2(\Omega)} + ||g_\alpha^{(P)}(\varepsilon)||_{H^{-\frac{1}{2}}(\Gamma_V)} = o(\varepsilon^{1+P}).$$

Proof: Assume first that $P > E(\alpha)$, owing to (21) and (28), we get

$$Y_{\alpha,\varepsilon} = \sum_{l=E(\alpha)+1}^{P} \varepsilon^{l-\alpha} p_\alpha^l[\ln\varepsilon] + \sum_{l=0}^{P-E(\alpha)} \sum_{\beta\in\Lambda\cap]\alpha-P+l-1,0[} \varepsilon^{l-\beta} p_{\alpha,\beta}^l[\ln\varepsilon] + v_{P,\varepsilon}.$$

Thanks to (20) and since $K(\alpha)$ is solution of problem (8), we have

$$(31)\qquad B_\varepsilon(\chi Y_{\alpha,\varepsilon}) = \chi\varepsilon\frac{\partial^2}{\partial\tau^2}S_{\alpha,\varepsilon}^{(E(\alpha))} - \varepsilon\left[\frac{\partial^2}{\partial\tau^2},\chi\right]Y_{\alpha,\varepsilon}.$$

The two above identities imply the third equation of (29) with $g_\alpha^{(P)}(\varepsilon) = -\varepsilon^{1+\alpha}\left[\frac{\partial^2}{\partial\tau^2},\chi\right]v_{P,\varepsilon}$.

Similarly as $\Delta w = \varepsilon^\alpha[\Delta,\chi]Y_{\alpha,\varepsilon}$, the above expression of $Y_{\alpha,\varepsilon}$ yields the first equation of problem (29) with $F_\alpha^{(P)}(\varepsilon) = -\varepsilon^\alpha[\Delta,\chi]v_{P,\varepsilon}$.

The estimate (30) is proved by the change of variable $s = \varepsilon r$ and a trace theorem.

Finally for $P \le E(\alpha)$, we apply the Lemma with $P = E(\alpha)+1$ and bring all the extra terms in the remainder. ■

At this stage, we can motivate the introduction of the sequence $(\delta_n)_{n\in\mathbf{N}}$. Indeed any power of ε in the right-hand side of (29), except the term $\varepsilon^{1+\alpha}\chi\frac{\partial^2}{\partial\tau^2}S_{\alpha,\varepsilon}^{(E(\alpha))}$, is an element of this sequence. The extension to some negative integer l_3 will be clarified in the next Corollary.

COROLLARY 4.3 *Assume that $w^n(\varepsilon)$ is defined by (26) with the coefficients given by (27). Then for all $m \in \mathbf{N}$, $m > n$, $w^n(\varepsilon)$ is solution of the following problem*

$$(32)\quad \begin{cases} \Delta w^n(\varepsilon) = \sum_{l=n+1}^{m} \varepsilon^{\delta_l} F_l^n[\ln\varepsilon] + F_n^{(m)}(\varepsilon) & \text{in } \Omega, \\ \gamma w^n(\varepsilon) = 0 & \text{on } \Gamma_D, \\ B_\varepsilon w^n(\varepsilon) = \sum_{\alpha\in\Lambda\cap]0,E(\delta_n)+1[} d_\alpha^n(\varepsilon)\varepsilon^{1+\alpha}\chi\frac{\partial^2}{\partial\tau^2}S_{\alpha,\varepsilon}^{(E(\alpha))} & \\ \quad - \sum_{l=n+1}^{m} \varepsilon^{\delta_l} g_l^n[\ln\varepsilon] - g_n^{(m)}(\varepsilon) & \text{on } \Gamma_V, \end{cases}$$

where the remainders $F_n^{(m)}(\varepsilon)$ and $g_n^{(m)}(\varepsilon)$ satisfy

$$(33)\qquad ||F_n^{(m)}(\varepsilon)||_{L^2(\Omega)} + ||g_n^{(m)}(\varepsilon)||_{H^{-\frac{1}{2}}(\Gamma_V)} = o(\varepsilon^{\delta_m}).$$

Proof: From Lemma 4.2 with $P \geq \delta_m$ and from the expressions of $w^n(\varepsilon)$ and its coefficients, we have

$$-\Delta w^n(\varepsilon) = \sum_{\alpha\in\Lambda\cap]0,E(\delta_n)+1[} \varepsilon^{\delta_n - E(\alpha)} p_\alpha^n(\ln\varepsilon)\Big\{ \sum_{l=E(\alpha)+1}^{P} \varepsilon^l F_\alpha^l[\ln\varepsilon]$$
$$+ \sum_{l=0}^{P-E(\alpha)} \sum_{\beta\in\Lambda\cap]\alpha-P+l-1,0[} \varepsilon^{l+\alpha-\beta} F_{\alpha,\beta}^l[\ln\varepsilon]\Big\} + F_n^{(m)}(\varepsilon) \text{ in } \Omega,$$

where $F_n^{(m)}(\varepsilon)$ satisfies (33) because $\delta_n - E(\alpha) \geq 0$ when $\alpha \in \Lambda\cap]0, E(\delta_n)+1[$. The two first terms of the right-hand side of the above expression may be written

$$\sum_{\alpha\in\Lambda\cap]0,E(\delta_n)+1[} \sum_{l=E(\alpha)+1}^{P} \varepsilon^{l+\delta_n-E(\alpha)} F_{\delta_n,\alpha}^l[\ln\varepsilon]$$
$$+ \sum_{\alpha\in\Lambda\cap]0,E(\delta_n)+1[} \sum_{l=0}^{P-E(\alpha)} \sum_{\beta\in\Lambda\cap]\alpha-P+l-1,0[} \varepsilon^{l+\alpha-\beta+\delta_n-E(\alpha)} F_{\delta_n,\alpha,\beta}^l[\ln\varepsilon].$$

For the first term, since $l - E(\alpha) \geq 1$, we remark that $l + \delta_n - E(\alpha)$ belongs to Δ_n. For the second term, from the expression of the singular exponent and since $E(\alpha) \leq E(\delta_n)$, we see that $l + \alpha - \beta + \delta_n - E(\alpha)$ belongs to Δ_n.

By the properties of the sequence $(\delta_n)_{n\in\mathbf{N}}$ stated before, we arrive at the first equation of (32). We argue similarly for the boundary condition on Γ_V. ∎

5 The asymptotic expansion

We first suppose that the data satisfy $f \in L^2(\Omega)$ and $g \in H^{\frac{1}{2}}(\Gamma_V)$. Under this assumption we will build the first terms in the expansion of the solution u_ε of (P_ε) in terms of ε. By Proposition 2.3, we deduce that a subsequence of (u_ε) tends weakly in $H^1(\Omega)$ to the solution u^0 of (P_0), therefore the first idea is to set

$$r_\varepsilon^0 = u_\varepsilon - u^0.$$

We then see that r_ε^0 is solution of

$$\begin{cases} -\Delta r_\varepsilon^0 = 0 & \text{in } \Omega, \\ \gamma r_\varepsilon^0 = 0 & \text{on } \Gamma_D, \\ B_\varepsilon r_\varepsilon^0 = \varepsilon \frac{\partial^2}{\partial \tau^2} u^0 & \text{on } \Gamma_V. \end{cases}$$

This problem is ill-posed since the right-hand side of the above problem does not belong to $H^{-\frac{1}{2}}(\Gamma_V)$ in general due to the singularities of u^0. That is the reason of the introduction of the corner layer term $w^0(\varepsilon)$ as the next result shows.

THEOREM 5.1 *Let us assume that $f \in L^2(\Omega)$ and $g \in H^{\frac{1}{2}}(\Gamma_V)$. Then the solution $u_\varepsilon \in V$ of problem (P_ε) admits the expansion*

$$u_\varepsilon = u^0 + \sum_{j=1,2} w_j^0(\varepsilon) + \tilde{r}_\varepsilon^0, \tag{34}$$

where for $j = 1$ or 2, $w_j^0(\varepsilon)$ is given by

$$w_j^0(\varepsilon) = \sum_{\alpha \in \Lambda_j \cap]0,1[} c_{j,\alpha}(u^0) \varepsilon^\alpha \chi_j Y_{j,\alpha,\varepsilon},$$

and the remainder $\tilde{r}_\varepsilon^0$ fulfils

$$||\tilde{r}_\varepsilon^0||_{H^1(\Omega)} \le C\varepsilon^{\delta_1}.$$

Proof: Let us set $\tilde{r}_\varepsilon^0 = u_\varepsilon - u^0 - \sum_{j=1,2} w_j^0(\varepsilon)$, with $w_j^0(\varepsilon)$ in the form (26) with coefficients $d_{j,\alpha}^0(\varepsilon)$ independent of ε. Using the expansion (2) of u^0 with $M = 1$ and Corollary 4.3 with $m = 1$, we can write

$$B_\varepsilon \tilde{r}_\varepsilon^0 = h_\varepsilon - \sum_{j=1,2} \varepsilon \chi_j \frac{\partial^2}{\partial \tau^2} \sum_{\alpha \in \Lambda_j \cap]0,1[} \left\{ d_{j,\alpha}^0(\varepsilon) \varepsilon^\alpha S_{j,\alpha,\varepsilon} - c_{j,\alpha}(u^0) S_{j,\alpha} \right\}. \tag{35}$$

where we have set

$$h_\varepsilon = \varepsilon \left(\frac{\partial^2}{\partial \tau^2} u^0_{reg(1)} + \sum_{j=1,2} \sum_{\alpha \in \Lambda_j \cap]0,1[} c_{j,\alpha}(u^0) \left[\frac{\partial^2}{\partial \tau^2}, \chi_j \right] S_{j,\alpha} \right) + g_0^{(1)}(\varepsilon).$$

Let us choose the coefficients $d^0_{j,\alpha}(\varepsilon)$ in (26) such that

$$d^0_{j,\alpha}(\varepsilon) = c_{j,\alpha}(u^0), \quad \forall \alpha \in \Lambda_j \cap]0,1[, \forall j = 1,2.$$

Since $S_{j,\alpha,\varepsilon} = \varepsilon^{-\alpha} S_{j,\alpha}$ the above choice cancels the last term of the identity (35). From Corollary 4.3 with $m = 1$, we then see that $\tilde{r}^0_\varepsilon$ is solution of the following problem

$$(36) \qquad \begin{cases} -\Delta \tilde{r}^0_\varepsilon = \varepsilon^{\delta_1} F^0_1 + F^{(1)}_0(\varepsilon) & \text{in } \Omega, \\ \tilde{r}^0_\varepsilon = 0 & \text{on } \Gamma_D, \\ B_\varepsilon \tilde{r}^0_\varepsilon = h_\varepsilon & \text{on } \Gamma_V, \end{cases}$$

where, as one easily shows, F^0_1 is a smooth function. Now the right-hand sides of this problem are suited to use Proposition 2.3. Therefore a unique solution $\tilde{r}^0_\varepsilon$ exists in V with the estimate

$$||\tilde{r}^0_\varepsilon||_{H^1(\Omega)} \leq C \left\{ ||\Delta \tilde{r}^0_\varepsilon||_{L^2(\Omega)} + ||B_\varepsilon \tilde{r}^0_\varepsilon||_{H^{-\frac{1}{2}}(\Gamma_V)} \right\}.$$

Since F^0_1 is a smooth function and using the estimate (33), we get

$$||\Delta \tilde{r}^0_\varepsilon||_{L^2(\Omega)} \leq C \varepsilon^{\delta_1}.$$

As $\left[\frac{\partial^2}{\partial \tau^2}, \chi_j\right] S_{j,\alpha}$ is smooth, $\frac{\partial^2}{\partial \tau^2} u^0_{reg(1)}$ belongs to $H^{-\frac{1}{2}}(\Gamma_V)$ (see Remark 1.4.4.7 of [4]) and using the estimate (33), we obtain

$$||B_\varepsilon \tilde{r}^0_\varepsilon||_{H^{-\frac{1}{2}}(\Gamma_V)} \leq C \varepsilon^{\delta_1},$$

which ends the proof. ∎

REMARK 5.2 The decomposition (34) may be equivalently written as

$$u_\varepsilon = u^0_{reg(1)} + \sum_{j=1,2} \sum_{\alpha \in \Lambda_j \cap]0,1[} c_{j,\alpha}(u^0) \chi_j \varepsilon^\alpha K_j(\alpha)_\varepsilon + \tilde{r}^0_\varepsilon.$$

As the expansion (20) and the estimate (23) imply that

$$\varepsilon^\alpha K_j(\alpha)_\varepsilon \to S_{j,\alpha}, \text{ as } \varepsilon \to 0,$$

the first singular terms in the expansion of u_ε tend to the first singular terms of u^0. In a certain sense, the singularities $S_{j,\alpha}$ of problem (P_0) are hidden in u_ε. ∎

As the remainder in the decomposition (34) is $o(1) = o(\varepsilon^{\delta_0})$, the second step consists in obtaining a remainder which is $o(\varepsilon^{\delta_1})$. To hit this goal we have to go back to problem (36) and to display in the right-hand sides all terms of powers of ε smaller or equal to δ_1. These terms will then contribute to a term in the asymptotic expansion of u_ε up to the order δ_1. This means that we rewrite problem (36) as follows

$$\begin{cases} -\Delta \tilde{r}_\varepsilon^0 = \varepsilon^{\delta_1} F_1^0 + F_0^{(1)}(\varepsilon) & \text{in } \Omega, \\ \gamma \tilde{r}_\varepsilon^0 = 0 & \text{on } \Gamma_D, \\ B_\varepsilon \tilde{r}_\varepsilon^0 = \varepsilon^{\delta_1} \dfrac{\partial^2}{\partial \tau^2} u_{reg(1)}^{\delta_1 - 1} & \\ +\varepsilon^{\delta_1} \sum\limits_{j=1,2} \sum\limits_{\alpha \in \Lambda_j \cap]0,1[} c_{j,\alpha}(u^{\delta_1 - 1}) \left[\dfrac{\partial^2}{\partial \tau^2}, \chi_j \right] S_{j,\alpha} + \tilde{g}_0^{(1)}(\varepsilon) & \text{on } \Gamma_V, \end{cases}$$

with the convention $u^{\delta_1 - 1} = 0$ if $\delta_1 - 1 < 0$, where $F_0^{(1)}(\varepsilon)$ and $\tilde{g}_0^{(1)}(\varepsilon)$ satisfy

$$\|F_0^{(1)}(\varepsilon)\|_{L^2(\Omega)} + \|\tilde{g}_0^{(1)}(\varepsilon)\|_{H^{-\frac{1}{2}}(\Gamma_V)} = o(\varepsilon^{\delta_1}).$$

Let then $u^{\delta_1} \in V_0$ be the unique solution of

$$\begin{cases} -\Delta u^{\delta_1} = F_1^0 & \text{in } \Omega, \\ \gamma u^{\delta_1} = 0 & \text{on } \Gamma_D, \\ \dfrac{\partial}{\partial \nu} u^{\delta_1} = \dfrac{\partial^2}{\partial \tau^2} u_{reg(1)}^{\delta_1 - 1} + \sum\limits_{j=1,2} \sum\limits_{\alpha \in \Lambda_j \cap]0,1[} c_{j,\alpha}(u^{\delta_1 - 1}) \left[\dfrac{\partial^2}{\partial \tau^2}, \chi_j \right] S_{j,\alpha} & \text{on } \Gamma_V. \end{cases}$$

(37)

At this stage, we define r_ε^1 by

$$r_\varepsilon^1 = \tilde{r}_\varepsilon^0 - \varepsilon^{\delta_1} u^{\delta_1}.$$

By the construction of u^{δ_1}, we clearly have

$$\text{(38)} \qquad \begin{cases} -\Delta r_\varepsilon^1 = F_0^{(1)}(\varepsilon) & \text{in } \Omega, \\ r_\varepsilon^1 = 0 & \text{on } \Gamma_D, \\ B_\varepsilon r_\varepsilon^1 = \varepsilon^{1+\delta_1} \frac{\partial^2}{\partial \tau^2} u^{\delta_1} + \tilde{g}_0^{(1)}(\varepsilon) & \text{on } \Gamma_V. \end{cases}$$

As in Theorem 5.1, the problem comes from the fact that $\frac{\partial^2}{\partial \tau^2} u^{\delta_1}$ do not belong to $H^{-\frac{1}{2}}(\Gamma_V)$ due to its singular behaviour near the corners. Indeed if $\delta_1 < 1$, applying Proposition 2.2 with $M = 1$ and if $\delta_1 = 1$, applying Proposition 2.4 with the assumption $f \in H^1(\Omega)$ and $g \in H^{\frac{3}{2}}(\Gamma_V)$, u^{δ_1} admits the decomposition

$$u^{\delta_1} = u_{reg(1)}^{\delta_1} + \sum_{j=1,2} \sum_{\gamma \in \Lambda_j \cap]0, E(\delta_1)+1[} c_{j,\gamma}^{\delta_1} \chi_j S_{j,\gamma}^{(E(\gamma))}.$$

Using this decomposition into (38), we arrive at

$$(39)\quad B_\varepsilon r_\varepsilon^1 = \sum_{j=1,2}\ \sum_{\gamma\in\Lambda_j\cap]0,E(\delta_1)+1[} c_{j,\gamma}^{\delta_1}\varepsilon^{1+\delta_1}\chi_j\frac{\partial^2}{\partial\tau^2}S_{j,\gamma}^{(E(\gamma))}+\breve{g}_0^{(1)}(\varepsilon) \text{ on } \Gamma_V,$$

with the estimate $\|\breve{g}_0^{(1)}(\varepsilon)\|_{H^{-\frac{1}{2}}(\Gamma_V)} = o(\varepsilon^{\delta_1})$. We then take the corner layers $w_j^1(\varepsilon)$ in the form (26) with coefficients $d_{j,\alpha}^1(\varepsilon)$ such that $B_\varepsilon(\sum_{j=1,2} w_j^1(\varepsilon))$ eliminates the first term of the right-hand side of (39). More precisely we impose for $j=1,2$ that

$$B_\varepsilon\left(\sum_{\kappa\in\Lambda_j\cap]0,E(\delta_1)+1[} d_{j,\kappa}^1(\varepsilon)\varepsilon^\kappa Y_{j,\kappa,\varepsilon}\right) = \sum_{\gamma\in\Lambda_j\cap]0,E(\delta_1)+1[} c_{j,\gamma}^{\delta_1}\varepsilon^{1+\delta_1}\frac{\partial^2}{\partial\tau^2}S_{j,\gamma}^{(E(\gamma))}.$$

By (31) and (28), the solution is

$$\left\{\begin{array}{l} d_{j,\kappa}^1(\varepsilon) = c_{j,\kappa}^{\delta_1}\varepsilon^{\delta_1-E(\kappa)}, \forall\kappa\in]E(\delta_1),E(\delta_1)+1[, \\ d_{j,\kappa}^1(\varepsilon) = \varepsilon^{\delta_1}\left(c_{j,\kappa}^{\delta_1} - \sum_{\substack{\tilde\kappa\in\Lambda_j,\\ \tilde\kappa-1=\kappa}} c_{j,\tilde\kappa}^{\delta_1}p_1(\ln\varepsilon)\right), \forall\kappa\in]0,1[, \text{ if } E(\delta_1)=1. \end{array}\right.$$

(40)

Note that the coefficients $d_{j,\kappa}^1(\varepsilon)$ have the form announced in (27). Consequently from Corollary 4.3 with $m=2$, we get

$$\left\{\begin{array}{lr} -\Delta w_j^1(\varepsilon) = \hat{F}_{j,1}^{(1)}(\varepsilon) & \text{in } \Omega, \\ w_j^1(\varepsilon) = 0 & \text{on } \Gamma_D, \\ B_\varepsilon w_j^1(\varepsilon) = \sum_{\alpha\in\Lambda_j\cap]0,E(\delta_1)+1[} d_{j,\alpha}^1(\varepsilon)\varepsilon^{1+\alpha}\chi_j\frac{\partial^2}{\partial\tau^2}S_{j,\alpha,\varepsilon}^{(E(\alpha))} + \hat{g}_{j,1}^{(1)}(\varepsilon) & \text{on } \Gamma_V, \end{array}\right.$$

with

$$\|\hat{F}_{j,1}^{(1)}(\varepsilon)\|_{L^2(\Omega)} + \|\hat{g}_{j,1}^{(1)}(\varepsilon)\|_{H^{-\frac{1}{2}}(\Gamma_V)} = o(\varepsilon^{\delta_1}).$$

All these considerations yield that $\tilde{r}_\varepsilon^1 = r_\varepsilon^1 - \sum_{j=1,2} w_j^1(\varepsilon)$ satisfies the problem

$$\left\{\begin{array}{ll} -\Delta\tilde{r}_\varepsilon^1 = \widetilde{F}_0^{(1)}(\varepsilon) & \text{in } \Omega, \\ \tilde{r}_\varepsilon^1 = 0 & \text{on } \Gamma_D, \\ B_\varepsilon\tilde{r}_\varepsilon^1 = \widetilde{g}_0^{(1)}(\varepsilon) & \text{on } \Gamma_V, \end{array}\right.$$

with the estimate

$$\|\widetilde{F}_0^{(1)}(\varepsilon)\|_{L^2(\Omega)} + \|\widetilde{g}_0^{(1)}(\varepsilon)\|_{H^{-\frac{1}{2}}(\Gamma_V)} = o(\varepsilon^{\delta_1}).$$

Consequently Proposition 2.3 guarantees the existence and uniqueness of $\tilde{r}_\varepsilon^1 \in V$ of the above problem with the estimate

$$\|\tilde{r}_\varepsilon^1\|_{H^1(\Omega)} = o(\varepsilon^{\delta_1}). \tag{41}$$

Summarizing we have proved the following asymptotic expansion for u_ε.

THEOREM 5.3 *Let us assume that $f \in H^{E(\delta_1)}(\Omega)$ and $g \in H^{E(\delta_1)+\frac{1}{2}}(\Gamma_V)$. Then the solution $u_\varepsilon \in V$ of problem (P_ε) admits the expansion*

$$u_\varepsilon = u^0 + \sum_{j=1,2} w_j^0(\varepsilon) + \varepsilon^{\delta_1} u^{\delta_1} + \sum_{j=1,2} w_j^1(\varepsilon) + \tilde{r}_\varepsilon^1,$$

where $w_j^0(\varepsilon)$ is defined in Theorem 5.1, $w_j^1(\varepsilon)$ is defined by (26) with the coefficients given by (40), $u^{\delta_1} \in V_0$ is the unique solution of (37) and finally the remainder $\tilde{r}_\varepsilon^1$ fulfils (41).

Similarly under higher regularity assumptions on the data, we prove by induction the expansion of u_ε up to the order ε^{δ_n}, for any $n \in \mathbf{N}$.

THEOREM 5.4 *Let $N \in \mathbf{N}$ and let us assume that $f \in H^{E(\delta_N)}(\Omega)$ and $g \in H^{E(\delta_N)+\frac{1}{2}}(\Gamma_V)$. Then the solution $u_\varepsilon \in V$ of problem (P_ε) admits the expansion*

$$u_\varepsilon = \sum_{n=0}^{N} \left\{ \varepsilon^{\delta_n} u^{\delta_n} + \sum_{j=1,2} w_j^n(\varepsilon) \right\} + \tilde{r}_\varepsilon^N,$$

where the remainder satisfies the estimate

$$\|\tilde{r}_\varepsilon^N\|_{H^1(\Omega)} = o(\varepsilon^{\delta_N}).$$

The outer expansion terms u^{δ_n} are the unique solutions in V_0 of

$$\begin{cases} -\Delta u^{\delta_n} = F^n[\ln\varepsilon] & \text{in } \Omega, \\ \gamma u^{\delta_n} = 0 & \text{on } \Gamma_D, \\ \frac{\partial}{\partial\nu} u^{\delta_n} = \frac{\partial^2}{\partial\tau^2} u_{reg(1)}^{\delta_n - 1} + g^n[\ln\varepsilon] & \text{on } \Gamma_V, \end{cases}$$

with the convention that $u^{\delta_n - 1} = 0$ if $\delta_n - 1 \notin \{\delta_0, \cdots, \delta_{n-1}\}$. The right-hand sides $F^n[\ln\varepsilon]$ and $g^n[\ln\varepsilon]$ come from the problem (32) with

$m = N$ solved by the corner layer terms $w_j^n(\varepsilon)$ and from the singular decomposition (42) of $u^{\delta_n - 1}$ and are the sum of the contributions corresponding to ε^{δ_n}, more precisely

$$F^n[\ln \varepsilon] = \sum_{j=1,2} \sum_{m=0}^{n-1} F_{j,n}^m[\ln \varepsilon],$$

$$g^n[\ln \varepsilon] = \sum_{j=1,2} \left(\sum_{m=0}^{n-1} g_{j,n}^m[\ln \varepsilon] + \sum_{\kappa \in \Lambda_j \cap]0, E(\delta_n)[} a_{j,\kappa}^{\delta_n - 1} \left[\frac{\partial^2}{\partial \tau^2}, \chi_j \right] S_{j,\kappa}^{(E(\kappa))} \right).$$

Finally the corner layer terms $w_j^n(\varepsilon)$ are defined by (26) with the coefficients $d_{j,\alpha}^n(\varepsilon)$ such that

$$\sum_{\kappa \in \Lambda_j \cap]0, E(\delta_n)+1[} d_{j,\kappa}^n(\varepsilon) \varepsilon^{\kappa+1} S_{j,\kappa,\varepsilon}^{(E(\kappa))} = \sum_{\kappa \in \Lambda_j \cap]0, E(\delta_n)+1[} a_{j,\kappa}^{\delta_n} \varepsilon^{1+\delta_n} S_{j,\kappa}^{(E(\kappa))}, \forall j = 1, 2,$$

where the coefficients $a_{j,\kappa}^{\delta_n}$ are the coefficients of the singularities of u^{δ_n} (which are polynomials in $\ln \varepsilon$), i.e.,

$$u^{\delta_n} = u_{reg(1)}^{\delta_n} + \sum_{j=1,2} \sum_{\kappa \in \Lambda_j \cap]0, E(\delta_n)+1[} a_{j,\kappa}^{\delta_n} \chi_j S_{j,\kappa}^{(E(\kappa))}. \tag{42}$$

Moreover these coefficients $d_{j,\alpha}^n(\varepsilon)$ have the form (27).

REFERENCES

[1] M. Bourlard, A. Maghnouji, S. Nicaise and L. Paquet, *Asymptotic expansion of the solution of a mixed Dirichlet-Ventcel problem with a small parameter,* Asymptotic Analysis, to appear.

[2] M. Costabel and M. Dauge, *A singularly perturbed mixed boundary value problem,* Comm. in PDE, **21**, 1996, 1919-1949.

[3] M. Dauge, *Elliptic boundary value problems in corner domains. Smoothness and asymptotics of solutions,* L.N. in Math., **1341**, Springer Verlag, Berlin, 1988.

[4] P. Grisvard, Elliptic problems in nonsmooth domains, Monographs and Studies in Mathematics 24, Pitman, Boston, 1985.

[5] A. M. Il'in, *Matching of asymptotic expansions of solutions of boundary value problems,* Translations of Math. Monographs **102**, AMS, Providence, 1992.

[6] R. B. Kellogg, *Boundary layers and corner singularities for a self-adjoint problem,* in: M. Costabel, M. Dauge and S. Nicaise eds, Boundary value problems and integral equations in nonsmooth domains, L.N. in Pure and Applied Math., **167**, 121-149, Marcel Dekker, New York, 1994.

[7] V.A. Kondratiev, Boundary problems for elliptic equations in domains with conical or angular points , Trans. Moscow Math. Soc., 16 , 1967, p.227-313.

[8] D. Leguillon and E. Sanchez-Palencia, *Computation of singular solutions in elliptic problems and elasticity,* RMA **5**, Masson, Paris, 1987.

[9] K. Lemrabet, *Problème aux limites de Ventcel dans un domaine non régulier,* C. R. Acad. Sc. Paris, Série I, **300**, 1985, 531-534.

[10] J.-L. Lions, *Perturbations singulières dans les problèmes aux limites non homogènes et en contrôle optimal,* L.N. in Math., **323**, Springer Verlag, Berlin, 1973.

[11] V.G. Maz'ja and B.A. Plamenevskii, L^p-estimates of solutions of elliptic boundary value problems in domains with edges, Trans. Moscow Math. Soc., 1, 1980, p.49-97.

[12] S. Nicaise, *Polygonal interface problems,* Series "Methoden und Verfahren der Mathematischen Physik", **39**, Peter Lang Verlag, 1993.

[13] S. Nazarov, *Vishik-Lyusternik method for elliptic boundary value problems in region with conical points I: The problem in a cone,* Siberian Math. J., **22**, 1981, 594-611.

[14] S. Nazarov, *Vishik-Lyusternik method for elliptic boundary value problems in region with conical points II: The problem in a bounded region,* Siberian Math. J., **22**, 1981, 753-769.

[15] S. Nazarov, *The spatial structure of the stress field in the neighbourhood of the corner point of a thin plate,* J. Appl. Math. Mechanics, **55**, 1991, 523-530.

On instantaneous control of singularly perturbed hyperbolic equations on graphs

T. FISCHER Institute of Mathematics, University of Bayreuth, D-95440 Bayreuth, Germany.

G. LEUGERING Institute of Mathematics, TU Darmstadt, Schloßgartenstr. 7, D-64289 Darmstadt, Germany.

1 ABSTRACT

We consider a system of $1-d$–diffusion-convection equations on graphs as being representative of flow or transport problems in channel, pipeline or root-systems appearing in many applications. Such systems are subject to controls the effect of which one wants to optimize. In these notes we discuss the mathematical framework of typical optimal control problems of such systems on graphs. We focus on singular perturbations with respect to diffusion and on the numerical implementation.

2 INTRODUCTION

We consider diffusion-convection (or sometimes advection) equations on a graph G consisting of a set $V = \{v_J | J \in \mathcal{J}\}$ of vertices (nodes, joints) and a set $E = \{\Omega_i | i \in \mathcal{I}\}$ of edges which are one-dimensional manifolds represented as simple C^2-curves in $\mathbb{R}^n$ with arclength function π_i, respectively. Functions defined on Ω_i are consequently expressed on the interval $(0, l_i)$, with l_i denoting the length of link i. l_i has to be finite. We consider finite graphs and use the following notation: for each $i \in \mathcal{I}$ there are indices $J_i^-, J_i^+ \in \mathcal{J}$, $\mathcal{J}_i = \{J_i^-, J_i^+\}$, such that $\pi_i(0) = v_{J_i^-}, \pi_i(l_i) = v_{J_i^+}$, i.e. the edge i starts at the node $v_{J_i^-}$ and ends at $v_{J_i^+}$. This convention turns G into a directed graph. Conversely, to each vertex v_J with $J \in \mathcal{J}$ we associate the set $\mathcal{I}_J^{\rightarrow} := \{i \in \mathcal{I} | J = J_i^-\}$ of edges starting at that node, and the set $\mathcal{I}_J^{\leftarrow} := \{i \in \mathcal{I} | J \in J_i^+\}$ ending there. The set $\mathcal{I}_J := \mathcal{I}_J^{\leftarrow} \dot\cup \mathcal{I}_J^{\rightarrow}$ then denotes the set of edges incident at v_J. Consequently, the edge-degree d_J of the node v_J is given by $d_J := |\mathcal{I}_J|$. On each of the edges we consider a scalar $1-d$ diffusion-advection equation of the following type

$$\dot{u}_i - \epsilon_i u_i'' + \beta_i u_i' + \alpha_i u_i = f_i \quad \text{on } \Omega_i \times (0,T), \tag{2.1}$$

where $u_i(x,t)$ represents the state (concentration, velocity of a fluid etc.) at the location x and time $t, \epsilon_i > 0$ is a diffusion parameter (considered small), $\beta_i \in C^1(\bar{\Omega}_i), \beta_i \geq 0$ a convection or transport parameter, and $\alpha_i \in C(\bar{\Omega}_i), \alpha_i \geq 0$, is a reaction coefficient. The sign condition on β_i is not restrictive, as, whenever β_i changes sign only a finite number of times, we can introduce an edge at each zero of β_i and reverse the orientation. Later, however, sign-conditions will apply, as we consider a "flow" through the network. We have denoted in (2.1) by a dot the time-derivative and by a prime the spatial derivative.

In order to keep matters as simple as possible and still representative of the underlying ideas, we restrict ourselves to distributed control actions which are represented by functions f_i acting along Ω_i. Boundary controls can also be handled. We, therefore, assume without loss of generality that at all nodes we have homogeneous conditions, i.e. Dirichlet or Neumann-conditions. We consider Dirichlet conditions only at simple nodes $v_J \in \mathcal{J}_D(d_j = 1)$ while Neumann conditions apply at the other simple or multiple ($\mathcal{J}_N$) nodes. Thus, $\mathcal{J} = \mathcal{J}_D \dot{\cup} \mathcal{J}_N$. In the case of a multiple node $v_J \in \mathcal{J}_N$ we define $d_{iJ}u_i'(v_J) := \pm u_i'(v_J)$ where the plus sign applies for $i \in \mathcal{I}_J^{\leftarrow}$ and the minus sign for $i \in \mathcal{I}_J^{\rightarrow}$.

With this notation and $\epsilon := \{\epsilon_i\}_{i\in\mathcal{I}}$ we can state the underlying evolutionary system in $G \times (0,T)$.

$$\begin{cases} \dot{u}_i - \epsilon_i u_i'' + \beta_i u_i' + \alpha_i u_i = f_i, \ (x,t) \in \Omega_i \times (0,T), \\ u_i(v_J) = 0, \ J \in \mathcal{J}_D, \ t \in (0,T), \\ u_i(v_J) = u_j(v_J), \ \forall i,j \in \mathcal{I}_J, \ t \in (0,T), \\ \sum\limits_{i\in\mathcal{I}_J} \epsilon_i d_{iJ} u_i'(v_J) =: \rho_J^\epsilon(u) = 0, \ J \in \mathcal{J}_N, \ t \in (0,T), \\ u_i(\cdot,0) = u_{i0}, \ x \in \Omega_i. \end{cases} \tag{2.2$^\epsilon$}$$

Only the statements $(2.2)^\epsilon{}_{3,4}$ need an explanation. Indeed, $(2.2)^\epsilon{}_3$ ensures continuity of u at the vertices, while $(2.2)^\epsilon{}_4$ is a balance condition analogous to the classical Kirchhoff-condition in circuit-theory. Systems as $(2.2)^\epsilon$ arise in a variety of applications including the linearized flow of water in sewer-systems, diffusive transport of ions along roots of plants in structured soil and other important models. From a mathematical point of view a complete treatment of such systems has been given J. von Below [4], [5], also for quasi-linear equations. Many papers and some monographs, notably by F. Ali-Mehmeti [2] and S. Nicaise [20] have been published also for elliptic and hyperbolic problems; we refer to bibliographies of these monographs for further references. The literature is sparse, however, for control problems related to systems like $(2.2)^\epsilon$. For hyperbolic or Petrovski-type problems on graphs or more general multi-domains see J.E. Lagnese, G. Leugering and E.J.P.G. Schmidt [13] and the references therein, see also the recent work of B. Dekoninck and S. Nicaise [8] and S. Nicaise [20], [21]. To the knowledge of the authors, control problems for parabolic problems on graphs have not yet received much attention. However, null-controllability results can be achieved, via Russell's equivalence principle from corresponding results concerning hyperbolic equations.

Let us pose the following simple optimal control problem related to $(2.2)^\epsilon$.

$$\begin{cases} (OCP)_\epsilon : \quad \min J_\epsilon(f) := \dfrac{1}{2}\displaystyle\sum_{i\in\mathcal{I}} \Big\{ \int_0^T \int_{\Omega_i} (u_i - z_i)^2 \, dx\, dt \\ \qquad\qquad + \sigma_i \displaystyle\int_0^T \int_{\Omega_i} f_i^2 \, dx\, dt \Big\} \\ \text{subject to } (2.2)^\epsilon, \end{cases} \tag{2.3}$$

where $f_i \in L^2(\Omega_i)$. In order to recast (2.3) into the framework of abstract optimization we introduce the following spaces and operators:

$$\begin{aligned} H_i &:= L^2(\Omega_i), \\ V_i &:= \{u_i \in H^1(\Omega_i) |\, u_i(v_J) = 0 \quad \forall\, J \in \mathcal{J}_i \cap \mathcal{J}_D\}, \\ A_i^\epsilon u_i &:= -\epsilon_i u_i'' + \beta_i u_i' + \alpha_i u_i, \\ D(A_i^\epsilon) &= \{u_i \in V_i |\, A_i^\epsilon u_i \in H_i\} = H^2(\Omega_i) \cap V_i, \\ V &:= \{u \in \prod_{i\in\mathcal{I}} V_i |\, u_i(v_J) = u_j(v_J) \quad \forall i, j \in \mathcal{I}_J, J \in \mathcal{J}_N, d_J > 1\}, \\ H &:= \prod_{i\in\mathcal{I}} H_i. \end{aligned}$$

Note that the spaces like H, V are examples of ramification spaces in the sense of G. Lumer (see [19], [1]).

It is plain that the system $(2.2)^\epsilon$ exhibits canonical well posedness properties in this space-framework. It is further clear that for controls $f \in \mathcal{U} = L^2(0,T;H)$ the optimal control problem $(OCP)_\epsilon$ has a unique solution which can be characterized by optimality conditions and in feed-back form, even though such results have not yet been reported in the literature. However, in a diploma thesis J. Geisler [10] discussed a numerical procedure based on multigrid-techniques. Computing optimal controls of systems like $(2.2)^\epsilon$, (2.3) in the context of real-time applications, as e.g. the optimal control of sewer-systems, is prohibitive with respect to computing time, if the computations are based on the entire time horizon and the entire graph.

One procedure in order to reduce the complexity of the problem has been introduced by R. Temam et. al. [7] who also coined the notion of instantaneous controls. The procedure which was originally designed for Navier-Stokes flow in channels or Burger's equation is as follows: One performs a semidiscretization in time for (2.1), ${(2.2)^\epsilon}_1$ such that, at each time step $t_k = k\Delta t$, the problem is elliptic. Further, the cost-function $(2.3)_1$ is discretized with respect to the time and then taken at a given time level, i.e. the new cost function depends on the actual time only. The resulting optimal control problem is thus of elliptic type, and the optimization is carried out for each consecutive time step.

The concept of instantaneous controls is a special case of the method of receding horizons known to control theorist in finite dimensional systems for finite some time. Rather than to compute optimal controls corresponding to a large time horizon, which necessitates the solution of Riccati algebraic (or differential) equations, one considers a small horizon, computes the control 'cheaply' but applies it at the current time, only. In the next step the horizon is moved by time unit and the procedure is repeated. In this respect, instantaneous control consists of looking ahead only one time step. The time horizon can be viewed as a number of moves a chess player looks ahead while applying the resulting strategy at a given move, only. The further moves depend on the moving horizon.

This strategy was taken up by K. Kunisch et. al. [6] for two-dimensional Navier-Stokes problems, see also [11]. R. Hundhammer and G. Leugering [12] have considered wave equations on graphs subjected to such instantaneous controls. The point in that analysis and the corresponding numerical work is that the underlying sequence of elliptic optimal control problems is treated by domain decomposition techniques for problems on graphs developed by G. Leugering [14], [15], [16].

The concern of this short paper is to outline the applicability of this approach to systems of the type $(2.2)^\epsilon$, (2.3), and, in particular, to give first results towards the resolution of the problem of singular perturbation of such optimal control problems.

Singular perturbation analysis for the problem above is very important in various respects. First of all it pares the way to a general 'vanishing viscosity'-type solution approach to optimal control problems for hyperbolic conservation laws on graphs. Secondly, even the linear theory is still silent when it comes to equations about boundary layers, in fact nodal layers for such systems on graphs. This is true for just the forward dynamics and even more evident for the optimality systems to be considered. The third aspect can be characterized by saying that one is interested in regularizing numerical schemes by the introduction of viscosity terms.

3 INSTANTANEOUS CONTROLS

We consider an implicit Euler scheme applied to (2.1). We set $u_i^k \approx u_i(\cdot, k\Delta t)$ as an approximation to the true state u_i at time t. We obtain

$$\frac{1}{\Delta t}(u_i^{k+1} - u_i^k) - \epsilon_i(u_i^{k+1})'' + \beta_i(u_i^{k+1})' + \alpha_i u_i^{k+1} = f_i^{k+1}.$$

We absorb $\dfrac{1}{\Delta t} + \alpha_i$ into α_i and rewrite the latter equation as $\left(\gamma_i = \frac{1}{\Delta t}\right)$

$$-\epsilon_i(u_i^{k+1})'' + \beta_i(u_i^{k+1})' + \alpha_i u_i^{k+1} = f_i^{k+1} + \gamma_i u_i^k =: g_i^k.$$

Therefore, instead of $(2.2)^\epsilon$ we consider the elliptic problem on G:

$$\begin{cases} -\epsilon_i u_i'' + \beta_i u_i' + \alpha_i u_i = f_i + \gamma_i u_i^k, \ x \in \Omega_i, \\ u_i(v_J) = 0, \ J \in \mathcal{J}_D, \\ u_i(v_J) = u_j(v_J), \ \forall i, j \in \mathcal{I}_J, \\ \sum\limits_{i \in \mathcal{I}_J} \epsilon_i d_{iJ} u_i'(v_J) = 0, \ J \in \mathcal{J}_N, \end{cases} \tag{3.1}$$

for given u_i^k (from the previous time step).

The instantaneous control problem can then be formulated as

$$\begin{cases} (IOCP)_\epsilon : \quad \min\limits_f J_\epsilon^k(f) := \dfrac{1}{2}\sum\limits_{i\in\mathcal{I}}\left\{ \displaystyle\int_{\Omega_i} (u_i - z_i^{k+1})^2\,dx \right. \\ \qquad\qquad\qquad\qquad \left. + \sum\limits_{i\in\mathcal{I}} \sigma_i \displaystyle\int_{\Omega_i} f_i^2\,dx \right\} \\ \text{subject to 3.1.} \end{cases} \tag{3.2}$$

The minimal f^{k+1} is relabelled as f and the corresponding u^{k+1} as u. These updates re-enter system (3.1), the time step is increased by 1 and the procedure is repeated until the time horizon has been reached. This procedure is suboptimal in some sense. Indeed, it has been shown by Hinze and Kauffmann [11] that, on the fully discrete level, the resulting optimality system can be reinterpreted as a stabilized discrete-time dynamical system.

In order to devise a strategy for solving (3.2) numerically, we need gradient information.

$$\delta J_\epsilon^k(f)(g) = \sum_{i\in\mathcal{I}} \int_{\Omega_i} (u_i - z_i)y_i + \sigma_i f_i g_i \, dx \tag{3.3}$$

gives the directional derivative of J_ϵ^k at f in the direction of g, where y solves (3.1) with $f_i + \gamma_i u_i^k$ replaced with g_i. We have

$$\begin{aligned} 0 &= \sum_{i\in\mathcal{I}} \int_{\Omega_i} (-\epsilon_i y_i'' + \beta_i y_i' + \alpha_i y_i - g_i)p_i \, dx \\ &= \sum_{J\in\mathcal{J}_N} \sum_{i\in\mathcal{I}_J} \Big[-\big(\epsilon_i d_{iJ} y_i' p_i\big)(v_J) + \big(\epsilon_i y_i d_{iJ} p_i'\big)(v_J) \Big] \\ &\quad + \sum_{J\in\mathcal{J}_N} \Big[\sum_{i\in\mathcal{I}_J^{\leftarrow}} (\beta_i y_i p_i)(v_J) - \sum_{i\in\mathcal{I}_J^{\rightarrow}} (\beta_i y_i p_i)(v_J) \Big] \\ &\quad + \sum_{i\in\mathcal{I}} \int_{\Omega_i} y_i(-\epsilon_i p_i'' - (\beta_i p_i)' + \alpha_i p_i) \, dx \\ &\quad - \sum_{i\in\mathcal{I}} \int_{\Omega_i} g_i p_i \, dx \end{aligned} \tag{3.4}$$

for sufficiently smooth functions p_i. We select these p_i such that p solves

$$\begin{cases} -\epsilon_i p_i'' - (\beta_i p_i)' + \alpha_i p_i = u_i - z_i^{k+1}, \quad x \in \Omega_i, \\ p_i(v_J) = 0, \; J \in \mathcal{J}_D, \\ p_i(v_J) = p_j(v_J), \quad \forall i,j \in \mathcal{I}_J, \; J \in \mathcal{J}_N, \\ \sum_{i\in\mathcal{I}_J} \epsilon_i d_{iJ} p_i'(v_J) + d_{iJ}(\beta_i p_i)(v_J) = 0. \end{cases} \tag{3.5}$$

Thus, (3.4) reads

$$\sum_{i\in\mathcal{I}} \int_{\Omega_i} y_i(u_i - z_i^{k+1}) \, dx = \sum_{i\in\mathcal{I}} \int_{\Omega_i} g_i p_i \, dx,$$

which inserted into (3.3) gives

$$\delta J_\epsilon^k(f)(g) = \sum_{i\in\mathcal{I}} \int_{\Omega_i} g_i(p_i + \sigma_i f_i) \, dx \tag{3.6}$$

for all $g \in H$. Thus, the gradient of J_ϵ^k at f is

$$\nabla J_\epsilon^k(f) = ((p + \sigma f)_i)_{i\in\mathcal{I}}. \tag{3.7}$$

The optimality condition $\nabla J_\epsilon^k(f) = 0$, amounts to

$$f_i = -\frac{1}{\sigma_i} p_i, \; i \in \mathcal{I}. \tag{3.8}$$

In order to solve (3.2) one can use (3.7) in a gradient method or directly solve the corresponding optimality system

$$\begin{cases} -\epsilon_i u_i'' + \beta_i u_i' + \alpha_i u_i + \dfrac{1}{\sigma_i} p_i = \gamma_i u_i^k, \\ -\epsilon_i p_i'' - (\beta_i p_i)' + \alpha_i p_i - u_i = -z_i^{k+1}, \\ u_i(v_J) = p_i(v_J) = 0, \ J \in \mathcal{J}_D, \\ u_i(v_J) = u_j(v_J), p_i(v_J) = p_j(v_J), \quad \forall i, j \in \mathcal{I}_J, \ J \in \mathcal{J}_N, \\ \sum\limits_{i \in \mathcal{I}_J} \epsilon_i d_{iJ} u_i'(v_J) = 0, \ J \in \mathcal{J}_N, \\ \sum\limits_{i \in \mathcal{I}_J} \epsilon_i d_{iJ} p_i'(v_J) + d_{iJ}(\beta_i p_i)(v_J) = 0, J \in \mathcal{J}_N. \end{cases} \tag{3.9}$$

We remark that it is in principle possible to let Δt approach 0. In finite dimensions, i. e. after spatial discretization, if there is a fixed time span for the cost (recall the receding horizon methodology mentioned in the introduction), then the corresponding solutions converge to the solution of the corresponding continuous problem. If, however, the time horizon shrinks to zero, then the corresponding optimal control problem is not properly posed. The infinite dimensional context is largely unexplored. Suffice it to say that in special cases one may interprete the resulting closed loop system as a discretization of stabilized dynamics. For the sake of brevity, we cannot dwell on this property further.

4 SINGULAR PERTURBATIONS

In this section we consider the following singular perturbation problem. Given

$$\begin{cases} \mathcal{A}^\epsilon := \prod\limits_{i \in \mathcal{I}} A_i^\epsilon, \\ D(\mathcal{A}^\epsilon) := \{u \in V \,|\, u \in \prod\limits_{i \in \mathcal{I}} D(A_i^\epsilon), \quad \rho_J^\epsilon(u) = 0 \quad \forall J \in \mathcal{J}_N\}, \end{cases} \tag{4.1}$$

we consider the problem

$$\mathcal{A}^\epsilon u = f, \ u \in D(\mathcal{A}^\epsilon) \tag{4.2}$$

for an arbitrary $f \in H$ or, in variational form, the weak problem

$$a_\epsilon(u, v) := \sum_{i \in \mathcal{I}} [\epsilon_i(u_i', v_i')_{\Omega_i} + (\beta_i u_i', v_i)_{\Omega_i} + (\alpha_i u_i, v_i)_{\Omega_i}] = (f, v), \ \forall v \in V. \tag{4.3}$$

The question is what happens as $\epsilon \longrightarrow 0$ ($\epsilon_i \longrightarrow 0 \ \forall i \in \mathcal{I}$). To our knowledge such a singular perturbation problem for an elliptic system on a graph with degeneration to an hyperbolic problem has not yet been considered in the literature.

Before we proceed with the analysis we want to emphasize that, of course, one has to expect boundary layers at out-flow boundaries. However, in this case we also have multiple nodes where, for instance, two edges end at that node and one starts from it. Therefore, such a node will be an out-flow node for two incoming links while it is an in-flow node the edge leaving there. Certainly, one has to expect discontinuities to develop at those nodes if $\epsilon \longrightarrow 0$. Even more important is the question what happens at circuits. After these preliminary remarks it seems clear that a straightforward application of well-established techniques, see for instance Eckhaus [9], will not suffice here.

We will follow the ideas of J. L. Lions [17]. Let us first introduce a flow condition for the graph.

DEFINITION 1 We say that the graph G with edges $\{\Omega_i\}_{i\in\mathcal{I}}$ and vertices $\{v_J\}_{J\in\mathcal{J}}$ satisfies the flow condition, if there is a reordering $i_1, \ldots, i_{|\mathcal{I}|} \in \mathcal{I}$ of the edges such that the following holds:
Define for each $k \in \{0, 1, \ldots, |\mathcal{I}|\}$ the sets

$$\begin{aligned}
\mathcal{I}_k &:= \{i_1, \ldots, i_k\}, \ (\mathcal{I}_0 = \emptyset), \\
\mathcal{J}_k^- &:= \{J \in \mathcal{J} \,|\, \exists i \in \mathcal{I}_k : J = J_i^-\}, \\
\mathcal{J}_k^+ &:= \{J \in \mathcal{J} \,|\, \exists i \in \mathcal{I}_k : J = J_i^+\}, \\
\mathcal{J}_k &:= J_k^- \cup \mathcal{J}_k^+.
\end{aligned}$$

Then for each $k \in \{1, \ldots, |\mathcal{I}|\}$ we have the implication:

$$J_{i_k}^+ \in \mathcal{J}_N \Longrightarrow J_{i_k}^+ \notin \mathcal{J}_{k-1}.$$

Taking into account condition (4.5), which makes it impossible for a Neumann vertex to have only outflowing edges incident, this definition essentially says that each vertex should be reachable from a Dirichlet node by a directed path, and that the graph does not contain circuits. We note that the flow condition is of technical nature as it guarantees the existence of a certain H^1-barrier function which is used in the proof of Theorem 1. In fact, we will discuss the case with circuits in another publication. Under the assumption that the flow condition is satisfied we can show the following uniform estimates.

THEOREM 1 Let G satisfy the flow condition. Let for each edge $\Omega_i, \theta_i, \delta_i \in \mathbb{R}, \delta_i > 0$ be constants such that

$$\theta_i \beta_i - \frac{1}{2}\beta_i' + \alpha_i \geq \delta_i, \tag{4.4}$$

and for each vertex v_J, $J \in \mathcal{J}_N$, let $\mu_J \in \mathbb{R}, \mu_J > 0$ be a constant such that

$$\sum_{i\in\mathcal{I}_J} d_{i_J}\beta_i(v_J) = \sum_{i\in\mathcal{I}_J^{\leftarrow}} \beta_i(v_J) - \sum_{i\in\mathcal{I}^{\rightarrow}} \beta_i(v_J) \geq \mu_J. \tag{4.5}$$

Then there exist positive constants ϵ_0, C such that all solutions u_i^ϵ of (4.2) with

$$\max_{i\in\mathcal{I}} |\epsilon_i| = \|\epsilon\| < \epsilon_0 \text{ satisfy}$$
$$\sum_{i\in\mathcal{I}} (\epsilon_i \|u_i'\|^2 + \|u_i\|^2) + \sum_{J\in\mathcal{J}_N} |u(v_J)|^2 \leq C \sum_{i\in\mathcal{I}} \|f_i\|^2. \tag{4.6}$$

Proof: We are able to construct a function ψ on G which is continuous on the entire graph and satisfies locally $\psi_i \in C^2(\bar{\Omega}_i) \hat{=} C^2([0, l_i])$. We also require $0 < m \leq \psi \leq M < \infty$, for suitable positive numbers m, M and

$$-\frac{1}{2}\psi_i' \geq \theta_i \psi_i \quad \forall i \in \mathcal{I}. \tag{4.7}$$

In particular, we set $\psi_i(x) = \lambda_i e^{\bar{\theta}_i x}$, with constants $\lambda_i > 0$ and $\bar{\theta}_i \in \mathbb{R}$ to be determined from these conditions.

We construct the function ψ step by step with respect to the reordering introduced in Definition 1. Assume, the functions ψ_i are already constructed for $i = 1, \ldots, i_{k-1}$. If

$$J_{i_k}^- \in \mathcal{J}_N \cap \mathcal{J}_k,$$

we have to choose

$$\lambda_{i_k} = \psi(v_{J_{i_k}^-}),$$

to ensure continuity at $v_{J_{i_k}^-}$. Otherwise we take

$$\lambda_{i_k} = 1.$$

The condition (4.7) is satisfied if and only if

$$\overline{\theta}_{i_k} \leq -2\theta_{i_k}. \tag{4.8}$$

Now, it is possible that the value of ψ is already fixed at the endpoint $v_{J_{i_k}^+}$ of the interval $\mathcal{I}_{i_k}$, i. e.

$$J_{i_k}^+ \in \mathcal{J}_N \cap \mathcal{J}_k.$$

In this case, there is no problem if

$$\psi(v_{J_{i_k}^+}) \leq \lambda_{i_k} e^{-2\theta_{i_k} l_{i_k}},$$

because the parameter $\overline{\theta}_{i_k}$ which makes the function continuous satisfies (4.8) automatically. If $\psi(v_{J_{i_k}^+}) > \lambda_{i_k} e^{-2\theta_{i_k} l_{i_k}}$, we take

$$\overline{\theta}_{i_k} = -2\theta_{i_k}$$

and make the parameters $\overline{\theta}_i$ of all the other inflow edges of $v_{J_{i_k}^+}$, i. e. the edges i with $J_i^+ = J_{i_k}^+$, smaller, in order to restore the continuity of ψ at $v_{J_{i_k}^+}$. It is obvious that neither condition (4.7) is violated at any adge nor the continuity of ψ, because $J_{i_k}^+ \notin \mathcal{J}_k^-$.

We use ψ as multiplier as follows:

$$\begin{aligned}
a_\epsilon^\psi(u,u) &:= a_\epsilon(u, \psi u) \\
&= \sum_{i\in\mathcal{I}} \Big(\epsilon_i \int_{\Omega_i} u_i'(\psi_i u_i)' dx + \int_{\Omega_i} \beta_i \psi_i u_i' u_i \, dx + \int_{\Omega_i} \alpha_i \psi_i u_i^2 \, dx\Big) \\
&\geq \sum_{i\in\mathcal{I}} \Big(\epsilon_i \int_{\Omega_i} \psi_i {u_i'}^2 \, dx + \int_{\Omega_i} u_i^2 (-\frac{1}{2}\epsilon_i \psi_i'' + \delta_i \psi_i) \, dx\Big) \\
&\quad + \frac{1}{2} \sum_{J\in\mathcal{J}_N} \Big(\sum_{i\in\mathcal{I}_J} \epsilon_i d_{iJ} \psi_i'(v_J) + \mu_J \psi(v_J)\Big) u(v_J)^2 \\
&\geq \sum_{i\in\mathcal{I}} \Big(\epsilon_i m \int_{\Omega_i} {u_i'}^2 \, dx + \delta_i^\star m \int_{\Omega_i} u_i^2 \, dx\Big) + \frac{1}{2} m \sum_{J\in\mathcal{J}_N} \mu_J^\star u(v_J)^2,
\end{aligned}$$

for some $\delta_i^\star \in (0, \delta_i), \mu_J^\star \in (0, \mu_J)$ and $\epsilon_0 > 0$ depending on the bounds of ψ_i and its first and second derivatives. As $\psi u^\epsilon \in V$ we conclude

$$a_\epsilon^\psi(u^\epsilon, u^\epsilon) \leq M\|f\|_H\|u^\epsilon\|_H,$$

from which the assertion follows. □

In order to proceed to a convergence result, let us consider a single edge first.

LEMMA 1 Let $\Omega = (a, b), \epsilon > 0, \beta, \varphi \in C^1(\bar{\Omega}), \alpha \in C(\bar{\Omega})$ be given. Further let $f \in L^2(\Omega)$ and $u \in H^2(\Omega)$ satisfy

$$-\epsilon u'' + \beta u' + \alpha u = f. \tag{4.9}$$

Then the following multiplier identity holds.

$$\begin{aligned} \frac{\epsilon}{2}(-\varphi\beta u'^2)\Big|_b^a + \frac{\epsilon}{2}\int_\Omega (\varphi\beta)' u'^2\,dx + \int_\Omega \varphi(\beta u')^2 \\ + \int_\Omega \varphi\beta\alpha u\,u'\,dx = \int_\Omega f\varphi\beta u'\,dx \end{aligned} \tag{4.10}$$

Proof: Multiply (4.9) by $\varphi\beta u'$ and integrate by parts. □

LEMMA 2 Let the assumptions of Lemma 1 be true. In addition assume $\beta \geq 0, \varphi \geq 0$ and $\varphi(b) = 0$. Then there exists a constant $C > 0$ depending on β and φ, such that for all $u \in H^2(\Omega), f \in L^2(\Omega)$ with (4.9) we have

$$\|\varphi^{1/2}\beta u'\|^2 \leq C(\epsilon\|u'\|^2 + \|f\|^2 + \|\alpha u\|^2). \tag{4.11}$$

Proof: By (4.10) and the assumptions made we have with $F = f - \alpha u$

$$\frac{\epsilon}{2}\int_\Omega (\varphi\beta)' u'^2\,dx + \int_\Omega \varphi(\beta u')^2\,dx \leq \int_\Omega F\varphi\beta u'\,dx,$$

from which we infer

$$\int_\Omega \varphi\Big(\beta u' - \frac{1}{2}F\Big)^2 dx \leq \frac{1}{4}\int_\Omega \varphi F^2\,dx + \frac{\epsilon}{2}\int_\Omega |(\varphi\beta)'|u'^2\,dx.$$

Upon using the triangle inequality and $2ab \leq \dfrac{a^2}{\sigma^2} + \sigma^2 b^2$ for $\sigma < 1$ we find a constant C depending on $\max_{\bar{\Omega}} |(\varphi\beta)'|$ such that (4.11) holds. □

In order to prepare for the limiting procedure we look at the first-order operator on $\Omega := (a, b)$

$$\begin{aligned} D_\beta u &:= (\beta u)' - \beta' u, \\ D(D_\beta) &= \{u \in L^2(\Omega) | \, \beta u \in H^1(\Omega)\}, \end{aligned} \tag{4.12}$$

which is a closed extension of the operator $\tilde{D}_\beta u = \beta u', \; D(\tilde{D}_\beta) = H^1(\Omega)$. The domain $D(D_\beta)$ can further be characterized by

$$D(D_\beta) = \{u \in L^2(\Omega) | \exists\, z \in L^2(\Omega) : (z, v) + (u, (\beta v)') = 0 \quad \forall v \in C_0^\infty(\Omega)\}.$$

Moreover, we can define closed restrictions of D_β by posing boundary conditions on either a or b. In fact, with $D(D_\beta^-) := \{u \in L^2(\Omega) |\ \beta u \in H^1(\Omega), (\beta u)(a) = 0\}$ for $\beta > 0$ and $D(D_\beta^+) := \{u \in L^2(\Omega) |\ \beta u \in H^1(\Omega), (\beta u)(b) = 0\}$ for $\beta < 0$ we have generators of C_0-semigroups in $L^2(\Omega)$. See Bardos [3] for the higher dimensional analogue.

LEMMA 3 Let the assumptions of Lemma 2 be satisfied. Consider for $\epsilon_n \searrow 0$ a sequence $(u_n) \subset H^2(\Omega)$ with

$$-\epsilon_n u_n'' + \beta u_n' + \alpha u_n = f. \tag{4.13}$$

Further assume that there are numbers ξ_-, ξ_+ and a function $\varphi \in C^1(\bar{\Omega})$ with $\varphi > 0$ on $[a, b)$ and $\varphi(b) = 0$. If

$$u_n \overset{w}{\rightharpoonup} u,\ \varphi\beta u_n' \overset{w}{\rightharpoonup} z,\ \epsilon_n u_n' \overset{w}{\rightharpoonup} 0,\ (\beta u_n)(a) \longrightarrow \xi_-, (\beta u_n)(b) \longrightarrow \xi_+$$

for some $u, z \in L^2(\Omega)$, then we have

$$\begin{aligned} &u \in D(D_\beta),\ D_\beta u + \alpha u = f, \\ &(\beta u)(a) = \xi_-,\ \epsilon_n u_n'(a) \longrightarrow 0,\ \epsilon_n u_n'(b) \longrightarrow \xi_+ - (\beta u)(b). \end{aligned} \tag{4.14}$$

Proof: By partial integration after multiplication by $v \in C_0^\infty(\Omega)$ we get

$$-\epsilon_n(u_n, v'') - (u_n, (\beta v)') + (u_n, \alpha v) = (f, v).$$

By our assumption we obtain in the limit

$$-(u, (\beta v)') + (u, \alpha v) = (f, v) \qquad \forall v \in C_0^\infty(\Omega),$$

i.e. $u \in D(D_\beta),\ D_\beta u + \alpha u = f$. Multiplying (4.13) by φv we obtain after integration by parts and then passing to the limit

$$(z, v) + (\alpha\varphi u, v) = (\varphi f, v) \qquad \forall v \in C_0^\infty(\Omega),$$

or $z = \varphi f - \alpha\varphi u = \varphi D_\beta u = D_{\varphi\beta} u,\ u \in D(D_{\varphi\beta})$. For $v \in H^1(\Omega)$ we have

$$(z, v) + (u, (\beta\varphi v)') = -\varphi(a)(\beta u)(a)v(a).$$

On the other side, since $u_n \in H^1(\Omega)$ we have

$$(\varphi\beta u_n', v) + (u_n, (\beta\varphi v)') = -\varphi(a)(\beta u_n)(a)v(a),$$

and passing to the limit we obtain

$$(z, v) + (u, (\beta\varphi v)') = -\varphi(a)\xi_- v(a),$$

which shows that $(\beta u)(a) = \xi_-$. As for the convergence of the derivatives at a and b, we choose $v \in H^1(\Omega)$ with $v(a) = 0$. Again, upon integration by parts

$$\epsilon_n(u_n', v') - \epsilon_n u_n'(b)v(b) - (u_n, (\beta v)') + (\beta u_n)(b)v(b) + (u_n, \alpha v) = (f, v).$$

Passing to the limit gives

$$\begin{aligned} \lim_{n\to\infty} \epsilon_n u_n'(b)v(b) &= -((u, (\beta v)') + (f, v) - (\alpha u, v)) + \xi_+ v(b) \\ &= (-(\beta u)(b) + \xi_+)v(b) \end{aligned}$$

and, hence, $\epsilon_n u_n'(b) \longrightarrow -(\beta u)(b) + \xi_+$. A similar argument and the fact that $(\beta u)(a) = \xi_-$ gives the other limit. □

Note that typically homogeneous boundary conditions are dealt with in the literature where the analysis concentrates on the inflow-boundary ($x = a$ for $\beta > 0$), while the outflow boundary is studied by boundary layer corrections, see Eckhaus [9]. For the application to networks, however, we need to include non-homogeneous conditions at multiple vertices.

We proceed to show weak or strong convergence as $\epsilon_i \longrightarrow 0$ of the solution of the network-system (2.2)$^\epsilon$ to corresponding solutions of a 'hyperbolic' system on the graph, depending on whether $m_i := -\frac{1}{2}\beta_i' + \alpha_i = 0$ or $m_i > 0$. To this end we first discuss the limiting problem.

$$\left\{\begin{array}{l} D_{\beta_i}u_i + \alpha_i u_i = f_i, \qquad x \in \Omega_i, \\ (\beta_i u_i)(v_{J_i^-}) = \beta_i u_{J_i^-} =: \beta_i u_J,\ i \in \mathcal{I}_J^{\rightarrow}, \\ u_J := \left\{\begin{array}{ll} 0 \ \text{ if } \ J \in \mathcal{J}_D & \\ \dfrac{\sum_{k\in\mathcal{I}_J^{\leftarrow}}(\beta_k u_k)(v_J)}{\sum_{k\in\mathcal{I}_J^{\leftarrow}}\beta_k(v_J)} & \text{if } \ J \in \mathcal{J}_N. \end{array}\right. \end{array}\right. \tag{4.15}$$

Before entering the analysis of (4.15) we remark that boundary/transmission conditions can only be assigned at inflow-vertices with respect to a given edge. For that reason assignments appear only at nodes $v_{J_i^-}$ $(i \in \mathcal{I}_J^{\rightarrow})$ in $(4.15)_2$. Furthermore, it is apparent that simple cancellation rules as in the one-link case do not apply, and that, therefore, some sort of averaging and mixing got to take place at multiple nodes. Now, multiply (4.15) by z_i and integrate by parts as follows

$$\begin{aligned} &\sum_{i\in\mathcal{I}}\int_{\Omega_i}(D_{\beta_i}u_i + \alpha_i u_i)z_i\,dx \\ &= \sum_J\sum_{i\in\mathcal{I}_J^{\leftarrow}}(\beta_i u_i)(v_J)z_i(v_J) - \sum_{i\in\mathcal{I}_J^{\rightarrow}}(\beta_i u_i)(v_J)z_i(v_J) \\ &\quad + \sum_{i\in\mathcal{I}}\int_{\Omega_i}u_i(-(\beta_i z_i)' + \alpha_i z_i)\,dx \\ &= \sum_J\sum_{i\in\mathcal{I}_J^{\leftarrow}}(\beta_i u_i)(v_J)z_i(v_J) \\ &\quad - \sum_{J\in\mathcal{J}_N}\frac{\sum_{k\in\mathcal{I}_J^{\leftarrow}}(\beta_k u_k)(v_J)}{\sum_{k\in\mathcal{I}_J^{\leftarrow}}\beta_k(v_J)}\sum_{k\in\mathcal{I}_J^{\rightarrow}}(\beta_i z_i)(v_J) \\ &\quad + \sum_{i\in\mathcal{I}}\int_{\Omega_i}u_i(-(\beta_i z_i)' + \alpha_i z_i)\,dx. \end{aligned} \tag{4.16}$$

We remark, that the previous computation can only be done, if the functions u_i and z_i are in $H^1(\Omega_i)$. Therefore we assume from now on that

$$\beta_i > 0 \quad \forall i \in \mathcal{I}, \tag{4.17}$$

because in that case $(\beta_i u_i) \in H^1(\Omega_i)$ is equivalent to $u_i \in H^1(\Omega_i)$. Define the 'boundary'-form at a node $J \in \mathcal{J}_N$

$$B_J(u,z) := \sum_{i\in\mathcal{I}_J^{\leftarrow}}(\beta_i u_i)(v_J)z_i(v_J) - \frac{\sum_{k\in\mathcal{I}_J^{\leftarrow}}(\beta_k u_k)(v_J)}{\sum_{k\in\mathcal{I}_J^{\leftarrow}}\beta_k(v_J)}\sum_{k\in\mathcal{I}_J^{\rightarrow}}(\beta_i z_i)(v_J) \tag{4.18}$$

and the differential expressions

$$\begin{array}{rcll} \mathcal{L} & := & (L_i)_{i\in\mathcal{I}}, & L_i u_i = D_{\beta_i} u_i + \alpha_i u_i, \\ \mathcal{L}^\star & := & (L_i^\star)_{i\in\mathcal{I}}, & L_i^\star u_i = -(\beta_i u_i)' + \alpha_i u_i = -L_i u_i + 2m_i u_i. \end{array} \tag{4.19}$$

Then in terms of the scalar product of H, (4.16) reads like

$$(\mathcal{L}u, z) = \sum_{J\in\mathcal{J}} B_J(u, z) + (u, \mathcal{L}^\star z). \tag{4.20}$$

Therefore, the adjoint boundary/transmission conditions at the Neumann-joints are given for inflow-edges as follows

$$z_i(v_J) = \frac{\sum_{k\in\mathcal{I}_J^\rightarrow}(\beta_k z_k)(v_J)}{\sum_{k\in\mathcal{I}_J^\leftarrow}\beta_k(v_J)}, \quad i \in \mathcal{I}_J^\leftarrow. \tag{4.21}$$

Let us discuss the case where $z = u$. Then for any $J \in \mathcal{J}_N$ we have

$$B_J(u,u) = \sum_{i\in\mathcal{I}_J^\leftarrow} \beta_i(v_J)u_i^2(v_J) - \frac{\sum_{i\in\mathcal{I}_J^\rightarrow}\beta_i(v_J)}{\left(\sum_{k\in\mathcal{I}_J^\leftarrow}\beta_i(v_J)\right)^2}\left(\sum_{i\in\mathcal{I}_J^\leftarrow}(\beta_i u_i)(v_J)\right)^2. \tag{4.22}$$

But

$$\left(\sum_{i\in\mathcal{I}_J^\leftarrow}(\beta_i^{1/2}\beta_i^{1/2}u_i)(v_J)\right)^2 \le \sum_{i\in\mathcal{I}_J^\leftarrow}\beta_i(v_J)\sum_{i\in\mathcal{I}_J^\leftarrow}\beta_i(v_J)u_i^2(v_J),$$

hence,

$$\begin{array}{rcl} B_J(u,u) & = & \displaystyle\sum_{i\in\mathcal{I}_J^\leftarrow}\beta_i(v_J)u_i^2(v_J)\left(1 - \frac{\sum_{i\in\mathcal{I}_J^\rightarrow}\beta_i(v_J)}{\sum_{i\in\mathcal{I}_J^\leftarrow}\beta_i(v_J)}\right) \\ & \ge & \mu_J\dfrac{\sum_{i\in\mathcal{I}_J^\leftarrow}\beta_i(v_J)u_i^2(v_J)}{\sum_{i\in\mathcal{I}_J^\leftarrow}\beta_i(v_J)} \ge 0, \end{array} \tag{4.23}$$

according to our flux condition $(4.5)_2$. Thus, by (4.23), (4.20)

$$(\mathcal{L}u, u) \ge \sum_{i\in\mathcal{I}}(m_i u_i, u_i) \tag{4.24}$$

for all u satisfying $(4.15)_{2,3}$. Indeed, the operator $\mathcal{A}^0$ defined by $\mathcal{A}^0 u = (A_i^0 u_i)_{i\in\mathcal{I}}$

$$\left\{\begin{array}{rcll} A_i^0 u_i & = & D_{\beta_i}u_i + \alpha_i u_i, & i \in \mathcal{I}, \\ D(\mathcal{A}^0) & = & \left\{u \in \prod_{i\in\mathcal{I}} D(D_{\beta_i}) \middle| \, u \text{ satisfies } (4.15)_{2,3}\right\} & \end{array}\right. \tag{4.25}$$

is a well-defined closed operator in H, with adjoint $(\mathcal{A}^0)^\star$ given by

$$\left\{\begin{array}{rcl} (A_i^0)^\star u_i & = & (\beta_i u_i)' + \alpha_i u_i \\ D((\mathcal{A}^0)^\star) & = & \Big\{u \in \prod_{i\in\mathcal{I}} D(D_{\beta_i}) \big| u \text{ satisfies } (4.21) \\ & & \qquad \text{for } J \in \mathcal{J}_N \text{ and } u_i(v_J) = 0 \text{ for } i \in \mathcal{I}_D^\leftarrow\Big\}. \end{array}\right. \tag{4.26}$$

Note that these results are true also for $m_i \equiv 0$. If all $m_i \geq m_0 > 0,\ i \in \mathcal{I}$, we obtain a positive definite operator with respect to the H-topology

$$(\mathcal{A}^0 u, u) \geq m_0^2 \|u\|_H. \tag{4.27}$$

For that situation one can apply the theory of C. Bardos [3] both for wellposedness of the corresponding time-dependent problem as well as for the limiting procedure when $\epsilon \longrightarrow 0$. Even though we discretize $(2.2)^\epsilon$ with respect to time and obtain (3.1), the constant α_i depends on the step-size. It is therefore reasonable to also handle the case $\alpha_i = 0$. The results are, however, also of independent interest in many flow problems on graphs. We, therefore, present a result which covers both cases $m_i > 0$ and $m_i = 0$ in a unified way, thereby revealing strong convergence in the first and weak convergence in the second case. This is in agreement with J.L. Lions [17] and C. Bardos [3].

THEOREM 2 Let the assumptions of Theorem 1 be satisfied. Then, as $\max_{i \in \mathcal{I}} \epsilon_i$ tends to zero, the solutions u^ϵ of $\mathcal{A}^\epsilon u^\epsilon = f$ tend to the solutions of the corresponding problem $\mathcal{A}^0 u = f$. If $m = (m_i)_{i \in I}$ is equal to zero the convergence is in the weak H-sense, if all $m_i > 0$, strong H-convergence holds. Moreover, we have

$$\begin{aligned} \epsilon_i (u_i^\epsilon(v_J))' &\longrightarrow 0 && \forall i \in \mathcal{I}_J^{\rightarrow}, \\ \epsilon_i (u_i^\epsilon(v_J))' &\longrightarrow \beta_i u_J - (\beta_i u_i)(v_J) && \forall i \in \mathcal{I}_J^{\leftarrow}. \end{aligned}$$

Proof: By Theorem 1 and Lemma 2, u^ϵ and $\varphi^{1/2}\beta(u^\epsilon)'$ as well as $u^\epsilon(v_J),\ J \in \mathcal{J}_N$ are all bounded. Hence, we can extract subsequences (which we again denote by the label ϵ) such that

$$\begin{array}{ll} u^\epsilon \overset{w}{\rightharpoonup} u, & \varphi\beta(u^\epsilon)' \overset{w}{\rightharpoonup} z \\ \epsilon(u^\epsilon)' \overset{w}{\rightharpoonup} 0, & u^\epsilon(v_J) \longrightarrow u_J \end{array} \tag{4.28}$$

for some $u, z \in H,\ u_J \in \mathbb{R}$, and the function φ constructed as in Theorem 1. The convergence results in (4.28) imply the convergence of the variational forms. As for the transmission conditions we consider

$$0 = \sum_{k \in \mathcal{I}_J^{\leftarrow}} \epsilon_k (u_k^\epsilon)'(v_J) - \sum_{k \in \mathcal{I}_J^{\rightarrow}} \epsilon_k (u_k^\epsilon)'(v_J), \qquad \forall J \in \mathcal{J}_N. \tag{4.29}$$

We apply Lemma 3 on each edge incident at $v_J,\ J \in \mathcal{J}_N$, and, passing to the limit in (4.29), we obtain

$$0 = \sum_{k \in \mathcal{I}_J^{\leftarrow}} (\beta_k(v_J) u_J - (\beta_k u_k)(v_J)),$$

hence $(4.15)_{2,3}$. The Dirichlet conditions at $v_J, J \in \mathcal{J}_D,\ i \in I_J^{\rightarrow}$ are preserved. Further, it is plain that the limiting functions are uniquely determined and that in turn the limit u satisfies $\mathcal{A}_u^0 = f$ weakly. As for strong convergence, we proceed along the lines of [17] Chap. II, Theorem 3.1: we know that for $m_i > 0\ \forall i \in \mathcal{I}$ (4.27) holds. We have

$$\begin{aligned} X_\epsilon &:= \sum_{i \in \mathcal{I}} (D_{\beta_i}(u_i^\epsilon - u_i) + \alpha_i(u_i^\epsilon - u_i), u_i^\epsilon - u_i) \\ &= \sum_{i \in \mathcal{I}} (D_{\beta_i} u_i^\epsilon + \alpha_i u_i^\epsilon, u_i^\epsilon - u_i) - (f_i, u_i^\epsilon - u_i). \end{aligned} \tag{4.30}$$

On the other hand

$$\begin{aligned}\sum_{i\in\mathcal{I}}(f_i,u_i^\epsilon) &= \sum_{i\in\mathcal{I}}(-\epsilon_i(u_i^\epsilon)'',u_i^\epsilon)+(D_{\beta_i}u_i^\epsilon+\alpha_i u_i^\epsilon,u_i^\epsilon)\\ &\geq \sum_{i\in\mathcal{I}}(D_{\beta_i}u_i^\epsilon+\alpha_i u_i^\epsilon,u_i^\epsilon),\end{aligned}$$

which inserted into (4.30) gives

$$X_\epsilon \leq \sum_{i\in\mathcal{I}}(f_i,u_i)-(D_{\beta_i}u_i^\epsilon+\alpha_i u_i^\epsilon,u_i).$$

Now we pass to the limit the right hand side of the previous inequality. For all terms it is clear what happens, except $-(D_{\beta_i}u_i^\epsilon,u_i)$. But we have

$$\sum_{i\in\mathcal{I}}-(D_{\beta_i}u_i^\epsilon,u_i)=\sum_{i\in\mathcal{I}}(u_i^\epsilon,(\beta_i u_i)')-\sum_{J\in\mathcal{J}_N}u^\epsilon(v_J)(\sum_{i\in\mathcal{I}_J^{\leftarrow}}(\beta_i u_i)(v_J)-\sum_{i\in\mathcal{I}_J^{\rightarrow}}(\beta_i u_i)(v_J)).$$

Taking into account that

$$\begin{aligned}\sum_{i\in\mathcal{I}_J^{\leftarrow}}(\beta_i u_i)(v_J) &= u_J\sum_{i\in\mathcal{I}_J^{\leftarrow}}\beta_i(v_J),\\ (\beta_i u_i)(v_J) &= \beta_i(v_J)u_J \quad \forall i\in\mathcal{I}_J^{\rightarrow},\end{aligned}$$

this is equal to

$$\sum_{i\in\mathcal{I}}(u_i^\epsilon,(\beta_i u_i)')-\sum_{J\in\mathcal{J}_N}u^\epsilon(v_J)u_J(\sum_{i\in\mathcal{I}_J^{\leftarrow}}\beta_i(v_J)-\sum_{i\in\mathcal{I}_J^{\rightarrow}}\beta_i(v_J)).$$

Passing to the limit, we obtain

$$\begin{aligned}\limsup X_\epsilon &\leq \sum_{i\in\mathcal{I}}(f_i,u_i)-(\alpha_i u_i,u_i)\\ &\quad +\sum_{i\in\mathcal{I}}(u_i,(\beta_i u_i)')-\sum_{J\in\mathcal{J}_N}u_J^2(\sum_{i\in\mathcal{I}_J^{\leftarrow}}\beta_i(v_J)-\sum_{i\in\mathcal{I}_J^{\rightarrow}}\beta_i(v_J))\\ &= \sum_{i\in\mathcal{I}}(f_i,u_i)-(\alpha_i u_i,u_i)-(D_{\beta_i}u_i,u_i)\\ &= 0.\end{aligned}$$

Using (4.26) gives the desired strong convergence. □

5 THE SINGULAR PERTURBATION OF THE OPTIMAL CONTROL PROBLEM

In this section we go back to the instantaneous optimal control problem (3.1), (3.2) at given ϵ-level. In particular, we can assume throughout this section that $m_i > 0, \quad \forall i \in \mathcal{I}$, and, hence, that the strong convergence result of Theorem 2 holds. We fix the time iteration at k and consider the input $F_i := f_i + u_i^k$ as in (3.1), but with u_i^k being the solution of the previous time step after passing to the limit with $\epsilon \longrightarrow 0$. Then, by Theorem 3 we know that u^ϵ, the solution of (4.2) converges strongly to a solution u of (4.15) with the same right hand side F_i. As the cost function itself does not depend on ϵ explicitly we take the same cost function for the limiting problem and denote it by $J^k(f)$.

By standard theory, there exists a unique optimal control for the problem

$$\left\{\begin{array}{l} (IOCP)_0: \quad \min_f J^k(f) \\ \text{subject to (4.15) with right hand side } f_i + \gamma_i u_i^k. \end{array}\right. \tag{5.1}$$

By continuity $J_\epsilon^k(f) \longrightarrow J^k(f)$, and, following the arguments in J.L. Lions [17] Chap. VII, Theorem 1.1, we conclude

THEOREM 3 Let f^ϵ, f be the solutions of the optimal control problems (3.1), (5.1), respectively. Denote by $u^\epsilon(f^\epsilon), u(f)$ the corresponding solutions of the differential equation (3.1), (4.15) (with proper right hand side). Then

$$f^\epsilon \longrightarrow f,\ u^\epsilon \longrightarrow u \text{ in } H \tag{5.2}$$

and

$$J_\epsilon^k(f^\epsilon) = \inf J_\epsilon^k(g) \longrightarrow J(f) = \inf J^k(g).$$

Moreover, let (u^ϵ, p^ϵ) be a solution of the optimality system (3.9). Then, as ϵ tends to zero, we obtain (u, p) solving the optimality system for $\epsilon = 0$

$$\left\{\begin{array}{l} D_{\beta_i} u_i + \alpha_i u_i + \dfrac{1}{\sigma_i} p_i = \gamma_i u_i^k \\ -(\beta_i p_i)' + \alpha_i p_i - u_i = -z_i^{k+1}, \\ (\beta_i u_i)(v_J) = 0,\ J \in \mathcal{J}_D,\ i \in \mathcal{I}_J^{\rightarrow}, \\ (\beta_i p_i)(v_J) = 0,\ J \in \mathcal{J}_D,\ i \in \mathcal{I}_J^{\leftarrow}, \\ (\beta_i u_i)(v_J) = \beta_i(v_J) \dfrac{\sum_{k \in \mathcal{I}_J^{\leftarrow}} \beta_k u_k(v_J)}{\sum_{k \in \mathcal{I}_J^{\leftarrow}} \beta_k(v_J)},\ J \in \mathcal{J}_N,\ i \in \mathcal{I}_J^{\rightarrow}, \\ (\beta_i p_i)(v_J) = \beta_i(v_J) \dfrac{\sum_{k \in \mathcal{I}_J^{\rightarrow}} \beta_k p_k(v_J)}{\sum_{k \in \mathcal{I}_J^{\leftarrow}} \beta_k(v_J)},\ J \in \mathcal{J}_N,\ i \in \mathcal{I}_J^{\leftarrow}. \end{array}\right. \tag{5.3}$$

Proof: The first part follows along the lines of [17]. As for (5.3) we have to show that it is in fact the optimality system associated with (5.1).

But, as in (3.3),

$$\delta J^k(f)(g) = \sum_{i \in \mathcal{I}} \int_{\Omega_i} (u_i - z_i^{k+1}) y_i + \sigma_i f_i g_i \, dx,$$

where now y solves (4.15) with right hand side g. Hence, according to (4.16), (4.18)

$$\begin{aligned} (g, p) &= \sum_{i \in \mathcal{I}} (g_i, p_i) = \sum_{i \in \mathcal{I}} \int_{\Omega_i} (D_{\beta_i} y_i + \alpha_i y_i) p_i \, dx \\ &= \sum_{J \in \mathcal{J}_N} B_J(y, p) + \sum_{i \in \mathcal{I}} \int_{\Omega_i} y_i(-D_{\beta_i} p_i + \alpha_i p_i) \, dx \\ &= \sum_{i \in \mathcal{I}} \int_{\Omega_i} y_i(u_i - z_i^{k+1}) \, dx, \end{aligned}$$

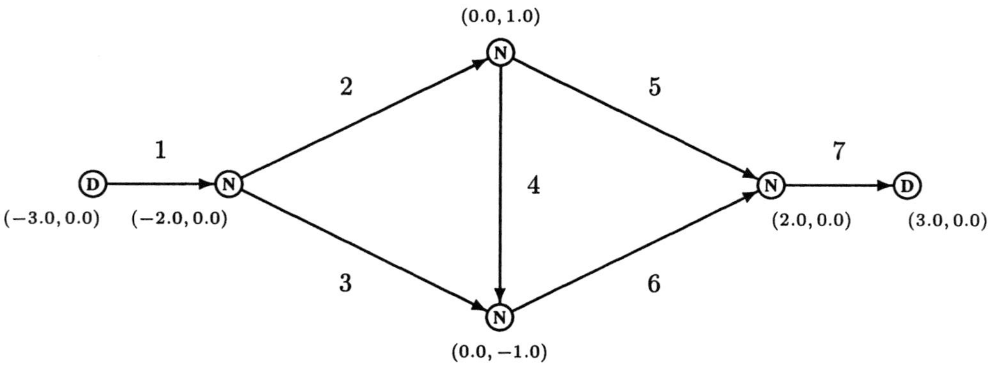

Figure 1: network

edge	$10^2\epsilon_i$	β_i
1	1.0	1.0
2	1.0	2.0
3	1.0	3.0
4	1.0	1.0
5	2.0	2.0
6	3.0	2.0
7	4.0	1.0

Table 1: values for ϵ_i and β_i

if p satisfies $(5.3)_{2,4,6}$. Therefore, the optimality condition $\delta J^k(f)(g) = 0 \; \forall g \in H$ gives again $f_i = -\dfrac{1}{\sigma_i} p_i$. The convergence of the solutions of the optimality system (3.9) to these of (5.3) follows by Theorem 2. □

6 NUMERICAL COMPUTATIONS

Finally, we give some numerical examples for instantaneous controls on a simple 2d-network. It is well known that the usual finite element method leads to oscillations if the diffusion parameter ϵ is taken small compared to the convection term β. It is also well known that the method can be stabilized by using bubble functions (see the SUPG-method), the analogon of upwind schemes for finite differences. Numerical studies along these lines for problems on graphs are currently under investigation.

For the purpose of demonstrating the effectiveness of our instantaneous control regime, however, it suffices to take sufficiently fine discretizations in order to represent the effects of the singular perturbation analysis outlined above, i. e. we take ϵ_i in the order of 10^{-2}, so that the averaging condition of the multiple nodes can be shown to be significant.

For our control problem we use the network shown by figure 1. The labels 'D' and 'N' at the nodes indicate Dirichlet-/Neumann-conditions with the constant parameters ϵ_i and β_i as shown in table 1 and $\alpha_i = 0$. As initial and target state we take the global zero function and a function with a bump on one edge. Figure 3 shows the uncontrolled time evolution of the bump function. As time discretization we take $\Delta t = 0.01$. We should mention here, that in all time evolution plots, the plots start with the the first time iteration step, not with

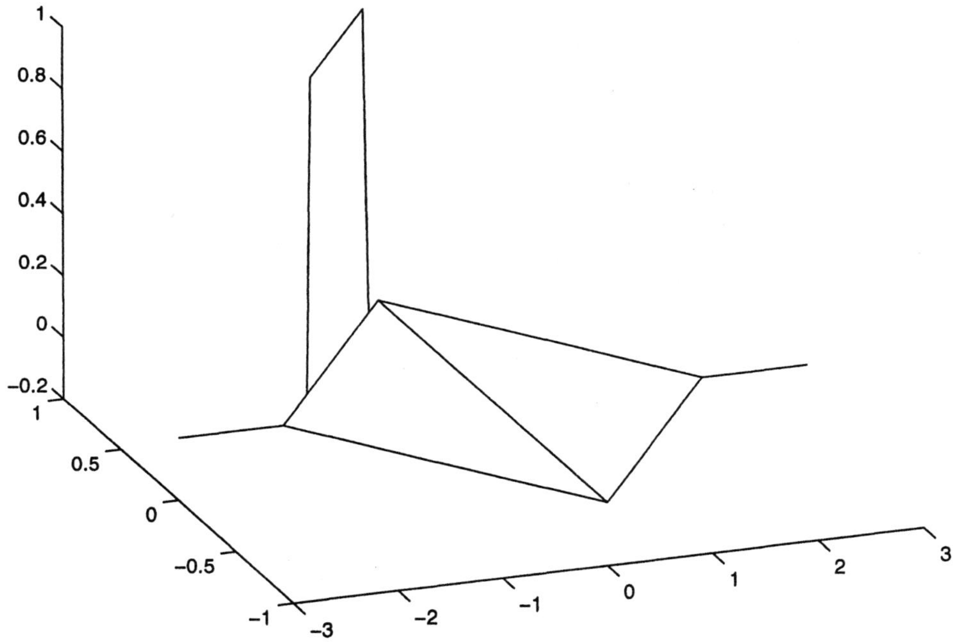

Figure 2: bump function

the initial state.

As for the controlled problems, we consider two scenarios. In the first set of experiments we are going to control the state to zero on the entire graph (B2Z). Unlike in the case of the heat equation, where boundary null-controllability is known to hold for a long time, the diffusion-convection case has only been handled very recently by using Carleman estimates, see [18]. For such problems on graphs no theoretical result is currently available, even though positive results along this line are likely to be true. We let the initial condition be given by a bump on edge number 2 and control everywhere on the graph. Boundary controls could be handled equally well. However, in these notes we confined ourselves to distributed controls. We take various choices of the penalty parameter $\sigma \in \{10^{-2}, 10^{-3}, 10^{-4}\}$. As σ is of the order $1, 10^{-1}$, not much is achieved by the controls when compared to the uncontrolled situation. If, however, σ is in the order of 10^{-2}, 10^{-3}, 10^{-4}, the effect of the controls becomes dramatic, as is easily seen from the self-explaining figures 4 to 7. In the second set of experiments (Z2B) we concentrate on the largely unexplored problem of attainability, which is of course much more interesting. Here we want to achieve a final distribution along the network. Speaking in terms of flow patterns, we are looking then at such patterns that can be achieved by our controls and that can be held by keeping controls active. Again, it is clearly seen that for moderate values of σ the diffusive properties and the penalty for the cost of the control drives the final state away from the desired configuration (the same bump again). For very small σ however, the bump is perfectly achieved, almost instantaneously. This is clearly seen in figures 8 to 11.

As a final remark, we wish to mention, that domain decomposition techniques for such optimal control problems are currently under investigation, such that, ultimately, problems on networks will be treated on parallel computers.

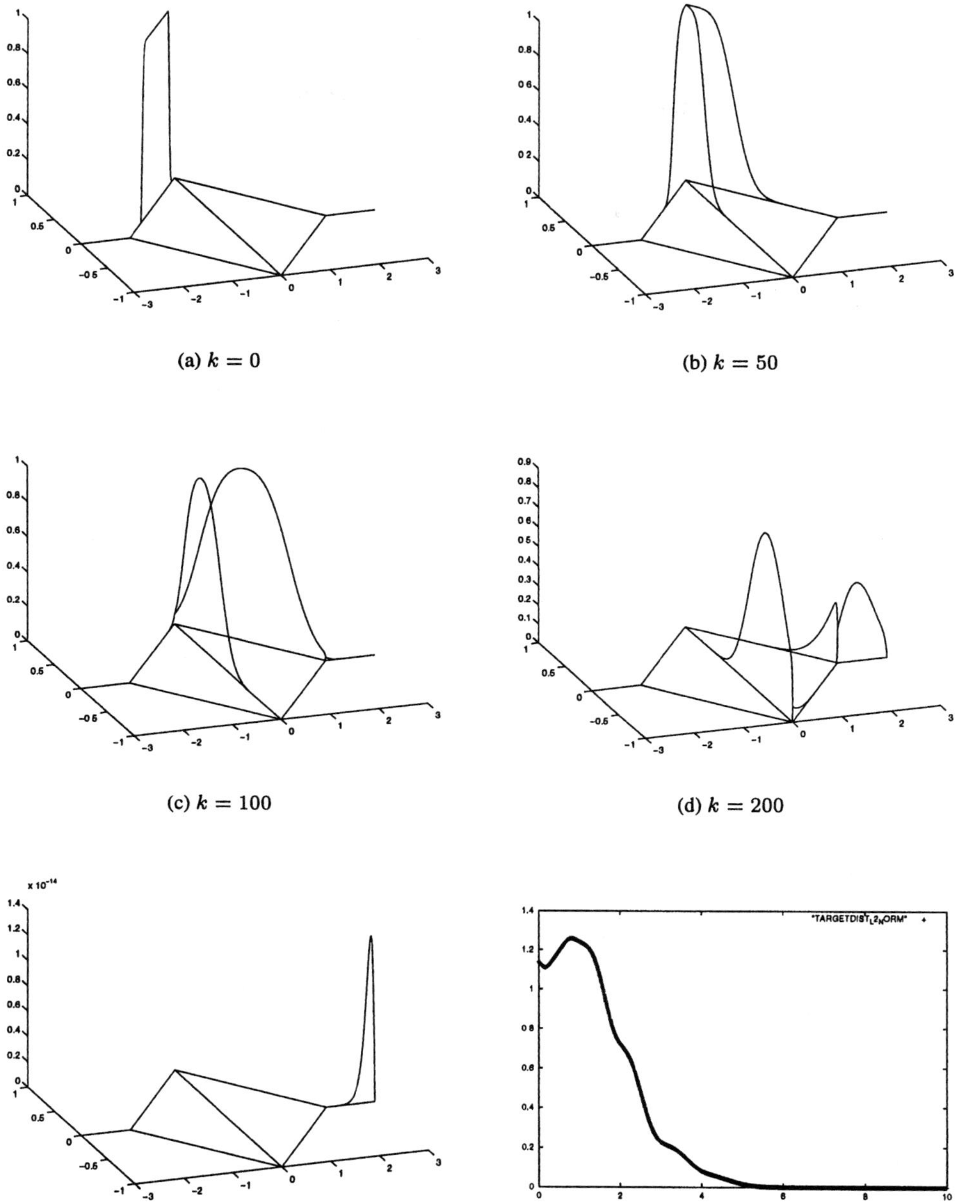

(a) $k = 0$ (b) $k = 50$

(c) $k = 100$ (d) $k = 200$

(e) $k = 1000$ (f) L^2-norm of the state

Figure 3: Time evolution of the bump function

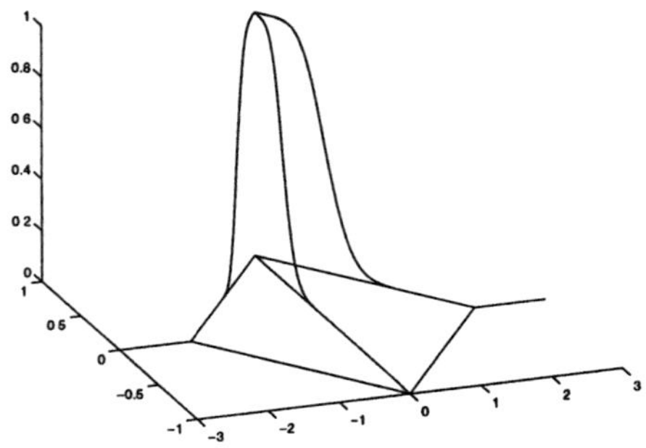

(a) $k = 50$

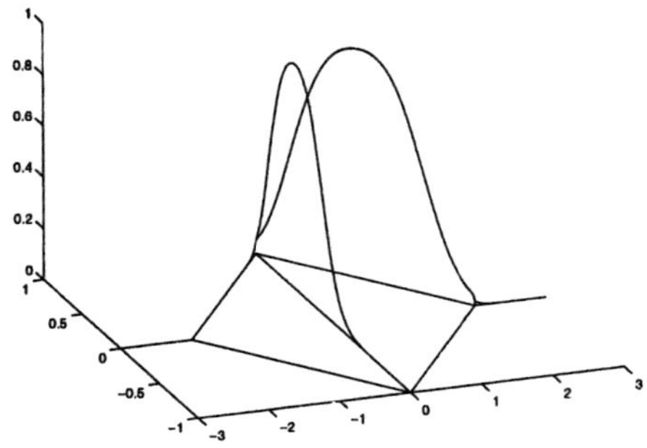

(b) $k = 100$

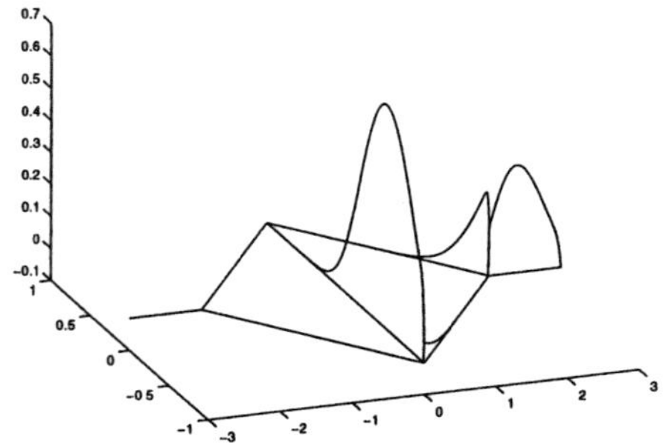

(c) $k = 200$

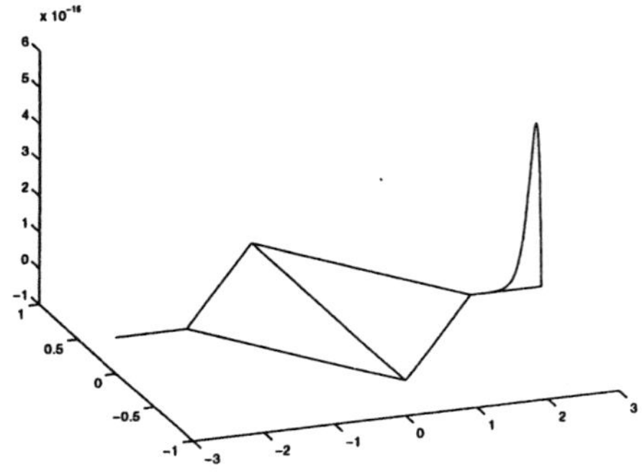

(d) $k = 1000$

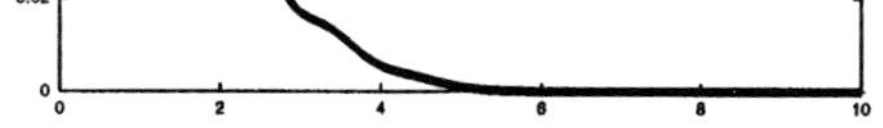

(e) L^2-norm of the control

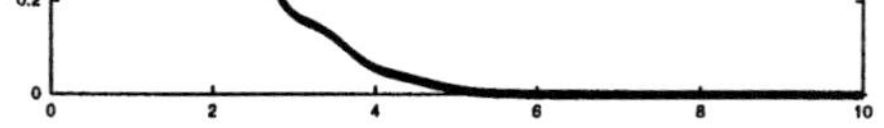

(f) L^2-norm of the distance from the target

Figure 4: (B2Z) with $\sigma = 0.1$

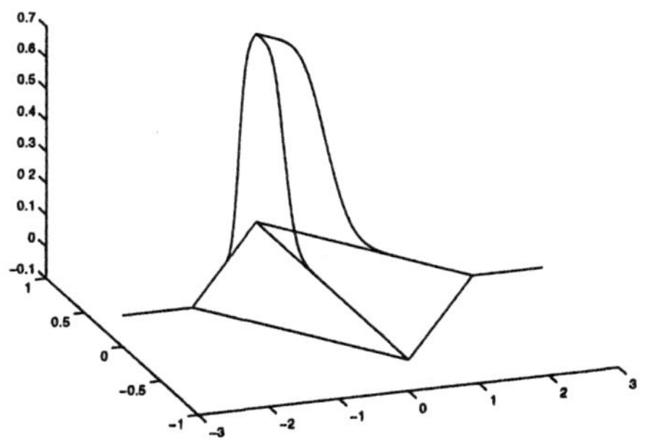

(a) $k = 50$

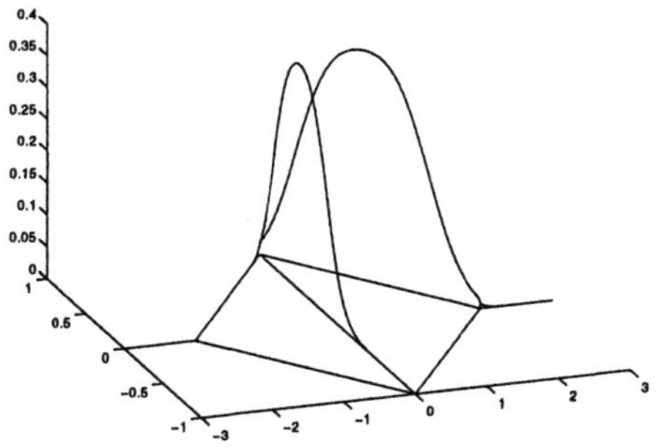

(b) $k = 100$

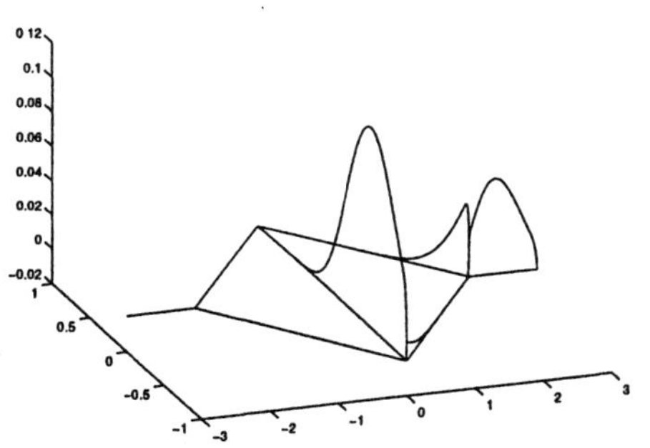

(c) $k = 200$

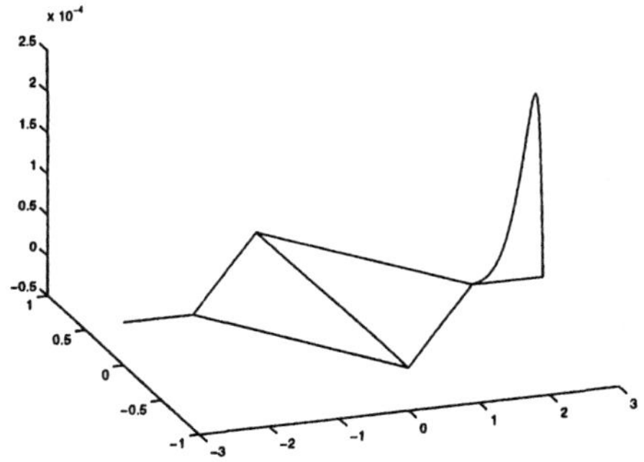

(d) $k = 500$

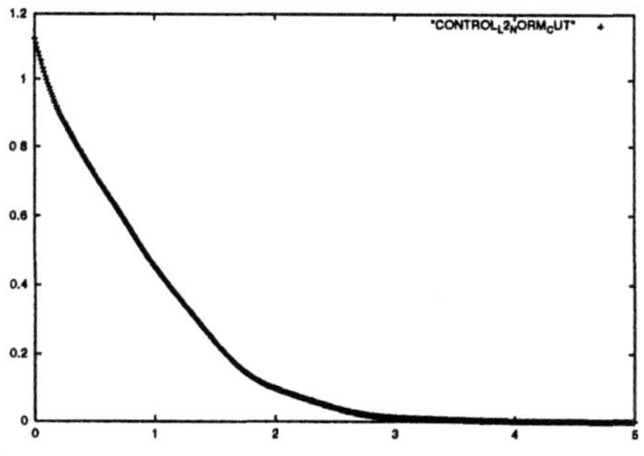

(e) L^2-norm of the control

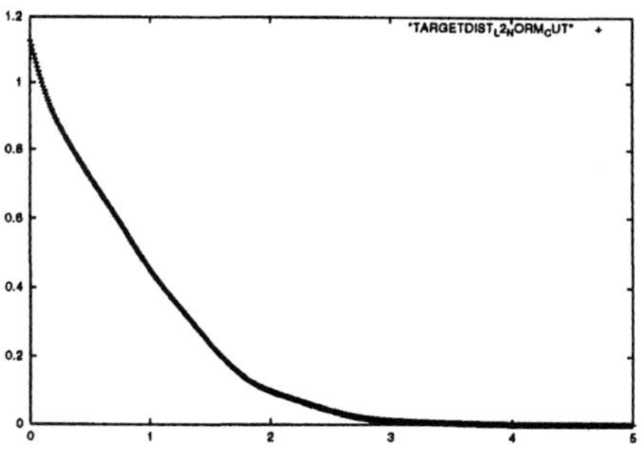

(f) L^2-norm of the distance from the target

Figure 5: (B2Z) with $\sigma = 0.01$

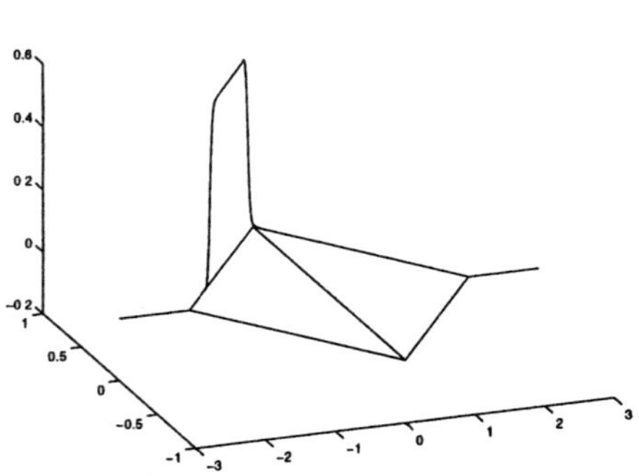

(a) $k = 5$

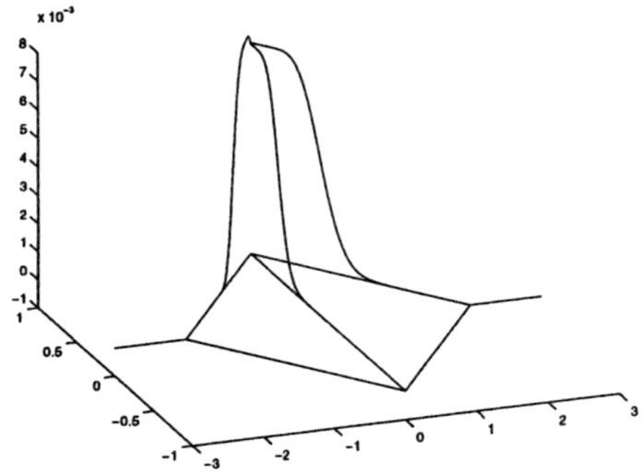

(b) $k = 50$

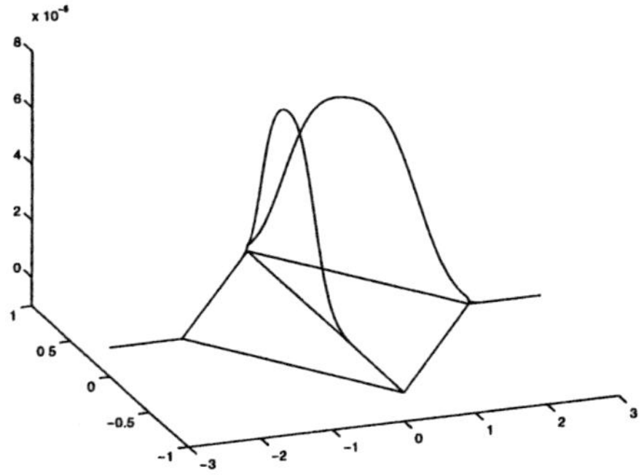

(c) $k = 100$

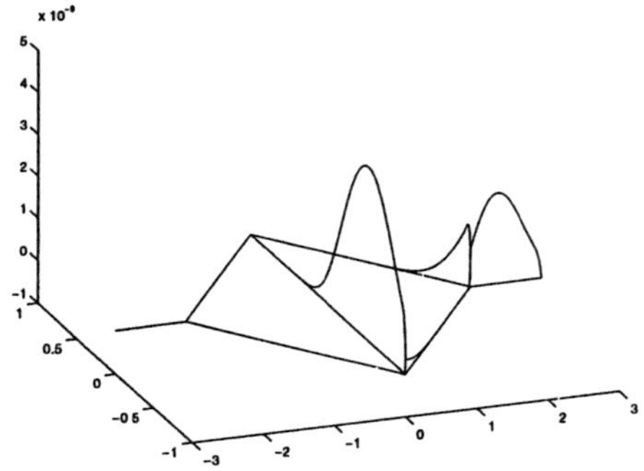

(d) $k = 200$

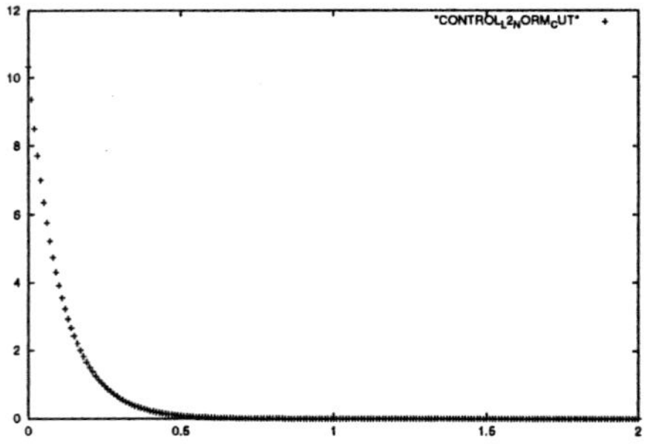

(e) L^2-norm of the control

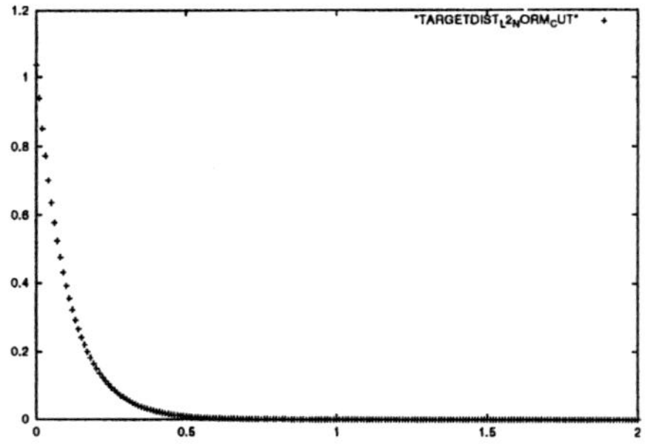

(f) L^2-norm of the distance from the target

Figure 6: (B2Z) with $\sigma = 0.001$

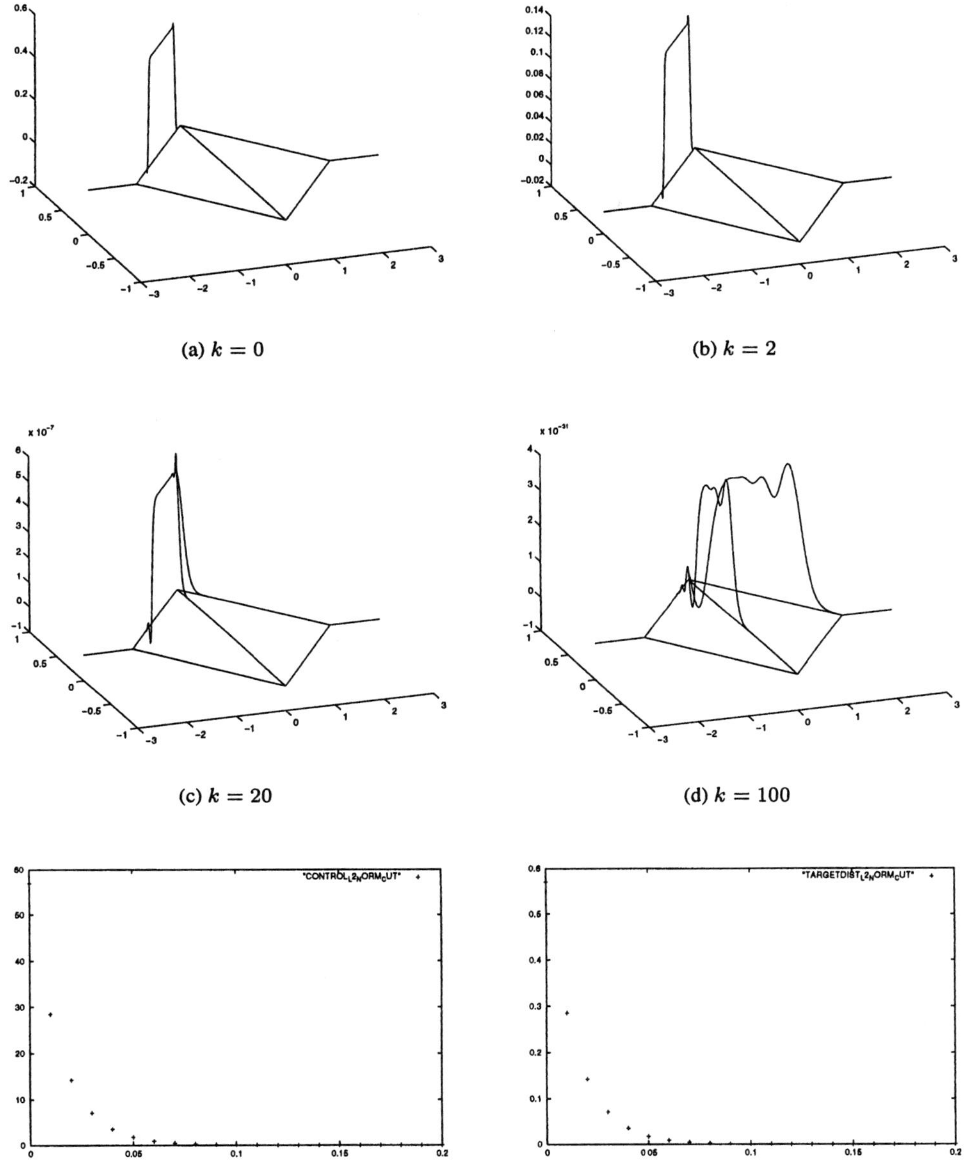

(a) $k = 0$

(b) $k = 2$

(c) $k = 20$

(d) $k = 100$

(e) L^2-norm of the control

(f) L^2-norm of the distance from the target

Figure 7: (B2Z) with $\sigma = 0.0001$

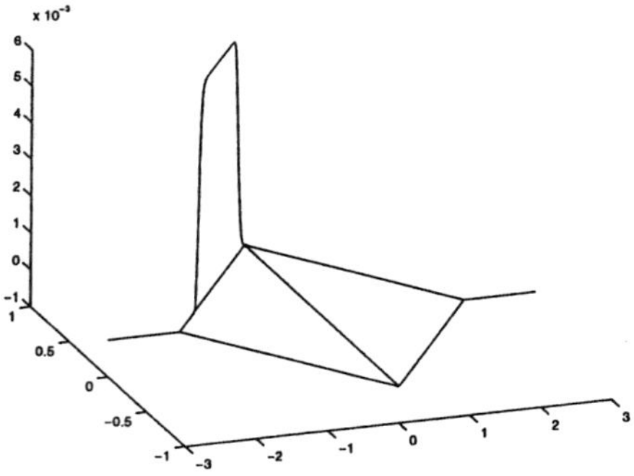

(a) $k = 5$

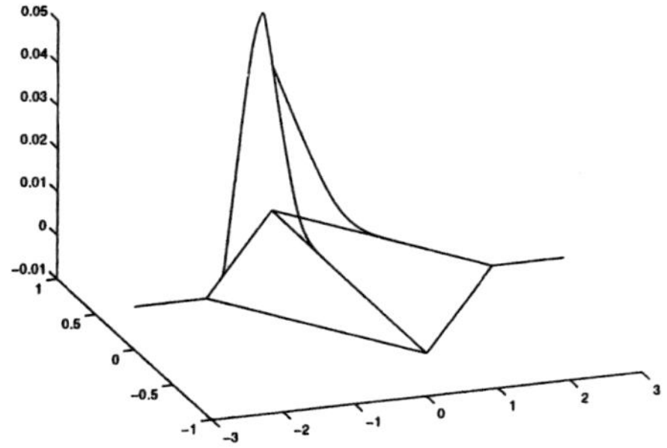

(b) $k = 50$

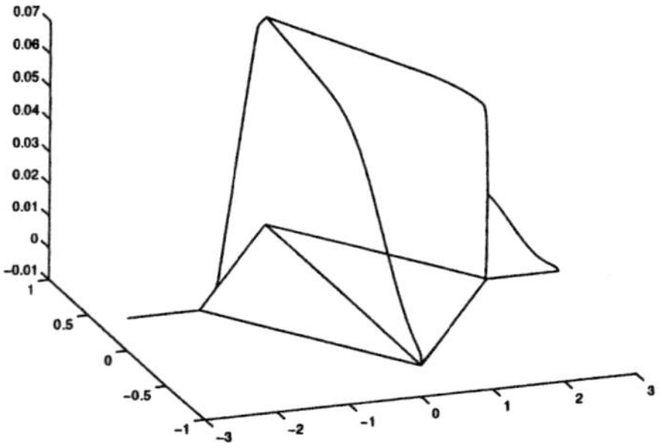

(c) $k = 200$

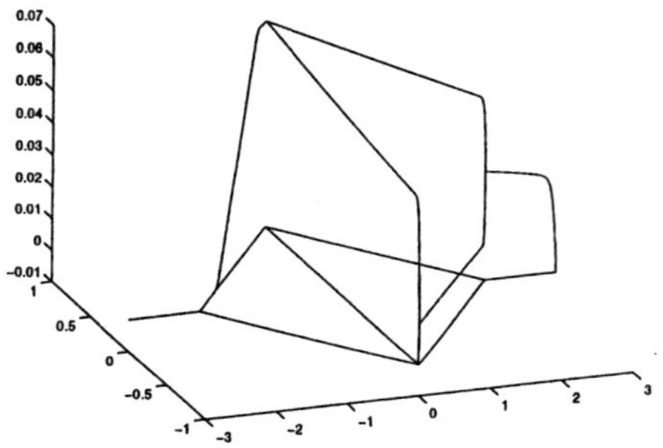

(d) $k = 500$

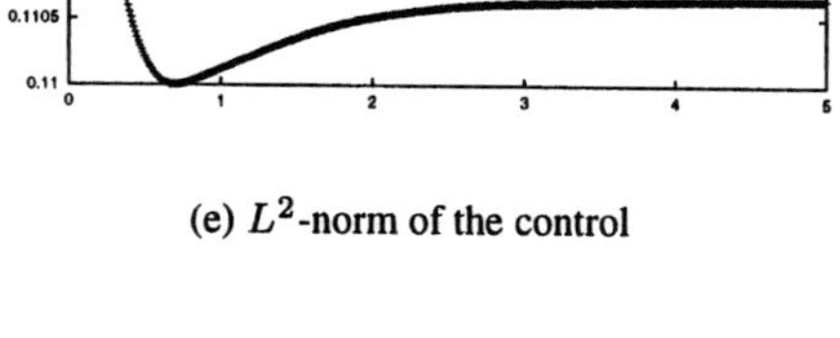

(e) L^2-norm of the control

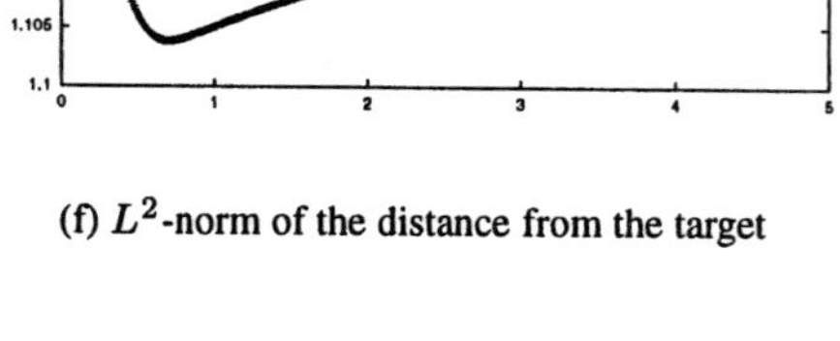

(f) L^2-norm of the distance from the target

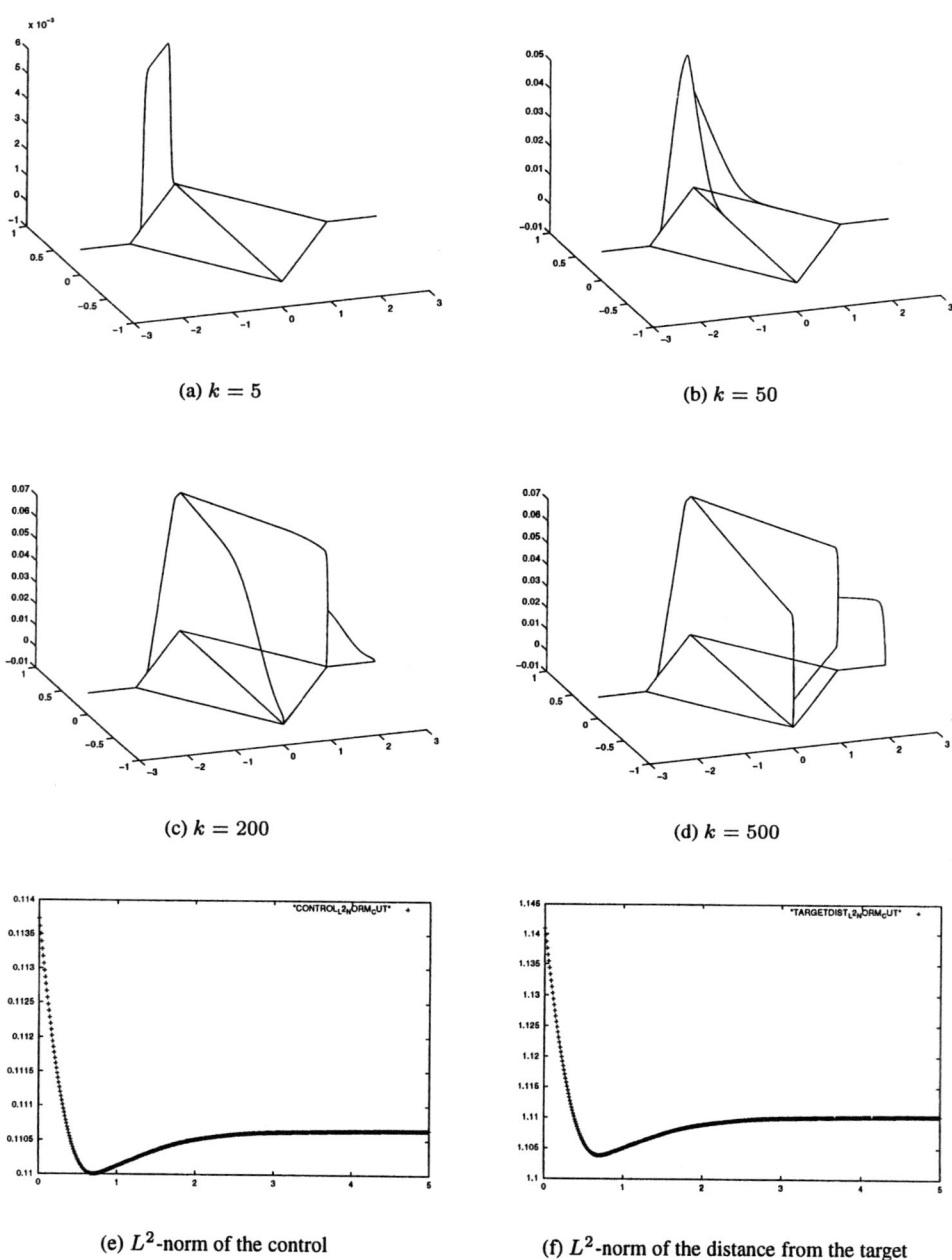

Figure 8: (Z2B) with $\sigma = 0.1$

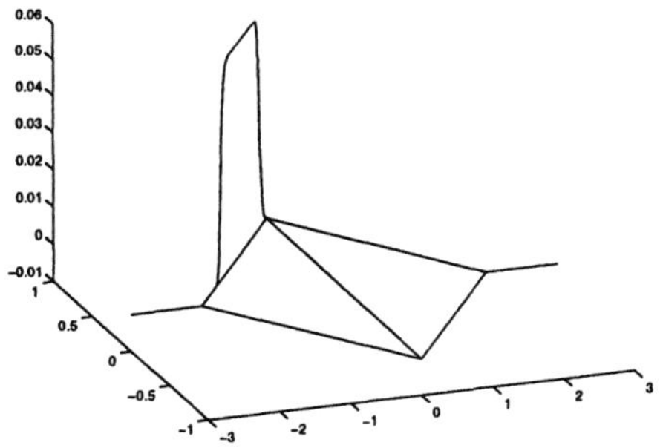

(a) $k = 5$

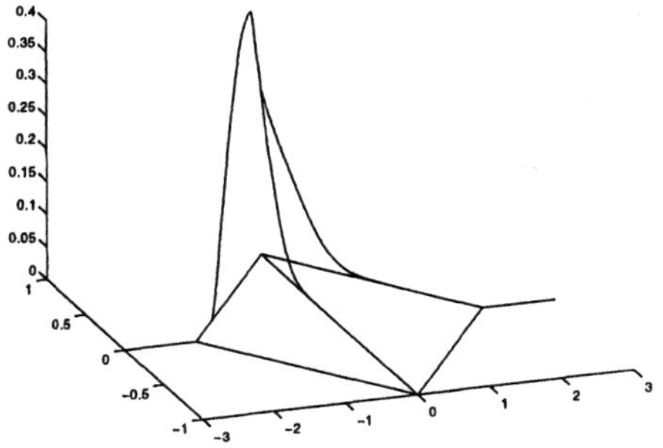

(b) $k = 50$

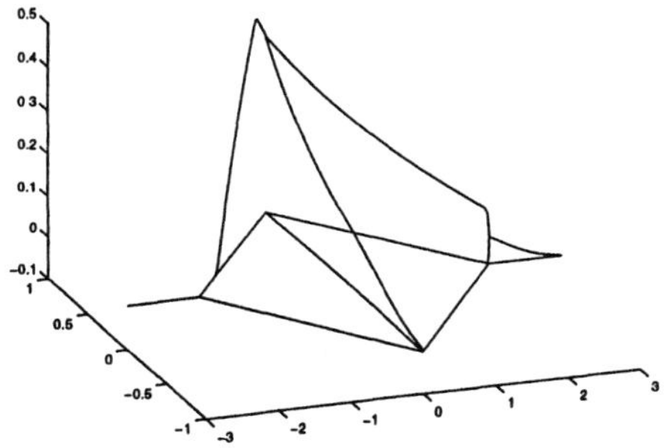

(c) $k = 200$

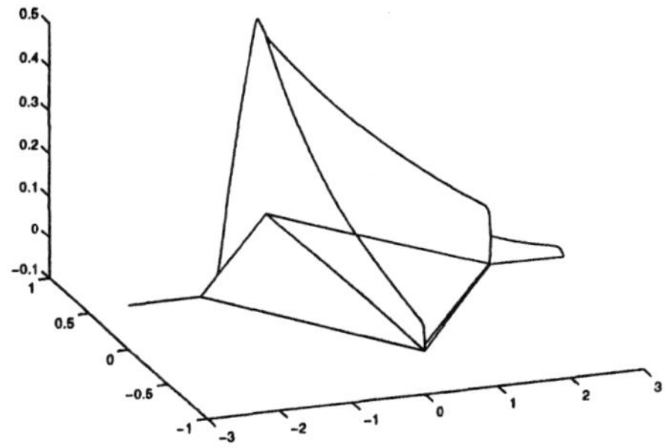

(d) $k = 500$

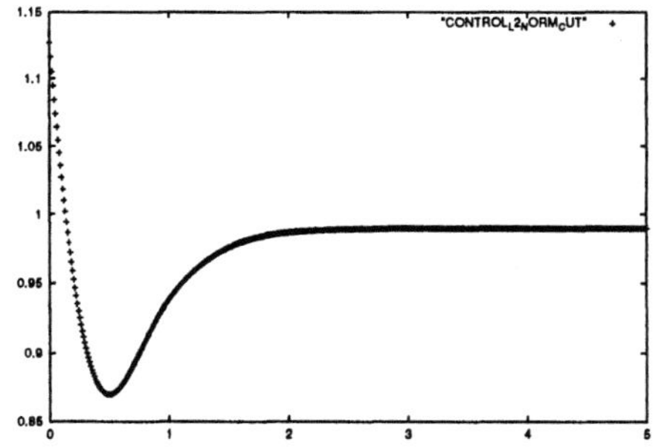

(e) L^2-norm of the control

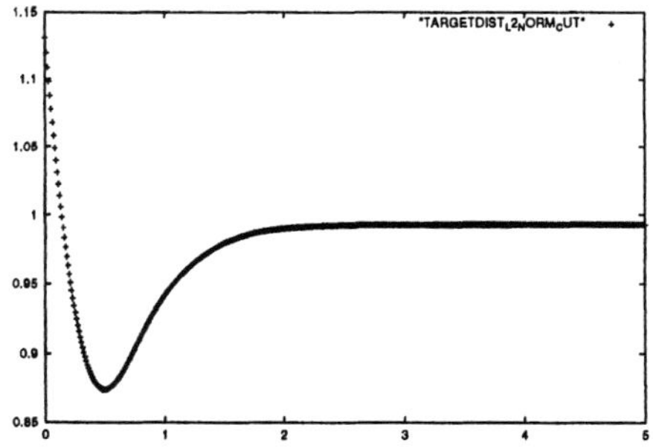

(f) L^2-norm of the distance from the target

Figure 9: (Z2B) with $\sigma = 0.01$

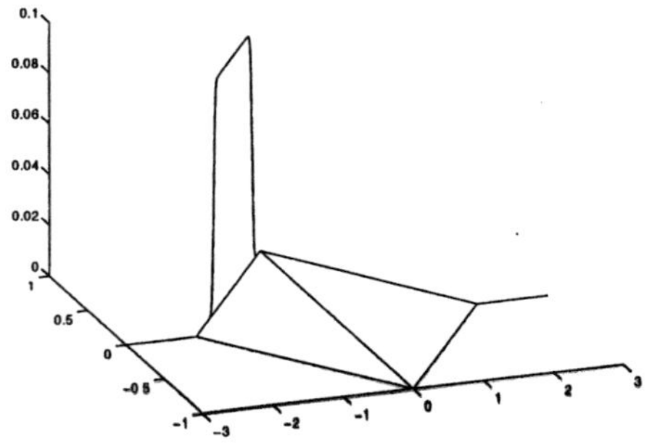

(a) $k = 0$

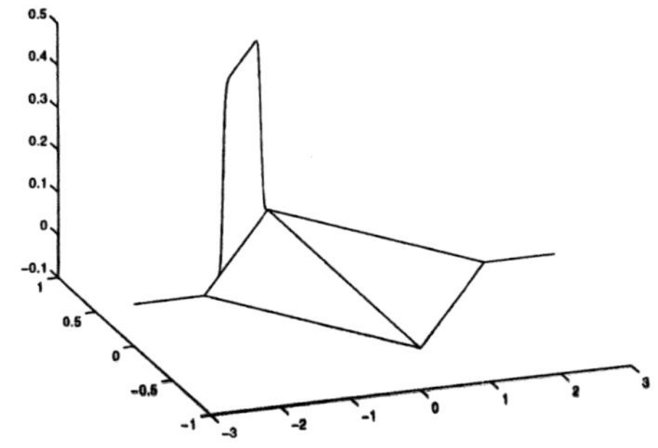

(b) $k = 5$

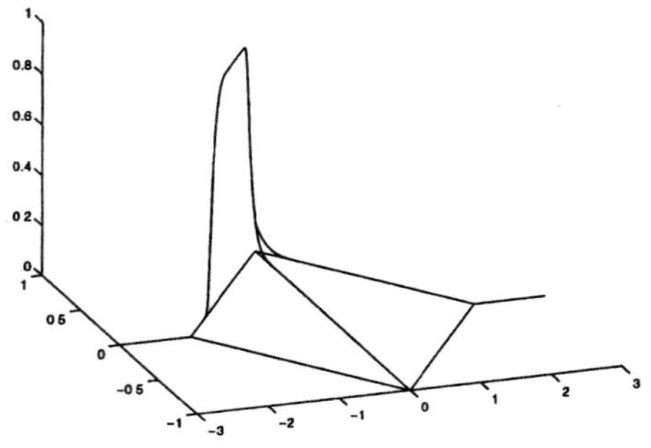

(c) $k = 20$

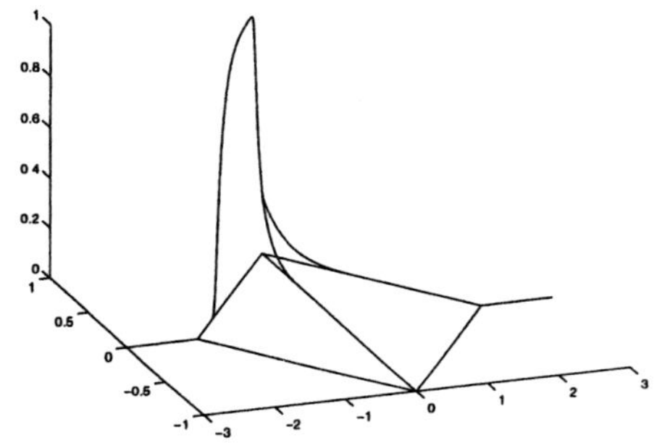

(d) $k = 50$

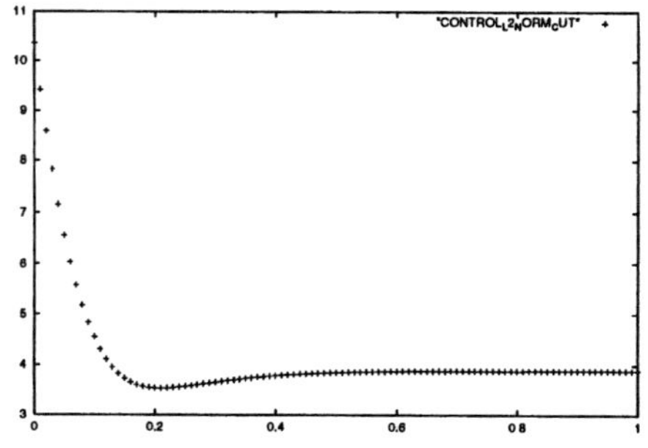

(e) L^2-norm of the control

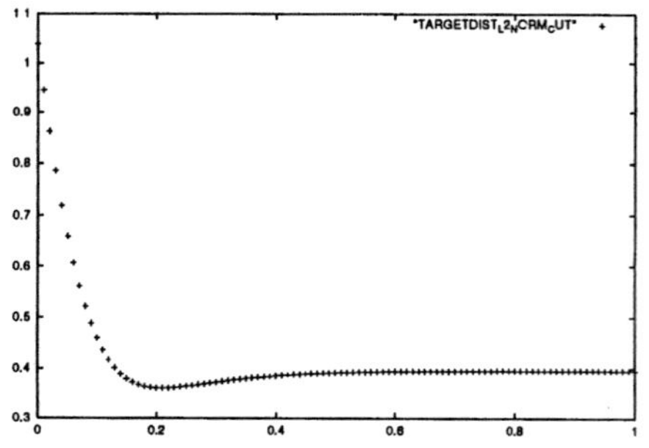

(f) L^2-norm of the distance from the target

Figure 10: (Z2B) with $\sigma = 0.001$

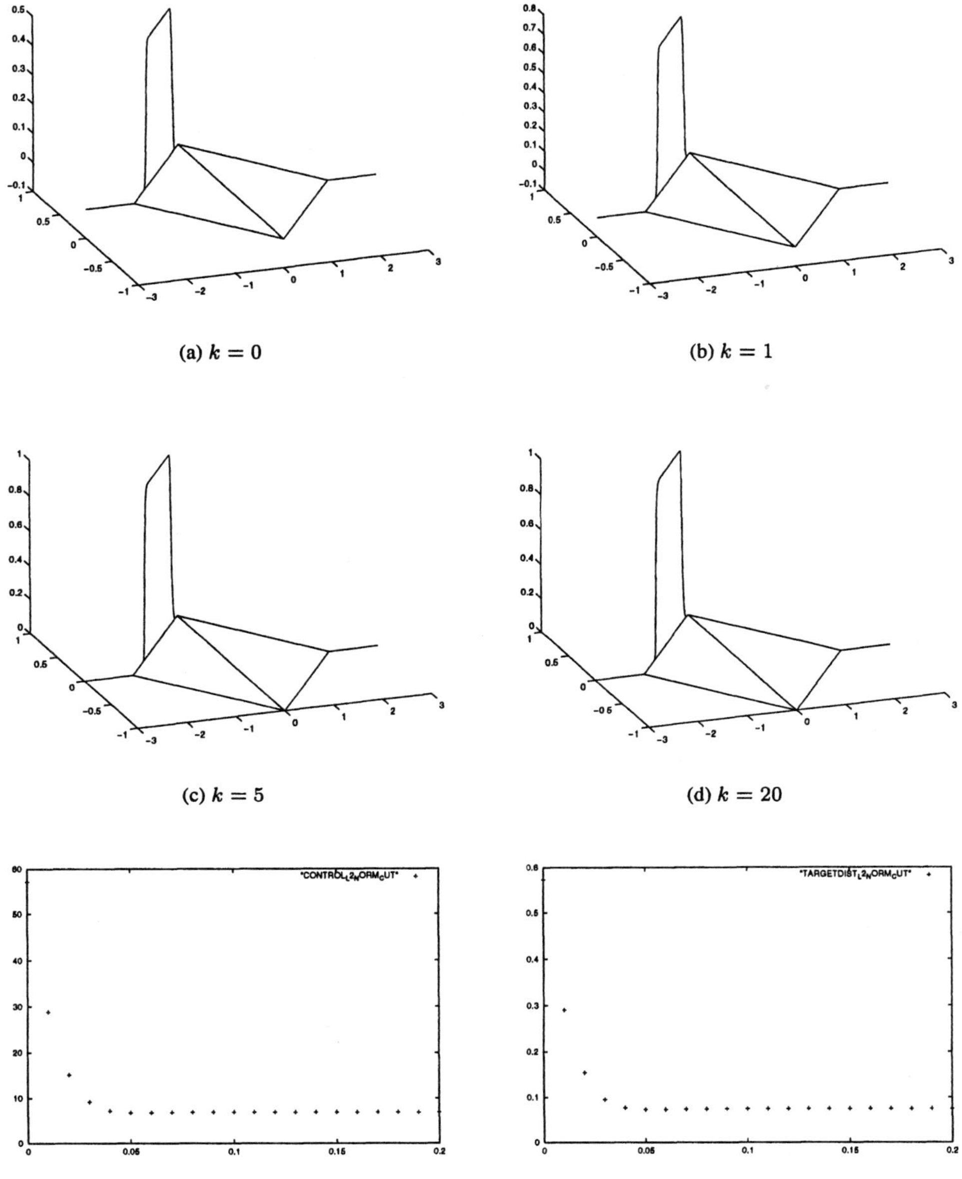

(a) $k = 0$

(b) $k = 1$

(c) $k = 5$

(d) $k = 20$

(e) L^2-norm of the control

(f) L^2-norm of the distance from the target

Figure 11: (Z2B) with $\sigma = 0.0001$

REFERENCES

1. F. Ali Mehmeti. Regular solutions of transmission and interaction problems for wave equation. Math. Meth. Appl. Sci, 11:665–685, 1989.
2. F. Ali Mehmeti. Nonlinear waves in networks. Akademie-Verlag, 1994.
3. C. Bardos. Problèmes aux limites pour les équations aux dérivées partielles du premier ordre a coefficients réels; théorèmes d'approximation; application a l'équation de transport. Ann. scient. Éc. Norm. Sup., 4(3):185–233, 1970.
4. J. von Below. Classical solvability of linear parabolic equations on networks. J. Diff. Eqns., 72:316–337, 1988.
5. J. von Below. Parabolic network equations, 1993. Habilitation, Tübingen.
6. H. Choi, M. Hinze, and K. Kunisch. Instantaneous control of backward-facing-step flows. Report 571/1997, Technische Universität Berlin, 1997.
7. H. Choi, R. Temam, P. Moin, and J. Kim. Feedback control for unsteady flow and its application to the stochastic Burgers equation. J. Fluid Mech., 253:509–543, 1993.
8. B. Dekoninck and S. Nicaise. Control of networks of Euler-Bernoulli beams. ESAIM-Control, Optimisation and Calculus of Variations, 4:57–81, 1999.
9. W. Eckhaus. Asymptotic analysis of singular Perturbations, volume 9 of North-Holland Mathematics Studies. North-Holland Publishing Company, Amsterdam–London, 1979.
10. J. Geisler. Dynamische Gebietszerlegung für Optimalsteuerungsprobleme auf vernetzten Gebieten unter Verwendung von Mehrgitterverfahren. Diplomarbeit, Universität Bayreuth, Bayreuth, 1999.
11. M. Hinze and A. Kauffmann. On a distributed control law with an application to the control of unsteady flow around a cylinder. In K.-H. Hoffmann, G. Leugering, and F. Tröltzsch, editors, Optimal Control of Partial Differential Equations, pages 177–190, Basel, 1999. International Series of Numerical Mathematics, 133.
12. R. Hundhammer and G. Leugering. Instantaneous control and domain decomposition for wave equations on graphs. In preparation.
13. J.E. Lagnese, G. Leugering, and E.J.P.G. Schmidt. Modeling, Analysis and Control of Dynamic Elastic Multi-Link Structures. Birkhäuser, Boston–Basel–Berlin, 1994.
14. G. Leugering. Domain decomposition of optimal control problems for networks of Euler-Bernoulli beams. In J. P. Puel and M. Tucsnak, editors, Control and Partial Differential Equations. ESAIM Paris, 1998. to appear.
15. G. Leugering. A domain decomposition of optimal control problems for dynamic networks of elastic strings. Computational Optimization and Applications, 1999. to appear.
16. G. Leugering. Domain decomposition of optimal control problems of networks of strings and Timoshenko-beams. SIAM Journal on Control and Optimization, 37(6):1649–1675, 1999.
17. J. L. Lions. Perturbations Singulières dans les Problèmes aux Limites et en Contrôle Optimale, volume 323 of Lecture Notes in Mathematics. Springer-Verlag, Berlin–Heidelberg–New York, 1973.
18. A. López. Control y perturbaciones singulares de sistemas parabólicos, 1999. Ph. D. Thesis, Universidad Complutense Madrid.
19. G. Lumer. Espaces ramifiés, et diffusions sur les réseaux topologiques. C. R. Acad. Sc. Paris, 291, Série A:627–630, 1980.
20. S. Nicaise. Exact controllability of a pluridimensional coupled problem. Revista

Matematica, 5:91–135, 1992.
21. S. Nicaise. Polygonal interface problems. Peter Lang, Frankfurt am Main, 1993.

Hadamard formula in nonsmooth domains and applications

G. FREMIOT and J. SOKOLOWSKI, Institut Elie Cartan, Laboratoire de Mathématiques, Université Henri Poincaré Nancy I, Vandoeuvre lès Nancy, France

ABSTRACT

In the present paper the Hadamard formula is derived for the differentiable shape functionals in nonsmooth domains. The structure theorem is applied to the shape sensitivity analysis of optimal control problems and nonlinear problems with unilateral conditions prescribed on the crack faces. The form of the first order shape derivatives of the associated shape functionals is obtained in terms of singularity coefficients of solutions to elliptic equations under considerations.

1 INTRODUCTION

In the present paper the so-called Hadamard formula [1] is derived for differentiable shape functionals defined in domains with geometrical singularities. We provide the proof of the representation theorem [2] and present two applications. First, the optimal value of the cost functional for an optimal control problem is considered as a shape functional. We derive the so-called Griffith formula for the shape functional in terms of four coefficients of singularity of solutions to elliptic equations associated with the control problem.

In the second part of the paper the energy functional for an elliptic equation defined in a domain with a crack is considered. The unilateral nonpenetration condition [3] is prescribed on the crack faces and mixed Dirichlet-Neumann boundary condition is given on the external boundary of the geometrical domain. We show that perturbations of the interface points on the external boundary lead to shape differentiable energy functional. The form of the shape derivative of the energy functional is obtained in terms of singularity coefficients of the solution to the elliptic boundary value problem.

The results presented in the paper for domains in $\mathbb{R}^2$ can be applied as well to the domains in $\mathbb{R}^3$, e.g. for the elasticity system with frictionless contact conditions prescribed on the crack faces [4].

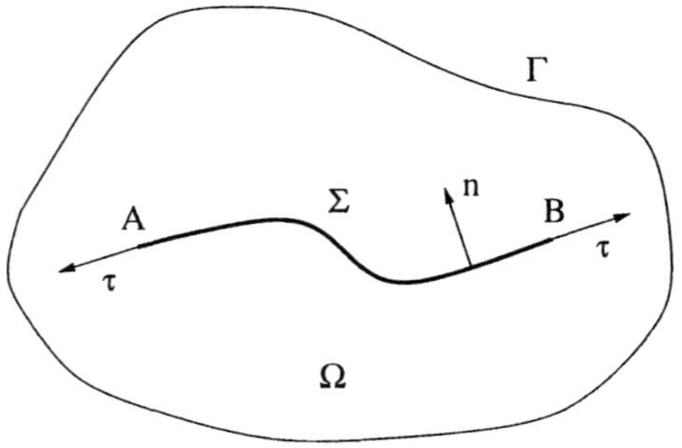

Figure 1 Domain Ω with the curved crack Σ

2 STRUCTURE THEOREM (HADAMARD FORMULA)

The structure theorem of shape derivatives of domain functionals is useful in shape optimization, because it allows to obtain the structure of the shape gradient of a specific domain functional by means of simple verifications of hypothesis, usually in the fixed domain setting, e.g. by an application of the material derivative method [1].

Let $D \subset \mathbb{R}^2$ be a bounded domain with smooth boundary Γ, and Σ be a part of a smooth curve. We assume that $\overline{\Sigma}$ belongs to the domain D. Therefore, we consider the domain $\Omega = D \setminus \overline{\Sigma}$ with crack Σ (see FIG.1). Let us denote by A and B the tips of $\overline{\Sigma}$, $\overline{\Sigma} = \{A\} \cup \Sigma \cup \{B\}$. Assume that J is a domain functional which is shape differentiable at Ω. For the convenience of the reader, we provide the definition of shape differentiability. The Eulerian derivative of the domain functional J at Ω in the direction of the vector field V is defined as the limit

$$dJ(\Omega; V) = \lim_{t \downarrow 0} \frac{J(\Omega_t) - J(\Omega)}{t},$$

where the velocity field V is used to construct a family of domains $\Omega_t = T_t(V)(\Omega)$, $t \in]-\delta, \delta[$, $\delta > 0$, using the technique described in [1]. In particular

$$\Omega_t = \{x \in \mathbb{R}^2 \mid x = x(t, X),\ X \in \Omega\}$$

where $x(t, X)$ denotes a solution to the system

$$\frac{dx}{dt}(t) = V(t, x(t)),\ x(0) = X.$$

The functional J is shape differentiable at Ω if there exists the Eulerian derivative $dJ(\Omega; V)$ in any direction V, and the mapping $V \mapsto dJ(\Omega; V)$ is linear and continuous.

Without losing the generality, we can consider the problem with autonomous vector fields. Let $\mathcal{D}^k(D; \mathbb{R}^2) = C_0^k(D; \mathbb{R}^2)$ be the space of k-times continuously differentiable transformations of $\mathbb{R}^2$ with compact support in D.

The following result on the structure of the Eulerian semiderivative $dJ(\Omega; V)$ of the shape functional $J(\Omega)$ is established in the present paper.

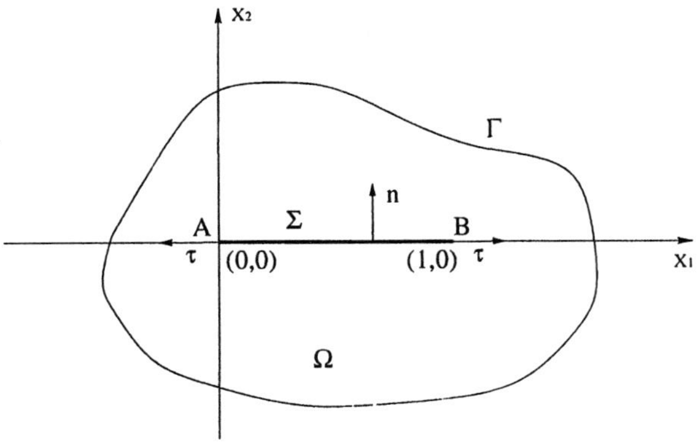

Figure 2 Domain Ω after changing of variables

THEOREM 1 Let k be a nonnegative integer. Assume that the mapping $\mathcal{D}^k(D;\mathbb{R}^2) \ni V \mapsto dJ(\Omega;V) \in \mathbb{R}$ is linear and continuous. Then there exist two real numbers α_A and α_B, and a linear form ϕ which is continuous on $C^k(\overline{\Sigma})$ $(\phi \in (C^k(\overline{\Sigma}))')$ such that:

$$dJ(\Omega;V) = \alpha_A(V.\tau)(A) + \alpha_B(V.\tau)(B) + \phi(V.n), \quad \forall V \in \mathcal{D}^k(D;\mathbb{R}^2),$$

where $V.\tau$ and $V.n$ denote the tangential and normal components of field V on $\overline{\Sigma}$, respectively.

Proof: We may assume that Σ is the set given by

$$\Sigma = \{(x_1,x_2) \mid 0 < x_1 < 1,\ x_2 = 0\} \tag{1}$$

otherwise, we can use an appropriate change of variables (see FIG.2).

We need the form of the tangent set

$$T_{\overline{\Sigma}}(x) = \left\{ v \in \mathbb{R}^2 \mid \liminf_{h\to 0+} \frac{d_{\overline{\Sigma}}(x+hv)}{h} = 0 \right\},$$

where $d_{\overline{\Sigma}}(x)$ denotes the distance between the crack $\overline{\Sigma}$ and the element x.

Evaluation of $T_{\overline{\Sigma}}(x)$ for $x \in \overline{\Sigma}$.

First case: $x = (x_1,x_2) \in \Sigma$, i.e. $0 < x_1 < 1$, $x_2 = 0$.

In this case, the normal $n(x)$ to $\overline{\Sigma}$ at $x \in \Sigma$ is well defined and, moreover we have

$$T_{\overline{\Sigma}}(x) = \mathcal{T}_x(\overline{\Sigma}), \text{ tangent space to } \overline{\Sigma} \text{ at } x,$$

so

$$V(x) \in T_{\overline{\Sigma}}(x) \quad \text{iff} \quad V(x).n(x) = 0.$$

Second case: $x = A = (0,0)$.

$$T_{\overline{\Sigma}}(x) = T_{\overline{\Sigma}}(A) = \left\{ v = (X_1,X_2) \in \mathbb{R}^2 \mid \liminf_{h\to 0+} \frac{d_{\overline{\Sigma}}(x+hv)}{h} = 0 \right\}$$

$$d_{\overline{\Sigma}}(x+hv) = \begin{cases} h\|v\| & \text{if } X_1 \le 0 \\ h|X_2| & \text{if } X_1 \ge 0 \end{cases}$$

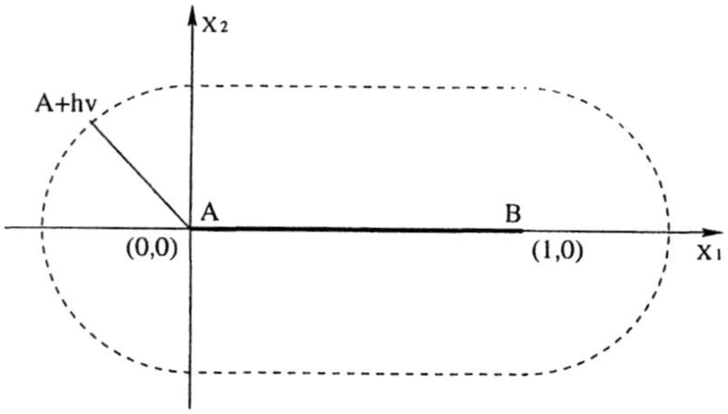

Figure 3 Evaluation of $T_{\overline{\Sigma}}(A)$

thus

$$\liminf_{h\to 0+} \frac{d_{\overline{\Sigma}}(x+hv)}{h} = \begin{cases} \|v\| & \text{if } X_1 \leq 0 \\ |X_2| & \text{if } X_1 \geq 0 \end{cases}$$

and consequently

$$T_{\overline{\Sigma}}((0,0)) = T_{\overline{\Sigma}}(A) = \{v = (X_1, X_2) \in \mathbb{R}^2 \mid X_2 = 0 \text{ and } X_1 \geq 0\}. \quad (2)$$

Third case: $x = B = (1,0)$.
In the same way as for $x = A = (0,0)$, we have

$$T_{\overline{\Sigma}}((1,0)) = T_{\overline{\Sigma}}(B) = \{v = (X_1, X_2) \in \mathbb{R}^2 \mid X_2 = 0 \text{ and } X_1 \leq 0\}. \quad (3)$$

For any $x \in \Sigma$, $T_{\overline{\Sigma}}(x)$ is a vector space thus $-T_{\overline{\Sigma}}(x) = T_{\overline{\Sigma}}(x)$. On the other hand

$$\{T_{\overline{\Sigma}}(A)\} \cap \{-T_{\overline{\Sigma}}(A)\} = \{T_{\overline{\Sigma}}(B)\} \cap \{-T_{\overline{\Sigma}}(B)\} = \{(0,0)\}, \quad (4)$$

so according to Nagumo's theorem [5] or to the double viability condition [6],[7], and in view of the relations (2)-(4), if the field $V \in \mathcal{D}^k(D;\mathbb{R}^2)$ satisfies the following conditions

$$\begin{cases} V(x).n(x) = 0, & \forall x \in \Sigma \\ V(A) = V(B) = (0,0) \end{cases} \quad (5)$$

then $\overline{\Sigma}$ is globally invariant by the associated transformation $T_t(V)$ (see [1] for the definition). The exterior boundary $\Gamma = \partial D$ is also invariant by the transformation $T_t(V)$, since the support of the field V is included in D. So the boundary of $\Omega = D \setminus \overline{\Sigma}$, i.e. $\partial\Omega = \Gamma \cup \overline{\Sigma}$, is globally invariant by the transformation $T_t(V)$. In consequence $\Omega_t \stackrel{def}{=} T_t(V)(\Omega) = \Omega$. Hence

$$dJ(\Omega; V) = 0 \quad (6)$$

for any vector field wich satisfies (5).
It is not difficult to see that it is possible to extend the notion of tangent

and normal vectors to the points A and B of $\overline{\Omega}$. We can formulate again the conditions (5) as follows

$$\begin{cases} V(x).n(x) = 0, \quad \forall x \in \overline{\Sigma} \\ (V.\tau)(A) = (V.\tau)(B) = 0. \end{cases} \tag{7}$$

So, we have established that for given $V \in \mathcal{D}^k(D;\mathbb{R}^2)$ which satisfies conditions (7), it follows that $dJ(\Omega;V) = 0$. Hence, it is natural to consider the following set

$$F(\Omega) = \{V \in \mathcal{D}^k(D;\mathbb{R}^2) \mid V.n = 0 \text{ on } \overline{\Sigma},\ (V.\tau)(A) = (V.\tau)(B) = 0\}. \tag{8}$$

According to the hypothesis that the mapping $V \mapsto dJ(\Omega;V)$ is linear continuous from $\mathcal{D}^k(D;\mathbb{R}^2)$ in $\mathbb{R}$, the set $F(\Omega)$ defined by (8) is included in its kernel. Consequently, we can prove the following lemma.

LEMMA 1 The mapping

$$\begin{array}{rcl} \psi : \mathcal{D}^k(D;\mathbb{R}^2)/F(\Omega) & \rightarrow & C^k(\overline{\Sigma}) \times \mathbb{R} \times \mathbb{R} \\ \{V\} & \mapsto & (V.n, (V.\tau)(A), (V.\tau)(B)) \end{array}$$

is an isomorphism.

Proof: The linear mapping $\psi : \{V\} \mapsto (V.n, (V.\tau)(A), (V.\tau)(B))$ is well defined since if $V_1 - V_2 \in F(\Omega)$ then

$$(V_1 - V_2).n = 0 \quad \text{on } \overline{\Sigma},\ ((V_1 - V_2).\tau)(A) = ((V_1 - V_2).\tau)(B) = 0.$$

Let $\{V\} \in \mathcal{D}^k(D;\mathbb{R}^2)/F(\Omega)$ be such that $\psi(\{V\}) = 0$ i.e.

$$V.n = 0 \quad \text{on } \overline{\Sigma},\ (V.\tau)(A) = (V.\tau)(B) = 0,$$

which means that $V \in F(\Omega)$ and then $\{V\} = \{0\}$. Consequently ψ is one-to-one.

Now let us show that ψ is onto. Let $(v, v_1, v_2) \in C^k(\overline{\Sigma}) \times \mathbb{R} \times \mathbb{R}$. We want to find $V \in \mathcal{D}^k(D;\mathbb{R}^2)$ such that $\psi(\{V\}) = (v, v_1, v_2)$.

For any $v \in C^k(\overline{\Sigma}) = C^k([0,1])$, by definition of the space $C^k([0,1])$, there exists $\tilde{v} \in C^k(\mathbb{R})$ such that $\tilde{v}_{|[0,1]} = v$. So we define $\widetilde{V_2}$ by

$$\widetilde{V_2}(x_1, x_2) = \tilde{v}(x_1), \quad \forall x_1,\ x_2 \in \mathbb{R}. \tag{9}$$

Then it is evident that $\widetilde{V_2} \in C^k(\mathbb{R}^2)$. Let $\theta \in \mathcal{D}(D;\mathbb{R}) = C_0^\infty(D;\mathbb{R})$ be such that $\theta \equiv 1$ in a sufficiently small neighbourhood of $\overline{\Sigma}$. Denote

$$V_2(x_1, x_2) = \theta(x_1, x_2)\widetilde{V_2}(x_1, x_2), \quad \forall x_1,\ x_2 \in \mathbb{R}. \tag{10}$$

Let us define the function μ by

$$\mu(x_1) = (1 - x_1)v_1 + x_1 v_2 \tag{11}$$

Then $\mu \in C^k(\mathbb{R})$ (extension by convex combination).

In the same way, we introduce

$$\widetilde{V_1}(x_1, x_2) = \mu(x_1), \quad \forall x_1,\ x_2 \in \mathbb{R}, \tag{12}$$

hence $\widetilde{V_1} \in C^k(\mathbb{R}^2)$. Then, we define V_1 by

$$V_1(x_1, x_2) = \theta(x_1, x_2)\widetilde{V_1}(x_1, x_2). \tag{13}$$

Let V the vector field whose components are V_1 and V_2 (V_1 and V_2 being successively given by (9),(10),(11),(12),(13)), so

$$V = (V_1, V_2) \in \mathcal{D}^k(D; \mathbb{R}^2),$$

moreover

$$\begin{aligned} V_1(x_1, x_2) &= \theta(x_1, x_2)\widetilde{V_1}(x_1, x_2) \\ &= \widetilde{V_1}(x_1, x_2) \quad \text{on } \overline{\Sigma} \\ &= \mu(x_1) \\ &= (1 - x_1)v_1 + x_1 v_2. \end{aligned}$$

In consequence $(V.\tau)(A) = \mu(0) = v_1$ and $(V.\tau)(B) = \mu(1) = v_2$

$$\begin{aligned} V_2(x_1, x_2) &= \theta(x_1, x_2)\widetilde{V_2}(x_1, x_2) \\ &= \widetilde{V_2}(x_1, x_2) \quad \text{on } \overline{\Sigma} \\ &= \tilde{v}(x_1) \\ &= v(x_1), \end{aligned}$$

i.e. $V.n = v$ on $\overline{\Sigma}$. We have defined $V \in \mathcal{D}^k(D; \mathbb{R}^2)$ such that $\psi(\{V\}) = (v, v_1, v_2)$. This completes the proof of lemma 1. □

LEMMA 2 There exists a linear, continuous mapping Φ

$$\Phi : C^k(\overline{\Sigma}) \times \mathbb{R} \times \mathbb{R} \to \mathbb{R}$$

such that for any vector field $V \in \mathcal{D}^k(D; \mathbb{R}^2)$,

$$dJ(\Omega; V) = \Phi\left(V.n, (V.\tau)(A), (V.\tau)(B)\right).$$

Proof: We define Φ by the following formula

$$\Phi(\{V\}) = dJ(\Omega; V). \tag{14}$$

Indeed, if $\{V'\} = \{V\}$, i.e. if $V' \in \{V\}$, we have $V' - V \in F(\Omega)$, since $F(\Omega)$ is included in the kernel of $dJ(\Omega; \cdot)$, it follows that

$$dJ(\Omega; V - V') = 0.$$

The Eulerian semiderivative $dJ(\Omega; \cdot)$ is linear by our assumption, therefore

$$dJ(\Omega; V) = dJ(\Omega; V'). \tag{15}$$

The relation (15) enables us to define Φ.
Using lemma 1 and the relation $\mathcal{D}^k(D; \mathbb{R}^2)/F(\Omega) \simeq C^k(\overline{\Sigma}) \times \mathbb{R} \times \mathbb{R}$, it follows that

$$\{V\} = (V.n, (V.\tau)(A), (V.\tau)(B)) \tag{16}$$

thus

$$dJ(\Omega; V) = \Phi(\{V\}) = \Phi(V.n, (V.\tau)(A), (V.\tau)(B)) . \tag{17}$$

Furthermore, $dJ(\Omega; \cdot)$ is linear and continuous which implies that Φ is linear and continuous.

Now, we can complete the proof of the structure theorem. Indeed, there exists a linear mapping Φ, which is continuous from $C^k(\overline{\Sigma}) \times \mathbb{R} \times \mathbb{R}$ in $\mathbb{R}$, such that

$$\forall V \in \mathcal{D}^k(D; \mathbb{R}^2), \quad dJ(\Omega; V) = \Phi(V.n, (V.\tau)(A), (V.\tau)(B))$$

with

$$\begin{aligned} \Phi \in \left(C^k(\overline{\Sigma}) \times \mathbb{R} \times \mathbb{R}\right)' &= \left(C^k(\overline{\Sigma})\right)' \times \mathbb{R}' \times \mathbb{R}' \\ &= \left(C^k(\overline{\Sigma})\right)' \times \mathbb{R} \times \mathbb{R}. \end{aligned}$$

We can conclude that there exist two real numbers α_A and α_B, and a linear form ϕ which is continuous on $C^k(\overline{\Sigma})$ such that

$$dJ(\Omega; V) = \phi(V.n) + \alpha_A (V.\tau)(A) + \alpha_B (V.\tau)(B), \quad \forall V \in \mathcal{D}^k(D; \mathbb{R}^2) \tag{18}$$

which completes the proof of structure theorem. □

3 OPTIMAL CONTROL PROBLEM

We apply the structure theorem in the case of domain functionals defined by the optimal values of cost functionals for control problems. The optimal control problem considered in this section is defined for the elliptic equation modelling the deflection of an elastic membrane with interior crack.

Let us consider the domain D in $\mathbb{R}^2$ with smooth boundary Γ. Let Σ_l be the set defined by $\{(y_1, y_2) \mid 0 < y_1 < l,\ y_2 = 0\}$, A and B denote its tips. We assume this set belongs to the domain D for $l > 0$ small enough. The domain with the crack is denoted by $\Omega = D \setminus \overline{\Sigma_l}$. Let K be an open subset of D (also with smooth boundary) and moreover we assume that $\overline{K} \cap \overline{\Sigma_l} = \emptyset$. The state equation for the control problem is of the form

$$\begin{cases} -\Delta q(u) = u\chi_K & \text{in} \quad \Omega, \\ q(u) = 0 & \text{on} \quad \Gamma = \partial D, \\ \dfrac{\partial q(u)}{\partial n} = 0 & \text{on} \quad {\Sigma_l}^{\pm}, \end{cases} \tag{19}$$

where χ_K denotes the characteristic function of K and ${\Sigma_l}^{\pm}$ denote the crack faces which correspond to the positive and negative directions of the normal vector n.

For given $u \in L^2(K)$, $q(u)$ represents the deflection of an elastic membrane loaded by the vertical force u concentrated on K. For the system (19), we define the cost functional

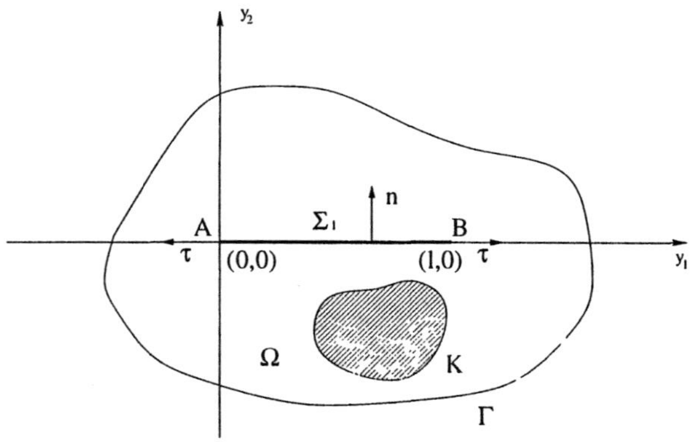

Figure 4 Domain Ω after changing of variables

$$I(u) = \frac{1}{2}\int_K [(q(u) - q_d)^2 + \alpha u^2] d\Omega, \tag{20}$$

which is minimized over the space of controls $u \in L^2(K)$, $\alpha > 0$, and $q_d \in L^2(K)$ is a given function. The minimization of the functional (20) with respect to u means approximation of a given function q_d in the region K by the deflection of an elastic membrane, using the smallest possible load u applied in K. The minimal value of the cost functional for this control problem defines the shape functional, depending on the geometrical domain Ω,

$$J(\Omega) = \min_{u \in L^2(K)} I(u).$$

It is not difficult to see that we have the following linear relation

$$q(u + sv) = q(u) + sq(v), \tag{21}$$

where $q(v)$ satisfies the equation

$$\begin{cases} -\Delta q(v) &= v\chi_K \quad \text{in } \Omega, \\ q(v) &= 0 \quad \text{on } \Gamma, \\ \dfrac{\partial q(v)}{\partial n} &= 0 \quad \text{on } \Sigma_l{}^{\pm}. \end{cases} \tag{22}$$

And in consequence $q(v)$ is the variation of the state corresponding to the variation of the control. According to (22), the variation $dI(u;v)$ of the cost functional is given by

$$dI(u;v) = \int_K [(q(u) - q_d)q(v) + \alpha uv] d\Omega. \tag{23}$$

Let us introduce the adjoint state p, which is defined by the following equation

$$\begin{cases} -\Delta p &= (q(u) - q_d)\chi_K \quad \text{in } \Omega, \\ p &= 0 \quad \text{on } \Gamma, \\ \dfrac{\partial p}{\partial n} &= 0 \quad \text{on } \Sigma_l{}^{\pm}. \end{cases} \tag{24}$$

The adjoint state p given by (24) allows us to have another expression for (23)

$$\begin{aligned} dI(u;v) &= \int_K [(q(u) - q_d)q(v) + \alpha uv] d\Omega \\ &= \int_\Omega \langle \nabla p, \nabla q(v) \rangle d\Omega + \int_K \alpha uv d\Omega, \end{aligned}$$

and using the equation (22) satisfied by $q(v)$, we obtain

$$dI(u;v) = \int_K [p + \alpha u] v d\Omega.$$

Thus the stationarity condition

$$dI(u;v) = 0, \quad \forall v \in L^2(K)$$

leads to the following equality

$$u(y) = -\frac{1}{\alpha} p(y), \text{ a.e. in } K. \tag{25}$$

And consequently, by using (25), the minimal value of the cost functional for the control problem takes the form

$$J(\Omega) = \frac{1}{2} \int_K [(q - q_d)^2 + \frac{1}{\alpha} p^2] d\Omega, \tag{26}$$

where p, q are given as a solution of the coupled system of equations:

$$\left\{ \begin{aligned} -\Delta q &= -\frac{1}{\alpha} p \chi_K && \text{in } \Omega, \\ -\Delta p &= (q - q_d)\chi_K && \text{in } \Omega, \\ q &= 0 && \text{on } \Gamma, \\ p &= 0 && \text{on } \Gamma, \\ \frac{\partial q}{\partial n} &= 0 && \text{on } \Sigma_l{}^{\pm}, \\ \frac{\partial p}{\partial n} &= 0 && \text{on } \Sigma_l{}^{\pm}. \end{aligned} \right. \tag{27}$$

Let $\theta_1,\ \theta_2 \in C_0^\infty(D) = \mathcal{D}(D)$. Then we consider the transformation (see [4]) defined by:

$$\left\{ \begin{aligned} y_1 &= x_1 - \delta\theta_1(x_1, x_2) \\ y_2 &= x_2 - \delta\theta_2(x_1, x_2), \quad \delta > 0. \end{aligned} \right. \tag{28}$$

The coordinates of a given point in open sets Ω, Ω_δ are denoted by $(y_1, y_2) \in \Omega$, $(x_1, x_2) \in \Omega_\delta$, respectively.
Let V be the vector field whose components are $\theta_1,\ \theta_2$, thus $V \in (\mathcal{D}(D))^2$, and moreover we assume that $\overline{K} \cap \text{supp}\{V\} = \emptyset$. The Jacobian of transformation (28) equals to:

$$\begin{aligned} q_\delta &= 1 - \delta(\theta_{1,x_1} + \theta_{2,x_2}) + \delta^2(\theta_{1,x_1}\theta_{2,x_2} - \theta_{1,x_2}\theta_{2,x_1}) \\ &= 1 - \delta \text{div} V + \delta^2 \det(DV). \end{aligned}$$

For $\delta > 0$, δ small enough, $q_\delta > 0$, so the transformation (28) is one-to-one and we denote $y = y(x, \delta)$, $x = x(y, \delta)$. Let Ω_δ be the image of Ω for the transformation (28).
For $\delta > 0$, the minimization problem is defined in the domain Ω_δ, with the optimal value of the cost functional

$$J(\Omega_\delta) = \frac{1}{2}\int_K [(q_\delta - q_d)^2 + \frac{1}{\alpha}p_\delta^2]d\Omega, \tag{29}$$

where p_δ, q_δ are the solutions of the following coupled equations (after the change of variables in order to transport the problem in Ω)

$$\begin{cases} \displaystyle\int_\Omega \langle C_\delta \cdot \nabla q_\delta, \nabla\varphi\rangle d\Omega &= \displaystyle -\frac{1}{\alpha}\int_K p_\delta \varphi d\Omega, \quad \forall\varphi \in H^1_\Gamma(\Omega), \\ \displaystyle\int_\Omega \langle C_\delta \cdot \nabla p_\delta, \nabla\psi\rangle d\Omega &= \displaystyle\int_K (q_\delta - q_d)\psi d\Omega, \quad \forall\psi \in H^1_\Gamma(\Omega), \end{cases} \tag{30}$$

where $C_\delta = \dfrac{1}{q_\delta} A_\delta^T \cdot A_\delta$ and the matrix function A_δ takes the following form

$$A_\delta = \begin{pmatrix} 1 - \delta\theta_{1,y_1} & -\delta\theta_{2,y_1} \\ -\delta\theta_{1,y_2} & 1 - \delta\theta_{2,y_2} \end{pmatrix}.$$

Then the domain functional $J(\Omega_\delta)$ is differentiable with respect to δ, namely we have the following result:

THEOREM 2 We have the following Griffith formula

$$\frac{dJ(\Omega_\delta)}{d\delta}|_{\delta=0} = -\frac{\pi}{2}(c_p c_\eta + c_q c_\xi)$$

where c_p, c_η, c_q, c_ξ are the coefficients of singularity of solutions p, η, q, ξ to the systems (27),(34).
Proof: Applying the implicite function theorem gives us the existence of the material derivatives $\dot{p}$, $\dot{q} \in H^1_\Gamma$. Moreover, we obtain the integral identities satisfied by $\dot{p}$ and $\dot{q}$:

$$\int_\Omega \langle C' \cdot \nabla q, \nabla\varphi\rangle d\Omega + \int_\Omega \langle \nabla\dot{q}, \nabla\varphi\rangle d\Omega = -\frac{1}{\alpha}\int_K \dot{p}\varphi d\Omega, \quad \forall\varphi \in H^1_\Gamma(\Omega), \tag{31}$$

$$\int_\Omega \langle C' \cdot \nabla p, \nabla\psi\rangle d\Omega + \int_\Omega \langle \nabla\dot{p}, \nabla\psi\rangle d\Omega = \int_K \dot{q}\psi d\Omega, \quad \forall\psi \in H^1_\Gamma(\Omega). \tag{32}$$

On the other hand, the previous results show that the cost functional is shape differentiable with the formula

$$dJ(\Omega; V) = \int_K [(q - q_d)\dot{q} + \frac{1}{\alpha}p\dot{p}]d\Omega. \tag{33}$$

In consequence, we can apply the structure theorem, which leads to the following formula for the derivative (33)

$$dJ(\Omega; V) = \alpha_A(V.\tau)(A) + \alpha_B(V.\tau)(B) + \phi(V.n),$$

where $\phi \in (C^1(\overline{\Sigma_l}))'$, α_A, $\alpha_B \in \mathbb{R}$.
It is possible to determine the coefficient α_A. In order to find the coefficient α_A, we can consider the perturbations of Ω such that the crack's length changes while the other part of the boundary does not move. In addition, we assume the crack grows without changing direction which implies that the vector field takes the form

$$V(y) = (\theta_1(y), 0),$$

where θ_1 is supported in D, $\overline{K} \cap \operatorname{supp}\{\theta_1\} = \emptyset$ and $\theta_1(y) = -1$ in the vicinity of the origin.
In order to express the Eulerian semiderivative $dJ(\Omega; V)$ by a formula without the material derivatives $\dot{p}$ and $\dot{q}$, we have to introduce the second level adjoint variables ξ, η (see [8]), defined by the following equations:

$$\left\{\begin{array}{rcll} -\Delta\xi - \eta\chi_K &=& (q - q_d)\chi_K & \text{in } \Omega, \\ -\Delta\eta + \dfrac{1}{\alpha}\xi\chi_K &=& \dfrac{1}{\alpha}p\chi_K & \text{in } \Omega, \\ \xi &=& 0 & \text{on } \Gamma, \\ \eta &=& 0 & \text{on } \Gamma, \\ \dfrac{\partial\xi}{\partial n} &=& 0 & \text{on } \Sigma^{\pm} \\ \dfrac{\partial\eta}{\partial n} &=& 0 & \text{on } \Sigma^{\pm}, \end{array}\right. \tag{34}$$

or, in the weak form:

$$\int_\Omega \langle\nabla\xi, \nabla\varphi\rangle d\Omega - \int_K \eta\varphi d\Omega = \int_K (q - q_d)\varphi d\Omega, \quad \forall\varphi \in H^1_\Gamma(\Omega), \tag{35}$$

$$\int_\Omega \langle\nabla\eta, \nabla\psi\rangle d\Omega + \frac{1}{\alpha}\int_K \xi\psi d\Omega = \frac{1}{\alpha}\int_K p\psi d\Omega, \quad \forall\psi \in H^1_\Gamma(\Omega). \tag{36}$$

Using (35) with $\varphi = \dot{q}$ leads to

$$\int_\Omega \langle\nabla\xi, \nabla\dot{q}\rangle d\Omega - \int_K \eta\dot{q} d\Omega = \int_K (q - q_d)\dot{q} d\Omega \tag{37}$$

and taking $\psi = \dot{p}$ in (36) gives us

$$\int_\Omega \langle\nabla\eta, \nabla\dot{p}\rangle d\Omega + \frac{1}{\alpha}\int_K \xi\dot{p} d\Omega = \frac{1}{\alpha}\int_K p\dot{p} d\Omega. \tag{38}$$

Moreover, we can substitute $\varphi = \xi$ and $\psi = \eta$ in the variational equalities (31),(32) respectively, so we obtain

$$\int_\Omega \langle C' \cdot \nabla q, \nabla\xi\rangle d\Omega + \int_\Omega \langle\nabla\dot{q}, \nabla\xi\rangle d\Omega = -\frac{1}{\alpha}\int_K \dot{p}\xi d\Omega, \tag{39}$$

and

$$\int_\Omega \langle C' \cdot \nabla p, \nabla\eta\rangle d\Omega + \int_\Omega \langle\nabla\dot{p}, \nabla\eta\rangle d\Omega = \int_K \dot{q}\eta d\Omega. \tag{40}$$

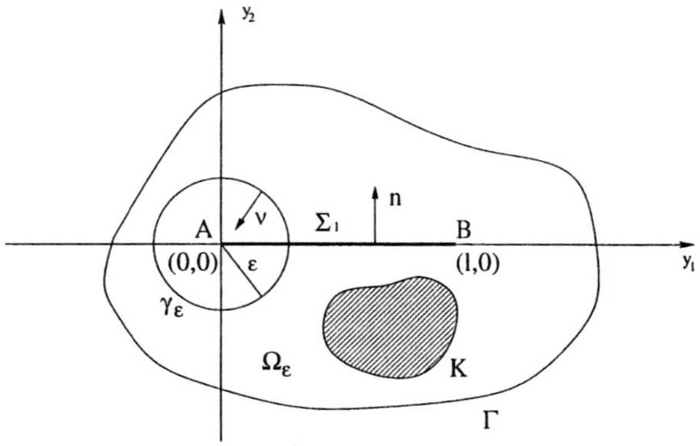

Figure 5 Domain Ω_ε with the crack Σ_l

Thus by combining the equalities (37),(38),(39),(40), it follows that

$$\begin{aligned} dJ(\Omega;V) &= \int_\Omega \langle\nabla\xi,\nabla\dot q\rangle d\Omega - \int_K \eta\dot q d\Omega + \int_\Omega \langle\nabla\eta,\nabla\dot p\rangle d\Omega + \frac{1}{\alpha}\int_K \xi\dot p d\Omega \\ &= -\int_\Omega \langle C'\cdot\nabla q,\nabla\xi\rangle d\Omega - \int_\Omega \langle C'\cdot\nabla p,\nabla\eta\rangle d\Omega \\ &= -\lim_{\varepsilon\to 0^+}\left(\int_{\Omega_\varepsilon}\langle C'\cdot\nabla q,\nabla\xi\rangle d\Omega + \int_{\Omega_\varepsilon}\langle C'\cdot\nabla p,\nabla\eta\rangle d\Omega\right) \end{aligned}$$

where Ω_ε is the subset of Ω defined by $r>\varepsilon$ (see FIG.5). Let γ_ε be the curve given in the polar coordinates by $r=\varepsilon$ and $0<\theta<2\pi$.
Let us introduce the notation

$$B_\varepsilon = \int_{\Omega_\varepsilon}\langle C'\cdot\nabla q,\nabla\xi\rangle d\Omega + \int_{\Omega_\varepsilon}\langle C'\cdot\nabla p,\nabla\eta\rangle d\Omega.$$

Moreover, we have

$$A_\delta = \begin{pmatrix} 1-\delta\theta_{1,y_1} & 0 \\ -\delta\theta_{1,y_2} & 1 \end{pmatrix},\quad q_\delta = 1-\delta\theta_{1,y_1}$$

which leads to

$$C_\delta = \frac{1}{1-\delta\theta_{1,y_1}}\begin{pmatrix} (1-\delta\theta_{1,y_1})^2+\delta^2\theta_{1,y_2}^2 & -\delta\theta_{1,y_2} \\ -\delta\theta_{1,y_2} & 1 \end{pmatrix},$$

and finally we obtain

$$C' = \frac{dC_\delta}{d\delta}|_{\delta=0} = \begin{pmatrix} -\theta_{1,y_1} & -\theta_{1,y_2} \\ -\theta_{1,y_2} & \theta_{1,y_1} \end{pmatrix}.$$

By using this relation, it follows that

$$\begin{aligned} B_\varepsilon &= \int_{\Omega_\varepsilon}\theta_{1,y_1}(-q_{y_1}\xi_{y_1}+q_{y_2}\xi_{y_2}-p_{y_1}\eta_{y_1}+p_{y_2}\eta_{y_2})d\Omega \\ &\quad + \int_{\Omega_\varepsilon}\theta_{1,y_2}(-q_{y_2}\xi_{y_1}-q_{y_1}\xi_{y_2}-p_{y_2}\eta_{y_1}-p_{y_1}\eta_{y_2})d\Omega. \end{aligned} \tag{41}$$

By integrating by parts in (41), we have

$$\begin{aligned} B_\varepsilon &= \int_{\gamma_\varepsilon} \theta_1(-q_{y_1}\xi_{y_1} + q_{y_2}\xi_{y_2} - p_{y_1}\eta_{y_1} + p_{y_2}\eta_{y_2})\nu_1 d\sigma \\ &+ \int_{\gamma_\varepsilon} \theta_1(-q_{y_2}\xi_{y_1} - q_{y_1}\xi_{y_2} - p_{y_2}\eta_{y_1} - p_{y_1}\eta_{y_2})\nu_2 d\sigma \\ &+ \int_{\Omega_\varepsilon} \theta_1(\xi_{y_1}\Delta q + q_{y_1}\Delta\xi + \eta_{y_1}\Delta p + p_{y_1}\Delta\eta)d\Omega. \end{aligned}$$

But, for ε small enough, $K \subset \Omega_\varepsilon$ and moreover $\overline{K} \cap \operatorname{supp}\{\theta_1\} = \emptyset$ that's why

$$\theta_1\Delta q = \theta_1\Delta\xi = \theta_1\Delta p = \theta_1\Delta\eta = 0 \quad \text{on } \Omega_\varepsilon$$

and in consequence

$$\begin{aligned} B_\varepsilon &= \int_{\gamma_\varepsilon} \theta_1(-q_{y_1}\xi_{y_1} + q_{y_2}\xi_{y_2} - p_{y_1}\eta_{y_1} + p_{y_2}\eta_{y_2})\nu_1 d\sigma \\ &+ \int_{\gamma_\varepsilon} \theta_1(-q_{y_2}\xi_{y_1} - q_{y_1}\xi_{y_2} - p_{y_2}\eta_{y_1} - p_{y_1}\eta_{y_2})\nu_2 d\sigma. \end{aligned} \tag{42}$$

For ε small enough, $\theta_1 \equiv -1$ on γ_ε, and in view of (42)

$$\begin{aligned} B_\varepsilon &= \int_{\gamma_\varepsilon} (q_{y_1}\xi_{y_1} - q_{y_2}\xi_{y_2} + p_{y_1}\eta_{y_1} - p_{y_2}\eta_{y_2})\nu_1 d\sigma \\ &+ \int_{\gamma_\varepsilon} (q_{y_2}\xi_{y_1} + q_{y_1}\xi_{y_2} + p_{y_2}\eta_{y_1} + p_{y_1}\eta_{y_2})\nu_2 d\sigma. \end{aligned} \tag{43}$$

Moreover, according to [9], we know that

$$\begin{cases} p = p^R + c_p S, \\ q = q^R + c_q S, \\ \eta = \eta^R + c_\eta S, \\ \xi = \xi^R + c_\xi S, \end{cases} \tag{44}$$

where $S = \sqrt{r}\cos\left(\frac{\theta}{2}\right)$ is the singular function for the boundary value problems under considerations, p^R, q^R, η^R, $\xi^R \in H^2(U)$ are the regular parts of the solutions, U is a neighbourhood of $(0,0)$ in Ω such that $\overline{U} \cap \Gamma = \emptyset$, and c_p, c_q, c_η, c_ξ denote the coefficients of singularity of solutions p, q, η, ξ to (27), (34), respectively.

Taking into account the decomposition (44) and developping in (43) we have

$$B_\varepsilon = B_\varepsilon^{(1)} + B_\varepsilon^{(2)} + B_\varepsilon^{(3)}$$

where $B_\varepsilon^{(1)}, B_\varepsilon^{(2)}, B_\varepsilon^{(3)}$ are defined by

$$B_\varepsilon^{(1)} = (c_p c_\eta + c_q c_\xi)\int_{\gamma_\varepsilon} [((S_{y_1})^2 - (S_{y_2})^2)\nu_1 + 2\nu_2 S_{y_1} S_{y_2}]d\sigma,$$

$$\begin{aligned} B_\varepsilon^{(2)} &= \int_{\gamma_\varepsilon} (c_\xi q_{y_1}^R + c_q \xi_{y_1}^R + c_\eta p_{y_1}^R + c_p \eta_{y_1}^R)(\nu_1 S_{y_1} + \nu_2 S_{y_2})d\sigma \\ &+ \int_{\gamma_\varepsilon} (c_\xi q_{y_2}^R + c_q \xi_{y_2}^R + c_\eta p_{y_2}^R + c_p \eta_{y_2}^R)(\nu_2 S_{y_1} - \nu_1 S_{y_2})d\sigma, \end{aligned}$$

$$
\begin{aligned}
B_\varepsilon^{(3)} &= \int_{\gamma_\varepsilon} \nu_1 (q_{y_1}^R \xi_{y_1}^R - q_{y_2}^R \xi_{y_2}^R + p_{y_1}^R \eta_{y_1}^R - p_{y_2}^R \eta_{y_2}^R) d\sigma \\
&+ \int_{\gamma_\varepsilon} \nu_2 (q_{y_2}^R \xi_{y_1}^R + q_{y_1}^R \xi_{y_2}^R + p_{y_2}^R \eta_{y_1}^R + p_{y_1}^R \eta_{y_2}^R) d\sigma.
\end{aligned}
$$

Moreover the form of singular function S is known in this case, therefore we have

$$
S_{y_1} = \frac{1}{2\sqrt{r}} \cos\left(\frac{\theta}{2}\right), \quad S_{y_2} = \frac{1}{2\sqrt{r}} \sin\left(\frac{\theta}{2}\right),
$$

and using polar coordinates, we have

$$
\begin{aligned}
B_\varepsilon^{(1)} &= (c_p c_\eta + c_q c_\xi) \int_0^{2\pi} \left(\frac{1}{4} \cos^2\left(\frac{\theta}{2}\right) - \frac{1}{4} \sin^2\left(\frac{\theta}{2}\right) \right) \cos\theta d\theta \\
&+ 2(c_p c_\eta + c_q c_\xi) \int_0^{2\pi} \frac{1}{4} \cos\left(\frac{\theta}{2}\right) \sin\left(\frac{\theta}{2}\right) \sin\theta d\theta \\
&= \frac{1}{4}(c_p c_\eta + c_q c_\xi) \int_0^{2\pi} (\cos^2\theta + \sin^2\theta) d\theta \\
&= \frac{\pi}{2}(c_p c_\eta + c_q c_\xi).
\end{aligned}
$$

It is not difficult to see that $B_\varepsilon^{(2)} \to 0$ and $B_\varepsilon^{(3)} \to 0$ as $\varepsilon \to 0^+$, in fact we have the estimations $B_\varepsilon^{(2)} = O(\sqrt{\varepsilon})$, $B_\varepsilon^{(3)} = O(\varepsilon)$. Finally,

$$
dJ(\Omega; V) = -\lim_{\varepsilon \to 0^+} B_\varepsilon = -\frac{\pi}{2}(c_p c_\eta + c_q c_\xi).
$$

We have identified the coefficient $\alpha_A = -\frac{\pi}{2}(c_p c_\eta + c_q c_\xi)$ in the expression of the Eulerian semiderivative $dJ(\Omega; V)$ given by the structure theorem.
We can remark that these results about the sensitivity of shape functionals in cracked domains are very similar to the results described in the paper [10] obtained by a direct approach. □

4 UNILATERAL CONDITIONS ON THE CRACK

In this section, we consider the Poisson equation, with mixed Dirichlet-Neumann boundary conditions, in a nonsmooth domain in $\mathbb{R}^2$, with an interior crack.
Let $D \subset \mathbb{R}^2$ be a bounded domain with smooth boundary ∂D. D plays the role of *hold-all*. Let E be a subset of D with the sufficiently smooth boundary $\Gamma = \partial E$. Let $\Gamma = \Gamma_0 \cup \Gamma_1 \cup \{A\} \cup \{B\}$ with $\overline{\Gamma_0} = \Gamma_0 \cup \{A\} \cup \{B\}$ and $\overline{\Gamma_1} = \Gamma_1 \cup \{A\} \cup \{B\}$.

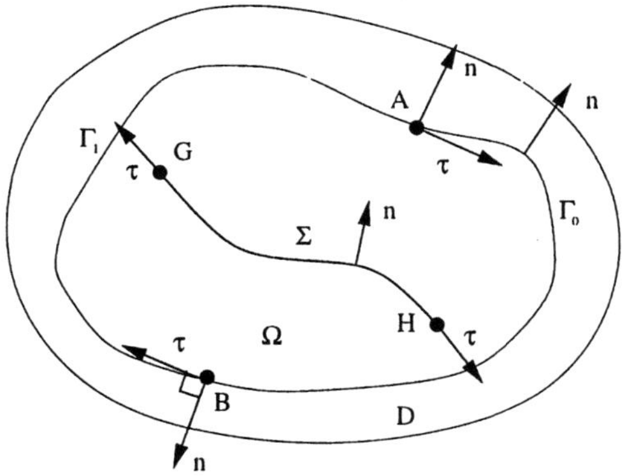

Figure 6 Domains D and Ω

Let Σ be a part of a smooth curve. Let us denote by G and H the tips of $\overline{\Sigma}$. We assume that $\overline{\Sigma}$ belongs to the domain E. The domain with the crack Σ is denoted by $\Omega = E \setminus \overline{\Sigma}$.

We consider the following boundary value problem with mixed conditions Dirichlet Neumann

$$-\Delta u = f \quad \text{in } \Omega, \tag{45}$$

$$u = 0 \quad \text{on } \Gamma_0, \ \frac{\partial u}{\partial n} = 0 \quad \text{on } \Gamma_1, \ [u] \geq 0 \quad \text{on } \Sigma. \tag{46}$$

Here $f \in C^1(\overline{D})$ is a given function, $[u] = u^+ - u^-$ is the jump of u accross Σ. The vector $n = (0, 1)$ is orthogonal to Σ, and $u^{\pm}$ denote the traces of u on the crack faces, corresponding to the positive and negative directions of n.

We consider the minimization of the functional

$$I(\phi) = \frac{1}{2} \int_{\Omega} |\nabla \phi|^2 - \int_{\Omega} f\phi$$

over the set of all functions of $H^1(\Omega)$ satisfying the conditions $u = 0$ on Γ_0, $[u] \geq 0$ on Σ. The solution of the minimization problem is unique and satisfies a variational inequality, and, in particular, u satisfies (45),(46) (there are additional relations verified on Σ). The energy functional for the problem (45),(46) is defined by the formula

$$J(\Omega) = \frac{1}{2} \int_{\Omega} |\nabla u|^2 - \int_{\Omega} fu \tag{47}$$

where u is the variational solution to (45), (46).

Let $\theta_1, \ \theta_2 \in C_0^{\infty}(D)$. We use the same notation as in section 3.

For $\delta > 0$, the minimization problem is defined in Ω_δ, with the energy functional

$$J(\Omega_\delta) = \frac{1}{2} \int_{\Omega_\delta} |\nabla u^\delta|^2 - \int_{\Omega_\delta} fu^\delta. \tag{48}$$

We derive the form of the shape derivative

$$\frac{dJ(\Omega_\delta)}{d\delta}|_{\delta=0} = \lim_{\delta \to 0+} \frac{J(\Omega_\delta) - J(\Omega)}{\delta} = dJ(\Omega; V) \tag{49}$$

in order to apply the structure theorem. Let $u^\delta(x)$ be the solution of minimization problem in Ω_δ, and $u^\delta(x) = u_\delta(y),\ x = x(y, \delta)$.
We have the following formula

$$\nabla_x u^\delta = A_\delta \cdot \nabla_y u_\delta \tag{50}$$

with

$$A_\delta = \begin{pmatrix} 1 - \delta\theta_{1,x_1} & -\delta\theta_{2,x_1} \\ -\delta\theta_{1,x_2} & 1 - \delta\theta_{2,x_2} \end{pmatrix}$$

thus $A_\delta = I + \delta B$, B is the matrix defined by

$$B = \begin{pmatrix} -\theta_{1,x_1} & -\theta_{2,x_1} \\ -\theta_{1,x_2} & -\theta_{2,x_2} \end{pmatrix}.$$

Consequently

$$\int_{\Omega_\delta} |\nabla u^\delta|^2 dx = \int_\Omega \frac{1}{q_\delta} |A_\delta \cdot \nabla u_\delta|^2 dy.$$

By the change of variables it follows that

$$\int_{\Omega_\delta} f u^\delta dx = \int_\Omega \frac{1}{q_\delta} f\left(x(y,\delta)\right) u_\delta(y) dy.$$

Denote $f^\delta(y) = \dfrac{f\left(x(y,\delta)\right)}{q_\delta}$, then

$$f'(y) = \frac{df^\delta(y)}{d\delta}|_{\delta=0} = \lim_{\delta \to 0+} \frac{f^\delta(y) - f^0(y)}{\delta}.$$

Assuming that y, δ are independent variables in (28), we have $x = x(y, \delta)$. Differentiation of (28) with respect to δ yields

$$\begin{cases} 0 = \dfrac{dx_1}{d\delta} - \theta_1 - \delta\theta_{1,x_1}\dfrac{dx_1}{d\delta} - \delta\theta_{1,x_2}\dfrac{dx_2}{d\delta} \\ 0 = \dfrac{dx_2}{d\delta} - \theta_2 - \delta\theta_{2,x_1}\dfrac{dx_1}{d\delta} - \delta\theta_{2,x_2}\dfrac{dx_2}{d\delta}, \end{cases}$$

thus

$$\begin{cases} \dfrac{dx_1}{d\delta} = \dfrac{\theta_1(1 - \delta\theta_{2,x_2}) + \delta\theta_2\theta_{1,x_2}}{q_\delta} \\ \dfrac{dx_2}{d\delta} = \dfrac{\theta_2(1 - \delta\theta_{1,x_1}) + \delta\theta_1\theta_{2,x_1}}{q_\delta}. \end{cases} \tag{51}$$

Consequently, by (51)

$$\frac{\partial f\left(x(y,\delta)\right)}{\partial\delta}|_{\delta=0} = f_{x_1}\frac{dx_1}{d\delta}|_{\delta=0} + f_{x_2}\frac{dx_2}{d\delta}|_{\delta=0} = f_{y_1}\theta_1 + f_{y_2}\theta_2. \tag{52}$$

Now, we are in a position to find the derivative $f'(y)$. Indeed, by (52)

$$\begin{aligned} f'(y) &= \lim_{\delta\to 0+}\left(\frac{f(x(y,\delta))}{q_\delta}-f(y)\right)\frac{1}{q_\delta}=\lim_{\delta\to 0+}\frac{f(x(y,\delta))-q_\delta f(y)}{\delta q_\delta}\\ &= \lim_{\delta\to 0+}\frac{f(x(y,\delta))-f(y)}{\delta}+\mathrm{div}V f(y)_{|\delta=0}\\ &= f_{y_1}\theta_1+f_{y_2}\theta_2+(\theta_{1,y_1}+\theta_{2,y_2})f \end{aligned}$$

i.e.

$$f'(y)=\frac{\partial}{\partial y_1}(\theta_1 f)+\frac{\partial}{\partial y_2}(\theta_2 f)=\mathrm{div}(Vf). \tag{53}$$

Since $f\in C^1(\overline{\Omega})$, we have the following convergence as $\delta\to 0^+$

$$\frac{f^\delta(y)-f^0(y)}{\delta}\to f'(y)\quad \text{in } L^\infty(\Omega). \tag{54}$$

The sets of admissible functions for the minimization problems under considerations are defined by

$$K_\delta=\{w\in H^1(\Omega_\delta)\mid [w]\geq 0 \text{ on } \Sigma_\delta;\ w=0 \text{ on } \Gamma_0\},$$

$$K_0=\{w\in H^1(\Omega)\mid [w]\geq 0 \text{ on } \Sigma;\ w=0 \text{ on } \Gamma_0\},$$

respectively.
In view of (28), let $x=x(y,\delta)$ denotes the inverse transformation. Then $w^\delta(x)=w_\delta(y)$. The inclusion $w^\delta\in K_\delta$ implies $w_\delta\in K_0$, and, conversely, $w_\delta\in K_0$ implies $w^\delta\in K_\delta$. This means that the transformation (28) maps K_δ into K_0, and it is one-to-one. Now we shall prove the continuity of u_δ with respect to δ

$$||u_\delta-u||_{H^1(\Omega)}\to 0\quad \text{as } \delta\to 0^+.$$

The function $u^\delta\in K_\delta$ is the solution of the variational inequality

$$\int_{\Omega_\delta}\langle\nabla u^\delta,\nabla v-\nabla u^\delta\rangle\geq\int_{\Omega_\delta} f(v-u^\delta),\quad \forall v\in K_\delta. \tag{55}$$

But by substituting $v=0$ in (55), it follows that

$$||u^\delta||_{H^1(\Omega_\delta)}\leq C \text{ uniformly in } \delta.$$

Consequently

$$||u_\delta||_{H^1(\Omega)}\leq C \text{ uniformly in } \delta.$$

By the change of variables in (55), it follows that

$$\int_\Omega\langle A_\delta\cdot\nabla u_\delta, A_\delta\cdot\nabla\tilde{v}-A_\delta\cdot\nabla u_\delta\rangle\frac{dy}{q_\delta}\geq\int_\Omega f^\delta(\tilde{v}-u_\delta)dy,\quad \forall\tilde{v}\in K_0 \tag{56}$$

or $A_\delta=I+\delta B$ so, according to (56) we have

$$\int_\Omega\langle\nabla u_\delta+\delta B\cdot\nabla u_\delta,\nabla\tilde{v}+\delta B\cdot\nabla\tilde{v}-\nabla u_\delta-\delta B\cdot\nabla u_\delta\rangle\frac{dy}{q_\delta}$$

$$\geq \int_\Omega f^\delta(\tilde{v} - u_\delta)dy, \quad \forall \tilde{v} \in K_0. \tag{57}$$

We can substitute $\tilde{v} = u$ in (57) and we obtain

$$\int_\Omega \langle \nabla u_\delta, \nabla u - \nabla u_\delta \rangle \frac{dy}{q_\delta} + P(\delta, u, u_\delta) \geq \int_\Omega f^\delta(u - u_\delta)dy, \tag{58}$$

$$\text{where } P(\delta, u, u_\delta) \to 0 \quad \text{as } \delta \to 0^+.$$

The solution of the problem (45),(46) is given by the solution of the variational inequality

$$u \in K_0 : \int_\Omega \langle \nabla u, \nabla v - \nabla u \rangle \geq \int_\Omega f(v - u), \quad \forall v \in K_0.$$

We can sustitute $v = u_\delta$

$$\int_\Omega \langle \nabla u, \nabla u_\delta - \nabla u \rangle \geq \int_\Omega f(u_\delta - u). \tag{59}$$

And suming up the relations (58),(59) implies that

$$||u_\delta - u||_{H^1(\Omega)} \to 0 \quad \text{as } \delta \to 0^+. \tag{60}$$

Denote,

$$\begin{aligned} J(\Omega) &= \pi(\Omega; u) \\ J(\Omega_\delta) &= \pi(\Omega_\delta; u^\delta). \end{aligned}$$

We have

$$\begin{aligned} \frac{J(\Omega_\delta) - J(\Omega)}{\delta} &= \frac{\pi(\Omega_\delta; u^\delta) - \pi(\Omega; u)}{\delta} = \frac{\pi_\delta(\Omega; u_\delta) - \pi(\Omega; u)}{\delta} \\ &\leq \frac{\pi_\delta(\Omega; u) - \pi(\Omega; u)}{\delta}, \end{aligned}$$

and consequently

$$\begin{aligned} \limsup_{\delta \to 0^+} \frac{J(\Omega_\delta) - J(\Omega)}{\delta} &\leq \limsup_{\delta \to 0^+} \frac{\pi_\delta(\Omega; u) - \pi(\Omega; u)}{\delta} \\ &\leq \int_\Omega \langle B \cdot \nabla u, \nabla u \rangle dy + \frac{1}{2}\int_\Omega \mathrm{div} V |\nabla u|^2 dy - \int_\Omega f' u dy. \end{aligned} \tag{61}$$

On the other hand

$$\begin{aligned} \frac{J(\Omega_\delta) - J(\Omega)}{\delta} &= \frac{\pi(\Omega_\delta; u^\delta) - \pi(\Omega; u)}{\delta} = \frac{\pi_\delta(\Omega; u_\delta) - \pi(\Omega; u)}{\delta} \\ &\geq \frac{\pi_\delta(\Omega; u_\delta) - \pi(\Omega; u_\delta)}{\delta}, \end{aligned}$$

therefore

$$\begin{aligned} \liminf_{\delta \to 0^+} \frac{J(\Omega_\delta) - J(\Omega)}{\delta} &\geq \liminf_{\delta \to 0^+} \frac{\pi_\delta(\Omega; u_\delta) - \pi(\Omega; u_\delta)}{\delta} \\ &\geq \int_\Omega \langle B \cdot \nabla u, \nabla u \rangle dy + \frac{1}{2}\int_\Omega \mathrm{div} V |\nabla u|^2 dy - \int_\Omega f' u dy \end{aligned} \tag{62}$$

taking into account (60).

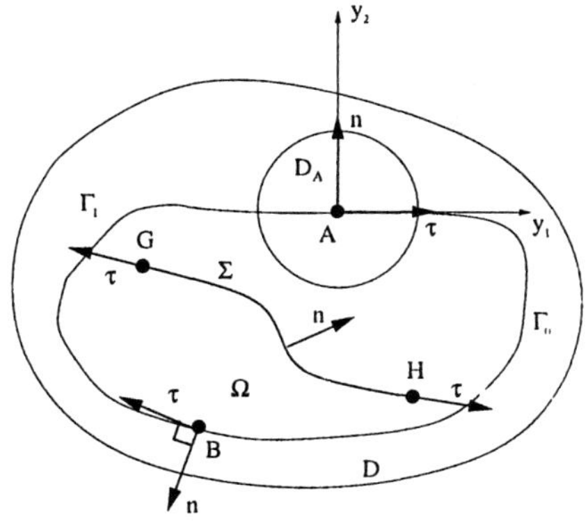

Figure 7 Domains Ω after changing of variables

Using (61), in view of (62), it follows that

$$\lim_{\delta\to 0+}\frac{J(\Omega_\delta)-J(\Omega)}{\delta}=\int_\Omega \langle B\cdot\nabla u,\nabla u\rangle dy+\frac{1}{2}\int_\Omega \mathrm{div}V|\nabla u|^2dy-\int_\Omega f'udy. \tag{63}$$

Substitution of B and f' in (63), compare to (53), leads to

$$\begin{aligned} dJ(\Omega;V) &= -\frac{1}{2}\int_\Omega \left((\theta_{1,y_1}-\theta_{2,y_2})\left((u_{y_1})^2-(u_{y_2})^2\right)+2(\theta_{1,y_2}+\theta_{2,y_1})u_{y_1}u_{y_2}\right) \\ &\quad -\int_\Omega ((\theta_1 f)_{y_1}+(\theta_2 f)_{y_2})\,u. \end{aligned} \tag{64}$$

The mapping $\mathcal{D}^k(D;\mathbb{R}^2)\ni V\mapsto dJ(\Omega;V)\in\mathbb{R}$ is linear and continuous for $k=1$. The functional $J(\Omega)$ is shape differentiable and therefore we can apply the structure theorem in order to obtain the following form of the shape derivative of $J(\Omega)$,

$$\begin{aligned} dJ(\Omega;V) &= \alpha_A(V.\tau)(A)+\alpha_B(V.\tau)(B)+\phi(V.n) \\ &+ \alpha_G(V.\tau)(G)+\alpha_H(V.\tau)(H)+\psi(V.n), \quad \forall V\in\mathcal{D}^1(D;\mathbb{R}^2), \end{aligned}$$

where α_A, α_B, α_G, $\alpha_H\in\mathbb{R}$, $\phi\in(C^1(\Gamma))'$, $\psi\in(C^1(\overline{\Sigma}))'$.
It is possible to determine the coefficient α_A. In the remainder of this section, using an appropriate change of variables if necessary, we may assume that there exists a neighbourhood $\mathcal{D}_A$ of A, $\mathcal{D}_A\subset D$, $\overline{\mathcal{D}_A}\cap\overline{\Sigma}=\emptyset$, such that $\Gamma\cap\mathcal{D}_A$ is rectilinear (see FIG.7). Moreover we use an appropriate coordinate system with origin A. In order to find the coefficient α_A, we can consider the deformations of Ω in a small neighbourhood of $\Gamma\cap\mathcal{D}_A$. The deformations are assumed to be such that $\Gamma\cap\mathcal{D}_A$ moves without changing direction, which implies that the vector field takes the form

$$V(y)=(\theta_1(y),0),$$

where θ_1 is supported in $\mathcal{D}_A$, and $\theta_1(y)=-1$ in the vicinity of the origin A. According to the relation (64), we have

$$dJ(\Omega;V)=-\frac{1}{2}\int_\Omega \left(\theta_{1,y_1}\left((u_{y_1})^2-(u_{y_2})^2\right)+2\theta_{1,y_2}u_{y_1}u_{y_2}\right)-\int_\Omega(\theta_1 f)_{y_1}u.$$

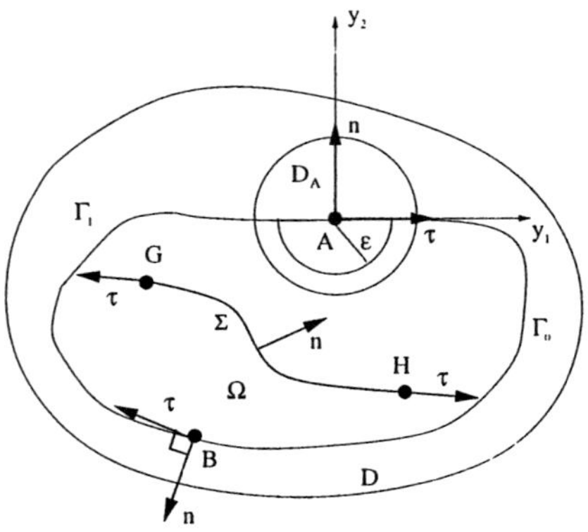

Figure 8 Domain Ω_ε

On the other hand,

$$dJ(\Omega; V) = \lim_{\varepsilon \to 0} \left(-\frac{1}{2} \int_{\Omega_\varepsilon} (\theta_{1,y_1} \left((u_{y_1})^2 - (u_{y_2})^2 \right) + 2\theta_{1,y_2} u_{y_1} u_{y_2}) \right) - \int_\Omega (\theta_1 f)_{y_1} u.$$

where Ω_ε is the subset of Ω defined by $r > \varepsilon$ (see FIG.8). Let γ_ε be the curve given in polar coordinates by $r = \varepsilon$ and $\pi < \theta < 2\pi$.
Let us denote

$$A_\varepsilon = -\frac{1}{2} \int_{\Omega_\varepsilon} \left(\theta_{1,y_1} \left((u_{y_1})^2 - (u_{y_2})^2 \right) + 2\theta_{1,y_2} u_{y_1} u_{y_2} \right). \tag{65}$$

By integrating by parts, we have

$$\begin{aligned} A_\varepsilon &= \int_{\Omega_\varepsilon} \theta_1 \left(\frac{1}{2} \left((u_{y_1})^2 - (u_{y_2})^2 \right)_{y_1} + (u_{y_1} u_{y_2})_{y_2} \right) dy \\ &\quad + \int_{\gamma_\varepsilon} \theta_1 \left(\frac{1}{2} \cos\theta \left((u_{y_1})^2 - (u_{y_2})^2 \right) + \sin\theta u_{y_1} u_{y_2} \right) d\sigma. \end{aligned}$$

Using the identity for the solution u,

$$\begin{aligned} \frac{1}{2} \left((u_{y_1})^2 - (u_{y_2})^2 \right)_{y_1} + (u_{y_1} u_{y_2})_{y_2} &= u_{y_1} (u_{y_1 y_1} + u_{y_2 y_2}) \\ &= u_{y_1} \Delta u = -f u_{y_1}, \end{aligned}$$

we obtain the formula

$$\begin{aligned} A_\varepsilon &= -\int_{\Omega_\varepsilon} \theta_1 f u_{y_1} dy \\ &\quad + \int_{\gamma_\varepsilon} \theta_1 \left(\frac{1}{2} \cos\theta \left((u_{y_1})^2 - (u_{y_2})^2 \right) + \sin\theta u_{y_1} u_{y_2} \right) d\sigma. \end{aligned} \tag{66}$$

Moreover, we have

$$\int_{\Omega_\varepsilon} \theta_1 f u_{y_1} dy \to \int_\Omega \theta_1 f u_{y_1} dy \quad \text{as } \varepsilon \to 0^+.$$

Let

$$B_\varepsilon = \int_{\gamma_\varepsilon} \theta_1 \left(\frac{1}{2} \cos\theta \left((u_{y_1})^2 - (u_{y_2})^2 \right) + \sin\theta u_{y_1} u_{y_2} \right) d\sigma. \qquad (67)$$

It is not difficult to obtain the following decomposition of the solution to (55) into regular and singular parts in the vicinity of A

$$u = u^R + cS, \qquad (68)$$

where $S = \sqrt{r} \sin\left(\frac{\theta}{2}\right)$ is the singular function, $u^R \in H^2(\mathcal{D}_A)$ and c denotes the coefficient of singularity.

For ε small enough, $\theta_1 \equiv -1$ on γ_ϵ and taking into account the decomposition (68) and developping in (67), we have

$$\begin{aligned} B_\varepsilon &= c^2 \int_\pi^{2\pi} \varepsilon \left(\frac{1}{2} \cos\theta \left((S_{y_1})^2 - (S_{y_2})^2 \right) + \sin\theta (S_{y_1} S_{y_2}) \right) d\theta \\ &\quad + c \int_\pi^{2\pi} \varepsilon \left(\cos\theta (u^R_{y_1} S_{y_1} - u^R_{y_2} S_{y_2}) + \sin\theta (u^R_{y_1} S_{y_2} + u^R_{y_2} S_{y_1}) \right) d\theta \\ &\quad + \int_\pi^{2\pi} \varepsilon \left(\frac{1}{2} \cos\theta \left((u^R_{y_1})^2 - (u^R_{y_2})^2 \right) + \sin\theta (u^R_{y_1} u^R_{y_2}) \right) d\theta \\ &= B^{(1)}_\varepsilon + B^{(2)}_\varepsilon + B^{(3)}_\varepsilon. \end{aligned}$$

The form of singular functions is known in this case,

$$S_{y_1} = -\frac{1}{2\sqrt{r}} \sin\left(\frac{\theta}{2}\right) \text{ and } S_{y_2} = \frac{1}{2\sqrt{r}} \cos\left(\frac{\theta}{2}\right).$$

The first integral in B_ε takes the form

$$\begin{aligned} B^{(1)}_\varepsilon &= c^2 \int_\pi^{2\pi} \left(\frac{\cos\theta}{2} \left(\frac{1}{4} \sin^2\left(\frac{\theta}{2}\right) - \frac{1}{4} \cos^2\left(\frac{\theta}{2}\right) \right) - \frac{\sin\theta}{4} \cos\left(\frac{\theta}{2}\right) \sin\left(\frac{\theta}{2}\right) \right) d\theta \\ &= -\frac{c^2}{8} \int_\pi^{2\pi} (\cos^2\theta + \sin^2\theta) d\theta = -\frac{\pi c^2}{8}. \end{aligned}$$

We have

$$B^{(1)}_\varepsilon = -\frac{\pi c^2}{8}. \qquad (69)$$

It is not difficult to see that $B^{(2)}_\varepsilon \to 0$ and $B^{(3)}_\varepsilon \to 0$ as $\varepsilon \to 0^+$(in fact we have the estimations $B^{(2)}_\varepsilon = O(\sqrt{\varepsilon})$, $B^{(3)}_\varepsilon = O(\varepsilon)$). Thus

$$B_\varepsilon \to -\frac{\pi c^2}{8}, \; A_\varepsilon \to -\int_\Omega \theta_1 f u_{y_1} dy - \frac{\pi c^2}{8} \quad \text{as } \varepsilon \to 0^+$$

hence

$$\begin{aligned} dJ(\Omega; V) &= -\int_\Omega ((\theta_1 f)_{y_1} u)\, dy - \int_\Omega \theta_1 f u_{y_1} dy - \frac{\pi c^2}{8} \\ &= -\lim_{\varepsilon \to 0^+} \underbrace{\left(\int_{\Omega_\epsilon} ((\theta_1 f)_{y_1} u + \theta_1 f u_{y_1})\, dy \right)}_{= \int_{\gamma_\varepsilon} f \theta_1 u \nu_1 d\sigma \to 0 \quad \text{as } \varepsilon \to 0^+} - \frac{\pi c^2}{8} \end{aligned}$$

and finally

$$dJ(\Omega; V) = -\frac{\pi c^2}{8} = -\alpha_A. \tag{70}$$

according to the structure theorem.

We have identified the coefficent $\alpha_A = \frac{\pi c^2}{8}$ in the expression of the Eulerian semiderivative $dJ(\Omega; V)$ given by the structure theorem.

REMARK Identification of coefficients α_G, α_H requires the knowledge of the singular functions associated to the crack Σ for the solutions to the variational inequality (55). □

References

[1] J. Sokołowski, J.-P. Zolésio. *Introduction to shape optimization: shape sensitivity analysis.* Springer Series in Computational Mathematics. 16. Berlin etc.: Springer-Verlag, (ISBN 3-540-54177-2). 250 p. (1992).

[2] G. Fremiot, J. Sokołowski. *The structure theorem for the Eulerian derivative of shape functionals defined in domains with cracks*, to appear in Siberian Mathematical Journal.

[3] A.-M. Khludnev, J. Sokołowski. *Griffith formula and Rice integral for an elliptic equation with unilateral conditions in nonsmooth domains*, European Journal of Applied Mathematics 10(1999), 379-394.

[4] A.-M. Khludnev, J. Sokołowski. *Griffith formula for elasticity system with unilateral conditions in domains with cracks*, European Journal of Elasticity/Solids. vol.19, No.1, 2000, 105-120.

[5] J.-P. Aubin. *Initiation à l'analyse appliquée. (Foundation of applied analysis).* Paris: Masson, (ISBN 2-225-84381-3/pbk). xxxi, 394 p. (1994).

[6] M.-C. Delfour. *Shape optimization and free boundaries.* Proceedings of the NATO Advanced Study Institute and Séminaire de mathématiques supérieures, held Montreal, Canada, June 25-July 13, 1990. NATO ASI Series. Series C. Mathematical and Physical Sciences. 380. Dordrecht: Kluwer Academic Publishers, (ISBN 0-7923-1944-3/hbk). xviii, 462 p. (1992).

[7] M.-C. Delfour, J.-P. Zolésio. *Structure of shape derivatives for nonsmooth domains.* J. Funct. Anal. 104, No.1, 1-33 (1992).

[8] J. Sokołowski, A. Zochowski. *Topological derivative for optimal control problems*, Les prépublications de l'Institut Élie Cartan. To appear in Control and Cybernetics, 3(1999).

[9] P. Grisvard. *Singularities in boundary value problems.* Recherches en Mathématiques Appliquées. 22. Paris: Masson, (ISBN 2-225-82770-2). Berlin: Springer-Verlag, (ISBN 3-540-55450-5). xiv, 198 p. (1992).

[10] M. Bochniak, A.-M. Sändig. *Sensitivity analysis of elastic structures in presence of stress singularities.* Arch. Mech. 51 (1999) pp. 155-171.

Singular Stress Field at the Tip of a Closed Interface Crack

DOMINIQUE LEGUILLON Laboratoire de Modélisation en Mécanique - CNRS UMR7607, Université Pierre et Marie Curie (Paris 6), 4 place Jussieu, F-75252 PARIS Cedex 05, France

1 INTRODUCTION – THE CLASSICAL OPENED INTERFACE CRACK

The generalized plane strain (GPS) linear elasticity model is considered. The displacement field $\underline{U}$ has three components U_1, U_2 and U_3 depending only on two space variables x_1 and x_2. In such a framework, the interface is a plane surface spaned by x_1 and x_3 and the interface crack front is a straight line parallel to x_3 (figure 1). It is implicitely assumed that the crack front ends are sufficiently far away to be ignored and then this model allows to study the three dimensional elastic fields in a vicinity of any interior point of the crack front. It embeds, as particular cases, the bidimensional plane strain model and the anti-plane scalar problem.
Let us first recall that the singular elastic displacement field near the tip of a crack lying in a homogeneous ans isotropic material (identical substrates on both sides of the interface) writes

$$\underline{U}(x_1, x_2) = \underline{U}(0,0) + k_I\sqrt{r}\ \underline{u}_I(\varphi) + k_{II}\sqrt{r}\ \underline{u}_{II}(\varphi) + k_{III}\sqrt{r}\ \underline{u}_{III}(\varphi) + O(r), \quad (1)$$

where r and φ are the polar coordinates with origin at a point of the crack front. The constant term $\underline{U}(0,0)$ has no mechanical role, it is present for consistency. The two first modes u_I and u_{II} are the in-plane opening and shear modes (the third component vanishes), whereas u_{III} is the out of plane shear mode (the two first components vanish). Coefficients k_I, k_{II} and k_{III} are the corresponding intensity factors. $O(r)$ stands for terms decreasing to 0 as r or faster.
This splitting still holds in case of an in-axis orthotropic material but disappears in case of pure anisotropy. The three modes have then three non zero components.
For an interface crack between different substrates, it is known, since the pioneering work by Williams [10, 14, 16], that the crack tip singularity is characterized by a

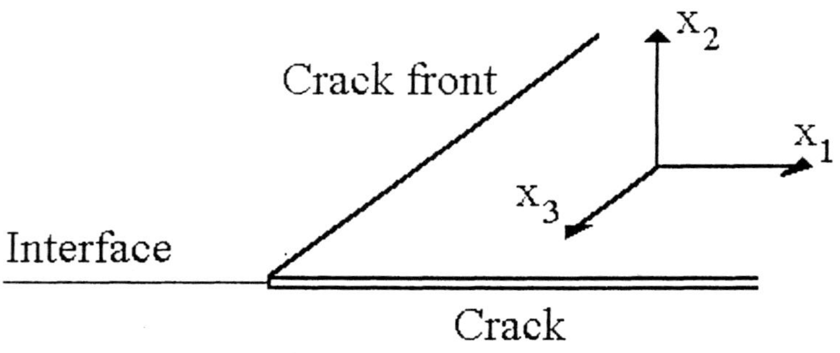

Figure 1: The plane interface crack

complex exponent

$$\underline{U}(x_1, x_2) = \underline{U}(0,0) + 2\mathcal{R}e\left(kr^{1/2+i\varepsilon}\underline{u}(\varphi)\right) + k_3\sqrt{r}\ \underline{u}_3(\varphi) + O(r). \qquad (2)$$

In the above expression k is a complex intensity factor and $\underline{u}$ the complex mode. The imaginary part ε of the exponent depends on the contrast between the two adjacent materials [14], practically it is small ($\varepsilon < 0.175$ with isotropic substrates [7]). The remaining part is real, k_3 is the intensity factor and $\underline{u}_3$ the associated mode.

In case of isotropic or in-axis orthotropic substrates, the complex part is an in-plane displacement and the reamining one is anti-plane. This property disappears if substrates are anisotrpic, each part (complex and real) has 3 non zero components. Two difficulties arise from this complex situation. First, it becomes impossible to split naturally the solution into an opening and a shear mode as in (1), the mode mixity (i.e. a ratio measuring the part of each mode) must be addressed [14, 12]. Second, the solution has an insuperable deficiency, the crack faces overlap in a vicinity of the front. This deficiency comes from the traction free condition (3) usually imposed on the crack faces which does not prevent from this troublesome feature. The crack opening is a posteriori checked. This condition writes

$$\sigma(\underline{U})\underline{n}^+ = 0\ , \qquad (3)$$

where $\sigma(\underline{U})$ is the stress field associated with the displacement $\underline{U}$ and $\underline{n}^+ = (0, -1, 0)$ the normal to the crack surface (see (19) in section 3.2 for the expression of σ in terms of $\underline{U}$). Whether this condition can be considered as admissible, since the crack opens and the overlaping zone is very small, or it exists a contact zone of finite length and one-sided contact conditions must be taken into account [2, 6]. These authors proved that the contact zone exists in any case but it is so small in the case of crack opening that it can be ignored. If contact conditions hold, a new question arises concerning the effect of friction on the two crack faces [3, 4, 11].

REMARK 1: For the sake of simplicity, this model is baptised opened interface crack because of condition (3) which means that no forces act on the crack faces just as if it was opened. Nevertheless, it is to be pointed out that, in case of a homogeneous isotropic body, $k_I = 0$ (i.e. no opening mode, see (1)) implies the crack closure. ■

REMARK 2 : Throughout this paper *in-plane* means in the plane (x_1, x_2) and *anti-plane* means orthogonal to this plane, i.e. parallel to x_3. The so-called *isotropic bimaterial case* is characterized by $E_1 = 1\,\text{GPa}$, $E_2 = 10\,\text{GPa}$, $\nu_1 = \nu_2 = 0.3$. Angle-ply laminates provide the anisotropic cases. Each ply is made of the carbon fiber reinforced composite T300/914 : $E_1 = E_2 = 9.4\,\text{GPa}$, $E_3 = 135\,\text{GPa}$, $G_{12} = G_{13} = G_{23} = 5.3\,\text{GPa}$, $\nu_{12} = \nu_{31} = \nu_{32} = 0.37$ (here fibers are parallel to x_3). The interface crack lies between two plies. The *in-axis orthotropic case* is obtained with a stacking sequence [0,90], the fibers are respectively parallel to x_3 and x_1 in two adjacent plies (angles between fibers and x_3 are respectively 0° and 90°). A stacking sequence [±45] provides the so-called *anisotropic case*, the fibers are in the x_1, x_3 plane and make an angle of +45°, resp. −45°, with x_3. ■

2 THE CLOSED INTERFACE CRACK WITH FRICTIONLESS CONTACT

In this section it is assumed that the two faces are in contact in a vicinity of the crack front, i.e. in a strip of finite and constant width to be consistent with the GPS model. Thus, the traction free condition (3) must be replaced by a condition of contact

$$[\![U_n]\!] = 0, \; \underline{F}_t = 0 \,, \tag{4}$$

where

$$[\![\underline{U}]\!] = [\![U_n]\!]\underline{n}^+ + [\![\underline{U}_t]\!] \text{ and } \sigma(\underline{U}).\underline{n}^+ = F_n\underline{n}^+ + \underline{F}_t \,. \tag{5}$$

The brackets denote a jump through the crack which sign is determined by the choice of the normal $\underline{n}^+$ (3), $[\![\underline{U}]\!] = \underline{U}^+ - \underline{U}^-$ where $\underline{U}^+$ and $\underline{U}^-$ are the values of $\underline{U}$ on the upper and lower crack faces. The index n stands for normal and t for tangential. Here the jump of the normal displacement must vanish (no opening). The second relation in (4) expresses the frictionless condition. Here $F_n \neq 0$ while (3) brings to $F_n = 0$.

2.1 Crack tip singularity

If one assumes the contact surface to be known, the problem remains a linear one and the usual tools extend [13]. The closure condition $(4)_1$ is treated as a particular kinematic condition and the frictionless condition $(4)_2$ is the homogeneous natural complement to the former.

If the crack lies along an interface between two isotropic or in-axis orthotropic substrates, the solution still splits into in-plane and anti-plane parts

$$\underline{U}(x_1, x_2) = \underline{U}(0,0) + k_2\sqrt{r}\;\underline{u}_2(\varphi) + k_3\sqrt{r}\;\underline{u}_3(\varphi) + O(r) \,. \tag{6}$$

The in-plane shear mode $\sqrt{r}\;\underline{u}_2(\varphi)$ has a non zero normal traction acting on the crack lips and thus differs from $\sqrt{r}\;\underline{u}_{II}$ in (1), whereas the anti-plane shear mode $\sqrt{r}\;\underline{u}_3(\varphi)$ fulfils the traction free condition (3), it is the same anti-plane component which is involved in the opened crack problem (1).

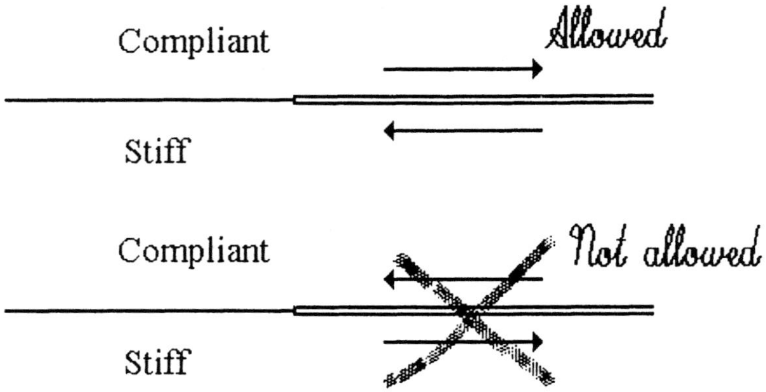

Figure 2: The allowed slip direction for a closed interface crack

In case of a crack lying between two anisotropic bodies, the situation is more entangled and the results derive of pure numerical ascertainements. There is a single complex mode

$$\underline{U}(x_1 . x_2) = \underline{U}(0,0) + 2\mathcal{R}e\left(k' r^{1/2+i\varepsilon} \underline{v}(\varphi)\right) + O(r). \qquad (7)$$

It involves oscillations but without faces overlap because of the closing condition $(4)_1$. It is numerically observed that the imaginary part ε of the exponent remains unchanged between this case and the classical opened crack one (as calculated in [14]).This particular case is not addressed in the following.

REMARK 3 : There is an additional condition to ensure the contact, the crack faces must undergo only compressive forces :

$$F_n < 0 \, . \qquad (8)$$

This condition brings to select one among the two opposite in-plane solutions. If $\sqrt{r}\ \underline{u}_2(\varphi)$ is solution. it fulfils (8) and by virtue of the same relation $-\sqrt{r}\ \underline{u}_2(\varphi)$ (which is an eigenmode as well) must be dismissed. There is a very important consequence on the displacement field at the tip of a closed interface crack : *only one in-plane slip direction is allowed.* as shown in figure 2. Thus if the applied loads were such that the other slip direction would be activated. the closure assumption would fail and the crack must open [1]. Note that the anti-plane solution $\sqrt{r}\ \underline{u}_3$ does not participate to the compression on the crack faces. ■

REMARK 4 : It is important to emphasize on another consequence of condition (8) concerning fracture mechanics. In in-plane problems ($k_3 = 0$), *the kinking of a closed interface crack out of the interface can occur only toward the most compliant material.* It is checked numerically in the isotropic bimaterial case for instance, that the admissible (8) mode $\underline{u}_2$ generates a compressive state parallel to the crack in the stiff material while the compliant one is in traction. ■

2.2 Intensity factors computation

The procedure is based on contour integrals [9], [13]. For any fields $\underline{U}$ and $\underline{V}$ satisfying the equilibrium equation in the substrates and the contact conditions (4) Ψ is a contour independent integral defined by

$$\Psi(\underline{U}, \underline{V}) = \int_{\Gamma} (\sigma(\underline{U})\underline{n}\ \underline{V} - \sigma(\underline{V})\underline{n}\ \underline{U})\ ds\ , \tag{9}$$

The integration line Γ is any contour surrounding the origin, starting and finishing at the same point of the crack faces, $\underline{n}$ is the unit normal to Γ pointing toward the origin. The main property of this integral is

$$\Psi(r^{\alpha}\underline{u}, r^{\beta}\underline{w}) \neq 0 \text{ only if } \beta = -\alpha\ . \tag{10}$$

where α and β denote generic real exponents, $\alpha = 1/2$ in the present singular frictionless contact case. Then the intensity factor k can be computed by

$$k = \frac{\Psi(\underline{U}, r^{-\alpha}\underline{v})}{\Psi(r^{\alpha}\underline{u}, r^{-\alpha}\underline{v})}\ . \tag{11}$$

The extraction function $r^{-\alpha}\underline{v}(\varphi)$ is the dual mode to $r^{\alpha}\underline{u}(\varphi)$ [13].
REMARK 5 : The procedure is a general one which can be used to compute the intensity factor of any term, singular or not, real or complex and corresponding to any kind of wedge (provided the edges undergo homogeneous conditions). On the other hand, it fails in case of friction as seen below in section 3. In case of complex exponent, the integral (9) must be extended to

$$\Psi(\underline{U}, \underline{V}) = \int_{\Gamma} [\sigma(\underline{U})\underline{n}\ \bar{\underline{V}} - \sigma(\bar{\underline{V}})\underline{n}\ \underline{U}]\ ds, \tag{12}$$

where the upper bar denotes the complex conjugate of a quantity. The counterpart to the property (10) is

$$\Psi(r^{\alpha}\underline{u}, r^{\beta}\underline{w}) \neq 0 \text{ only if } \beta = -\bar{\alpha}. \tag{13}$$

■

3 THE CLOSED INTERFACE CRACK WITH CONTACT AND FRICTION

This problem is more entangled than the frictionless contact case. In particular, the exponent is generally not 1/2 and the contour independent integral (9) no longer exists.
As in section 2 the contact with friction is assumed to occur in a strip of finite and constant width along the front. The closure condition in $(4)_1$ is still valid and must be completed by the Coulomb's law characterized by a friction coefficient μ

$$\| \underline{F}_t \| = \mu \mid F_n \mid \text{ and it exists } \lambda > 0 \text{ such that } [\![\underline{U}_t]\!] = -\lambda \underline{F}_t\ . \tag{14}$$

It is written here in the simplified incremental form [5] involving directly $[\![\underline{U}_t]\!]$ instead of the velocity $[\![\dot{\underline{U}}_t]\!]$. It implies monotonic and proportional applied loads. The slip discontinuity is an increasing function of the load intensity. Moreover, the additional closure condition (8) is taken into account a posteriori as in section 2.

3.1 The homogeneous isotropic case

In the homogeneous isotropic case [8], the asymptotics of the solution near the crack tip derive from the classical situation and as a particular case $\alpha = 1/2$. First, $k_I = 0$ in (1) ensures the closure condition to hold. Modes II and III fulfil trivially the contact and friction conditions. Thus in the homogeneous case, friction plays a role in the leading (most singular) terms only through the intensity factors k_{II} and k_{III}. New terms, homogeneous to r, appear at the next order. In addition to the classical so-called "non-singular" stresses involved in the Williams series [16] (or more precisely in its generalized counterpart in the GPS model) $\underline{T}_1 = r\ \underline{t}_1(\varphi)$ and $\underline{S}_{13} = r\ \underline{s}_{13}(\varphi)$ which are a traction and a shear mode parallel to the crack surface

$$\sigma_{11}(\underline{T}_1) = 1,\ \sigma_{13}(\underline{S}_{13}) = 1, \tag{15}$$

the other components vanishing, it comes a compression $\underline{T}_2 = r\ \underline{t}_2(\varphi)$ orthogonal to the crack surface and two shear modes $\underline{S}_{12} = r\ \underline{s}_{12}(\varphi)$ and $\underline{S}_{23} = r\ \underline{s}_{23}(\varphi)$

$$\sigma_{22}(\underline{T}_2) = 1,\ \sigma_{12}(\underline{S}_{12}) = 1,\ \sigma_{23}(\underline{S}_{23}) = 1, \tag{16}$$

the other components vanishing. Note that the 3D traction field $\underline{T}_3$ satisfying $\sigma_{33}(\underline{T}_3) = 1$ is out of the GPS model.

$$\begin{aligned}\underline{U}(x_1,x_2) = \ & \underline{U}(0,0) + k_{II}\sqrt{r}\ \underline{u}_{II}(\varphi) + k_{III}\sqrt{r}\ \underline{u}_{III}(\varphi) + \\ & c_1 r\ \underline{t}_1(\varphi) + c_{13} r\ \underline{t}_{13}(\varphi) + \\ & c_2 r\ \underline{t}_2(\varphi) + c_{12} r\ \underline{t}_{12}(\varphi) + c_{23} r\ \underline{t}_{23}(\varphi) + O(r^{3/2}).\end{aligned} \tag{17}$$

The constant $\underline{U}(0,0)$ as well as the two first non singular terms $\underline{T}_1$ and $\underline{S}_{13}$ are not involved in the friction mechanism, they do not produce slip discontinuities nor forces on the crack faces (as well as $\underline{T}_3$ in a complete 3D model). The three others are connected together by the friction law (14) through their intensity factors

$$c_2 < 0 \text{ and } \sqrt{c_{12}^2 + c_{23}^2} = \mu \mid c_2 \mid . \tag{18}$$

But curiously, these additional three terms do not trigger any slip discontinuity of the two crack faces. There is a kind of uncoupling between the singular terms and the non-singular ones.

REMARK 6 : It seems to be possible to extend this property to any term of the expansion. Terms corresponding to integer powers of r derive from an expansion for an interior point of a homogeneous material, they do not trigger any displacement discontinuity along the crack faces but produce stresses acting on the crack faces and are linked together by the friction law. While terms corresponding to "integer+1/2" powers of r derive from the Williams expansion for a crack in a homogeneous material, they fulfil a stress free condition (3) along the crack faces but allow slip discontinuities. Moreover, *there is no restriction on the slip direction* (see remark 3). ■

3.2 The interface crack tip singularity calculation

Before further investigations, it is necessary to extend the computational method used to get the characteristic elements (exponent and mode) of the singularities [13]. It is based on the variational formulation of the elasticity problems. Involving the most general form of elastic coefficients a_{ijkh}, the equilibrium equation

$$0 = -\nabla.\sigma(\underline{U}) = -\frac{\partial}{\partial x_\iota}\left(a_{i\iota k\kappa}\frac{\partial U_k}{\partial x_\kappa}\right), \tag{19}$$

where ι and κ on the one hand and i and k on the other hand run respectively from 1 to 2 and from 1 to 3, leads to

$$\int_\Phi a\nabla\underline{U}\nabla\underline{\phi}\; dx_1 dx_2 - \int_C \left(F_n[\![\phi_n]\!] + \underline{F}_t \cdot[\![\underline{\phi}_t]\!]\right) dx_1 = 0 \;\forall\; \underline{\phi}\,, \tag{20}$$

where a is the abridged form for the elastic coefficients. The domain Φ of the plane (x_1,x_2) is the support of the test function ϕ, embedding the crack tip and C is the part of the crack line included in Φ. The line C undergoes contact on its whole length. The test function $\underline{\phi}$ is admissible, it satisfies the closure condition $(4)_1$ in addition to the smoothness requirements. Its support Φ is strictly included in the structure and does not meet the boundaries. Then the variational formulation (20) reduces to

$$\int_\Phi a\nabla\underline{U}\nabla\underline{\phi}\; dx_1 dx_2 - \int_C \underline{F}_t\cdot[\![\underline{\phi}_t]\!] dx_1 = 0 \;\forall\; \underline{\phi}\,. \tag{21}$$

From (14), $\underline{F}_t$ is parallel to $[\![\underline{U}_t]\!]$, thus

$$[\![\underline{U}_t]\!] = [\![U_\tau]\!]\underline{\tau},\; \underline{F}_t = F_\tau\underline{\tau} \text{ and then } F_\tau = fF_n\,, \tag{22}$$

where $\underline{\tau}$ is the (tangential) slip direction and $f = \pm\mu$, the sign selection will be discussed in section 3.4). This direction being assumed as a kinematic condition, the test functions fulfil (22_1) which leads finally to

$$\int_\Phi a\nabla\underline{U}\nabla\underline{\phi}\; dx_1 dx_2 - f\int_C F_n(\underline{U})[\![\phi_\tau]\!] dx_1 = 0 \;\forall\; \underline{\phi}\,, \tag{23}$$

where the dependence of F_n on the unknown solution $\underline{U}$ has been highlighted. It is true for the exact solution which fulfils the equilibrium equation ensuring the second term to be meaningful and we consider formally the non symmetrical bilinear form defined on the space of admissible functions

$$a(\underline{\psi},\underline{\phi}) = \int_\Phi a\nabla\underline{\psi}\nabla\underline{\phi}\; dx_1 dx_2 - f\int_C F_n(\underline{\psi})[\![\phi_\tau]\!] dx_1\,. \tag{24}$$

Obviously the second integral is questionable, this is why the next reasoning is purely formal. Nevertheless, in this step the procedure already used in the classical elastic case [13] is put to work. The unknown solution is searched under the form

$$\underline{U}(x_1,x_2) = r^\alpha\underline{u}(\varphi)\,, \tag{25}$$

with test functions $\underline{\phi}(x_1,x_2) = g(r)\underline{\omega}(\varphi)$. The terms depending on r eliminate and then the functions $\underline{u}$ and $\underline{\omega}$ of the single variable φ are discretized by first order finite elements to lead to an implicit eigenvalue problem. The exponent α plays the role of the eigenvalue and $\underline{u}$ is the eigenvector [13].

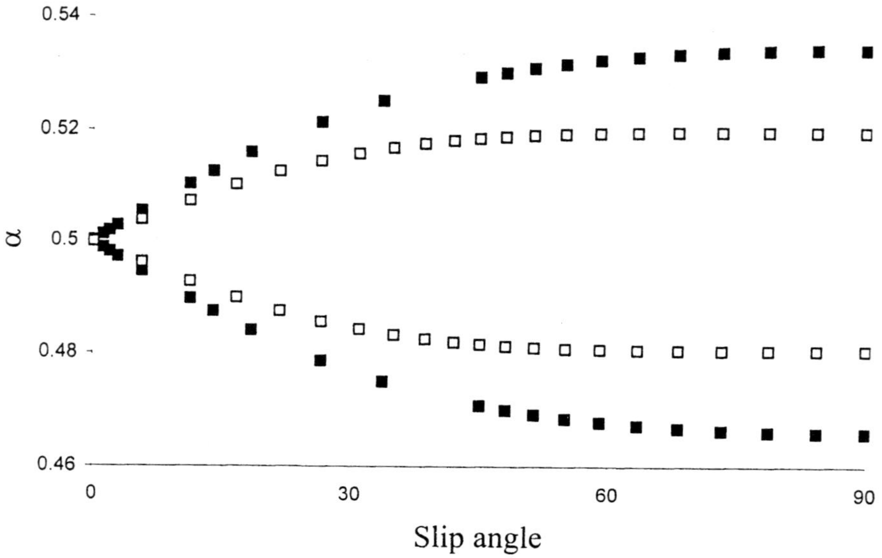

Figure 3: Singular exponents in the bimaterial isotropic case (black squares) and in the orthotropic case (white squares) vs. slip angle

3.3 Numerical results

In the above reasoning, it is recalled that the slip direction $\underline{\tau}$ is assumed, this is numerically obtained by imposing an additional constraint to the discretized system, the displacement discontinuity vanishes in the orthogonal direction to $\underline{\tau}$. It amounts to include some additional linear combinations of degrees of freedom. Thus the problem is solved for a prescribed slip direction (or angle) and a given friction coefficient μ. Such a framework is consistent with the GPS model.
In case of a crack lying between two isotropic substrates and a slip direction parallel to x_1 ($\underline{\tau} = (1, 0, 0)$, slip angle 90°), in-plane results [11] are in a very good agreement with the anlytical calculations of Comninou [3]. The advantage of the method is that computations can be worked out for any slip direction and any anisotropic situation in the GPS model. Figure 3 plots the singular exponent in the above bimaterial isotropic and orthotropic cases (remark 2) for different slip directions from a parallel to x_1 (angle = 90°, $\underline{\tau} = (1, 0, 0)$) to a parallel to x_3 (angle = 0°, $\underline{\tau} = (0, 0, 1)$). The symmetrical branches correspond to $f > 0$ and $f < 0$.
Table 1 exhibits some exponents corresponding to different anisotropic situations. The symmetrical stacking sequence [±45] (as any symmetric sequence [$\pm\gamma$]) gives $\alpha = 1/2$, it is numerically checked that it corresponds to a frictionless situation, $(4)_2$ holds true.
The role of the friction coefficient μ is examined in [11]. For smooth surfaces it lies between 0 (frictionless case) and 0.3 and roughness is simulated by higher values up to 1 and even more. Within this range the dependence is found almost linear.
REMARK 7 : Remarks 3 and 4 of the frictionless contact case concerning allowed slip directions and crack interface kinking are still valid. ■

Slip angle	Stacking sequence	f	α
90	[0,90]	+0.5	0.5195
90	[-15,+75]	+0.5	0.5187
90	[-30,+60]	+0.5	0.5098
90	[-45,+45]	+0.5	0.5001

Table 1: Singularity exponent vs. stacking sequence, the first and last cases are respectively the othortropic ans anisotropic examples of remark 2, the two others are different anisotropic situations

3.4 Discussion

It remains now to remove the sign ambiguity on f. It is observed (figure 3) that the exponent values are symmetrical with respect to 1/2 when changing the sign of f (it is shown by Comninou [3] in the plane isotropic bimaterial case but remains unproved in the general case) but there is not a one to one correspondance between the sign of f and the exponent α. In the isotropic bimaterial case for instance, if the compliant material is above (figure 1) $f > 0$ gives $\alpha > 1/2$ while if it is below the same $f > 0$ gives $\alpha < 1/2$ (and a symmetrical property for $f < 0$). Moreover this property holds in any case, condition (8) implies to change the sign of f if the materials are reversed.

Following the reasoning proposed by Comninou [3] in the bidimensional isotropic case, the situation leading to $\alpha > 1/2$ is to be kept. It is based on the analysis of the shear stress F_τ and slip discontinuity $[\![U_\tau]\!]$ signs, regarding the second part of the Coulomb's law (14).

However, in a numerical approach such a method cannot be used systematically, the shear stress is computed at Gauss points and must be extrapolated on the crack faces and since it is often small (zero in the frictionless case and small for small friction coefficients), it induces a prejudicial lack of accuracy.

Thus, another criterion must be developed. It is still the examination of the eigenvectors and a comparison with the frictionless case which brings to a selection. Let us denote

$$\begin{aligned} \underline{U}^+(r,\varphi) &= r^\alpha \underline{u}^+(\varphi) , \\ \underline{U}(r,\varphi) &= r^{1/2}\underline{u}(\varphi) , \\ \underline{U}^-(r,\varphi) &= r^\beta \underline{u}^-(\varphi) , \end{aligned} \tag{26}$$

the three modes corresponding respectively to the three cases : $\alpha > 1/2$, frictionless case and $\beta < 1/2$ (note that by virtue of the symmetry of the exponents with respect to 1/2, $\beta = 1 - \alpha$). The associated compressive components σ_{22} are

$$\begin{aligned} \sigma_{22}(\underline{U}^+) &= r^{\alpha-1} s_{22}^+(\varphi) , \\ \sigma_{22}(\underline{U}) &= r^{-1/2} s_{22}(\varphi) , \\ \sigma_{22}(\underline{U}^-) &= r^{\beta-1} s_{22}^-(\varphi) . \end{aligned} \tag{27}$$

In order to normalized them to -1 at a point of the crack faces located at a distance $r = r_0$ of the tip, first the components s_{22}^+, s_{22} and s_{22}^- are set to -1. Then the normalization amounts to multiplying the first vector by $r_0^{1-\alpha}$, the second one by $r_0^{1/2}$ and the last one by $r_0^{1-\beta}$. Finally, for a same unit compression on the crack

f	0	0.25	-0.25	0.50	-0.50	0.75	-0.75
α	0.500	0.510	0.489	0.519	0.478	0.529	0.468
normalized $[\![u_\tau]\!]$	1.000	0.989	1.011	0.977	1.023	0.966	1.035

Table 2: Normalized slip magnitude vs. friction coefficient and sign choice leading to $\alpha > 1/2$ or $\alpha < 1/2$ in the orthotropic case. The tangential displacement discontinuity is normalized by the same discontinuity in the frictionless case (28)

faces at the point $r = r_0$, the slip discontinuity reads

$$[\![U_\tau^+]\!] = r_0[\![u_\tau^+]\!] \ , \ [\![U_\tau]\!] = r_0[\![u_\tau]\!] \ , \ [\![U_\tau^-]\!] = r_0[\![u_\tau^-]\!] \ , \tag{28}$$

and it is easy to compare them by comparing $[\![u_\tau^+]\!]$, $[\![u_\tau]\!]$ and $[\![u_\tau^-]\!]$. It is observed that, as drawn before, the sign of f providing $\alpha > 1/2$ is to be kept because in that case the slip discontinuity is smaller than in the frictionless case, itself smaller than in the situation corresponding to $\beta = 1-\alpha < 1/2$. Reasonably, friction must reduce the amount of displacement discontinuity and not increase it. For a given friction coefficient μ, this criterion requires at most two (cheap) computations corresponding to $f = +\mu$ and $f = -\mu$, retaining the solution in which the tangential discontinuity of displacements is the smallest. This is illustrated in table 2. As a consequence, the meaningful exponent is $\alpha > 1/2$, *solutions exhibiting exponents smaller than 1/2 are mechanically unrealistic and must be dismissed.* This conclusion holds of course in the present framework of monotonic loads, history of loading is not addressed here. It is in agreement with [15] where the authors distinguish the "artificially" closed crack ($\alpha < 1/2$) and the "naturally" closed crack ($\alpha > 1/2$).

3.5 Intensity factors calculation

The intensity factor can no longer be computed exactly as in section 2.2 where the calculation is based on the existence of a dual mode and on a biorthogonality property (9) and (10) as well as on contour independent integrals. All these properties must be revisited.

Let us first consider a domain D surrounding the crack tip and bounded by two circular lines Γ_a (radius a) and Γ_b (radius $b > a$) (see figure 4).

In linear elasticity, for two displacement fields $\underline{U}$ and $\underline{V}$, the following integral vanishes

$$\int_D [\sigma(\underline{U})\nabla(\underline{V}) - \sigma(\underline{V})\nabla(\underline{U})]dx = 0 \ . \tag{29}$$

If $\underline{U}$ and $\underline{V}$ satisfy the equilibrium equation, (29) leads to

$$\Psi_a(\underline{U},\underline{V}) - \Psi_b(\underline{U},\underline{V}) + \int_C \left(\sigma(\underline{U})\underline{n}^+[\![\underline{V}]\!] - \sigma(\underline{V})\underline{n}^+[\![\underline{U}]\!]\right) ds = 0 \ , \tag{30}$$

where Ψ_a and Ψ_b denote the integral Ψ (9) computed respectively on Γ_a and Γ_b. The line C is the part of the crack located between the two contours (see section 2 for the other notations). Applying (30) to two solutions $\underline{U} = r^\alpha \underline{u}(\varphi)$ and $\underline{V} = r^{\alpha'} \underline{u}'(\varphi)$

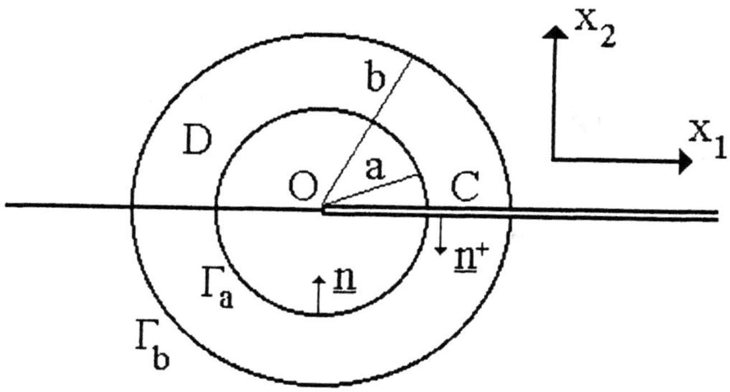

Figure 4: The domain D bounded by two circular contours Γ_a and Γ_b

gives

$$(a^{\alpha+\alpha'} - b^{\alpha+\alpha'})p(\underline{u},\underline{u}') + \int_a^b r^{\alpha+\alpha'-1}dr \times q(\underline{u},\underline{u}') = 0 \,. \tag{31}$$

The bilinear forms p and q are defined by

$$p(\underline{u},\underline{u}') = \int_0^{2\pi} [s(\varphi)\underline{n}\ \underline{u}'(\varphi) - s'(\varphi)\underline{n}\ \underline{u}(\varphi)]d\varphi \,, \tag{32}$$

$$q(\underline{u},\underline{u}') = s'_{12}(0)[\![u_1]\!] - s_{12}(0)[\![u'_1]\!] \,, \tag{33}$$

where s and s' are derived from the stress fields by

$$\sigma(\underline{U}) = r^{\alpha-1}s(\varphi),\ \sigma(\underline{U}') = r^{\alpha'-1}s'(\varphi) \,. \tag{34}$$

In this step two cases arise

$$\begin{aligned} \alpha + \alpha' = 0 \quad &\Rightarrow q(\underline{u},\underline{u}') = 0 \,, \\ \alpha + \alpha' \neq 0 \quad &\Rightarrow p(\underline{u},\underline{u}') = \frac{1}{\alpha+\alpha'}\, q(\underline{u},\underline{u}') \,. \end{aligned} \tag{35}$$

REMARK 8 : For a given α, the first case $(35)_1$ occurs if it exists $\alpha' = -\alpha$, i.e. a dual mode to the primary one, but this property has not yet been proved. Nevertheless it is equivalent to the symmetry with respect to 1/2 of the two exponents obtained with f and $-f$ (also not proved in the general case). It is the consequence of the invariance of the problem with respect to x_1 which implies that if $\underline{U}(x_1, x_2)$ is a solution then $\partial\underline{U}/\partial x_1(x_1, x_2)$ is also a solution. Thus, if α and its symmetric $1-\alpha$ are admissible exponents, then $\alpha - 1$ and $-\alpha$ are also admissible. *The dual mode to a primary one with an exponent α obtained with f is characterized by $-\alpha$ obtained with $-f$.* ∎

Let $\underline{U}$ be a solution of a structural problem which expands in a vicinity of a closed crack tip as

$$\underline{U}(x_1, x_2) = \underline{U}(0,0) + k\ r^{\alpha}\underline{u}(\varphi) + t\ r\ \underline{u}'(\varphi) + k'' r^{\alpha+1}\underline{u}''(\varphi) + O(r^2) \,. \tag{36}$$

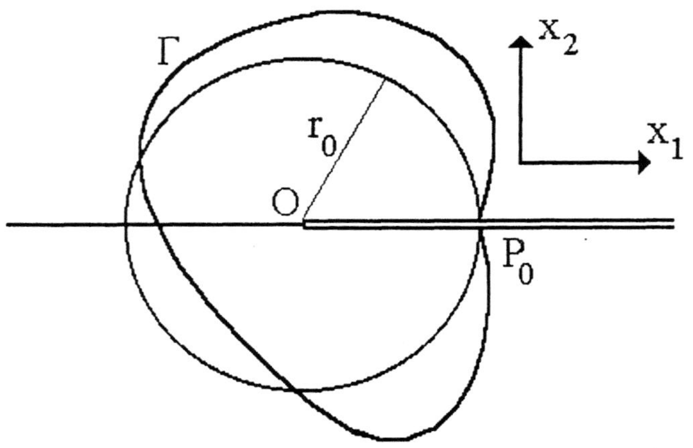

Figure 5: Two equivalent contours for the integral (37)

For simplicity, the first non singular term corresponding to r is written as a single term whereas multiplicty is respectively 2 and 4 in 2D elasticity and in the GPS model (plus one rigid rotation). The exponent $\alpha + 1$ of the next term is still a consequence of the invariance of the problem with respect to x_1.
We have to estimate k in (36). To this aim, let us consider the following integral I defined on a contour Γ surrounding the crack tip and starting and finishing at the same point P_0 of the crack lips, located at a distance $r = r_0$ of the tip. It involves the solution (36) and the dual mode $\underline{V} = r^{-\alpha}\underline{v}(\varphi)$ (see remark 8)

$$I = \int_\Gamma [\sigma(\underline{U})\underline{n}\ \underline{V} - \sigma(\underline{V})\underline{n}\ \underline{U}]ds - \frac{r_0}{1-\alpha}\{\sigma_{12}(\underline{V})[\![U_1]\!] - \sigma_{12}(\underline{U})[\![V_1]\!]\}_{|P_0}\ , \qquad (37)$$

where $\sigma_{12}(\underline{U})$ and $\sigma_{12}(\underline{V})$ and the jumps in the second term are computed on the crack lips at P_0 (i.e. for $r = r_0$).
REMARK 9 : A circular contour is not a necessary condition. By virtue of the equilibrium equation, any contour can be used provided it starts and finishes at the same point P_0 (fig. 5). ■
Replacing (36) into (37) leads to split the integral I into four parts. The first one associated with the constant term $\underline{U}(0,0)$ vanishes

$$I^0 = -\underline{U}(0,0)\int_\Gamma \sigma(\underline{V})\underline{n}\ ds = 0\ . \qquad (38)$$

Using the previous domain D allows to prove on the one hand that the integral (38) is contour independent and on the other hand it depends on $r_0^{-\alpha}$, thus it is zero.
The next term, corresponding to the singularity, satisfies $(35)_1$ then

$$I^1 = k\ p(\underline{u}, \underline{v})\ . \qquad (39)$$

The third term, associated with the first non singular term of the expansion, fulfils $(35)_2$, but a judicious choice of the multiplicative constant $r_0/(1-\alpha)$ in (37) gives

$$I^2 = 0\ . \qquad (40)$$

Finally, using once more (35) leads to conclude in the last case

$$I^3 = -k'' r_0 \frac{\alpha}{1-\alpha} p(\underline{u}'', \underline{v}) = O(r_0) \; . \tag{41}$$

As a consequence

$$k = \frac{I}{p(\underline{u}, \underline{v})} + O(r_0) \; . \tag{42}$$

As already anounced, it is only an approximate value which can be obtained by this way and computations must be carried out on small contours (i.e.small r_0) to be accurate. Unfortunately, the exact solution (36) is generally known from a finite element computation which reliability is poor in a vicinity of a singular point. By chance, $p(\underline{u}'', \underline{v}) = 0$ and then $I^3 = 0$ (and following terms) in the general and frictionless cases as a consequence of (10) and it can be reasonably expected that they remain small at least for small friction coefficients μ. Thus reasonable small r_0 can provide satisfying results.
Numerical tests have been performed to check these ascertainements. In the bimaterial isotropic case of remark 2 for instance, with $\mu = 0.5$, the agreement is quite good. Let us first consider the constant term $\underline{U}(0,0) = r^0 \underline{c} = \underline{c}$, it gives almost 0 trivially

$$p(\underline{c}, \underline{v}) < 0.01, \; q(\underline{c}, \underline{v}) \simeq 0 \; , \tag{43}$$

where p and q are defined by (32) and (33). For the second term, with $\alpha' = -\alpha$ and $(35)_1$, it comes

$$p(\underline{u}, \underline{v}) \simeq 219.11 \; , \; q(\underline{u}, \underline{v}) \simeq 0.17 \; . \tag{44}$$

The two next terms fulfil $(35)_2$, with respectivelly $\alpha + \alpha' = 1 - \alpha$ which leads to

$$\begin{aligned} p(\underline{u}', \underline{v}) \simeq 16.24 \; , \quad & q(\underline{u}', \underline{v}) \simeq 7.53 \; \Rightarrow \\ & p(\underline{u}', \underline{v}) - \frac{1}{1-\alpha} q(\underline{u}', \underline{v}) \simeq -0.07 \; , \end{aligned} \tag{45}$$

and $\alpha + \alpha' = 1$ which leads to

$$\begin{aligned} p(\underline{u}'', \underline{v}) \simeq 4.84 \; , \quad & q(\underline{u}', \underline{v}) \simeq 4.95 \; \Rightarrow \\ & p(\underline{u}', \underline{v}) - q(\underline{u}'', \underline{v}) \simeq -0.11 \; . \end{aligned} \tag{46}$$

Obviously $p(\underline{u}, \underline{v})$ is the governing term and as expected $p(\underline{u}'', \underline{v})$ and then I^3 are small compared to $p(\underline{u}, \underline{v})$ and I^1.

3.6 CONCLUSION

It is difficult to investigate more thoroughly the problem with regards to a fracture mechanism. There is a single real mode avoiding troublesome oscillations and the associated intensity factor can be computed but due to friction, no equivalence with an energy formulation can be settled. As a matter of fact, the definition of an energy release rate is doubtful, since a part of the supplied energy is consumed by friction.

References

[1] Audoly B. (1999) Asymptotic study of the interfacial crack with friction, to appear in J. Mech. Phys. Solids.

[2] Comninou M. (1977) The interface crack, J. of Applied Mech. : 44, 631-636.

[3] Comninou M. (1977) Interface crack with friction in the contact zone, J. of Applied Mech. : 44, 780-781.

[4] Deng X. (1994) An asymptotic analysis of stationary and moving cracks with frictional contact along bimaterial interfaces and in homogeneous solids, Int. J. Solids Structures : 31, 2407-2429.

[5] Duvaut G. & Lions J.L. (1972) *Les inéquations en mécanique et en physique*, Dunod, Paris.

[6] Gautesen A.K. & Dundurs J. (1987) The interface crack in a tension field, J. Appl. Mech. : 54, 93-98.

[7] Hutchinson J.W., Mear M.E. & Rice J.R. (1987) Crack paralleling an interface between dissimilar materials, J. Appl. Mech. : 54, 828-832.

[8] Leblond J.B. & Frelat J. (1999) Crack kinking from an initially closed crack, Int. J. Solids Stuctures : 37, 1595-1614.

[9] Leguillon D. (1993) Asymptotic and numerical analysis of a crack branching in non-isotropic materials, Eur. J. in Mech.- A/Solids : 12, 33-51.

[10] Leguillon D. (1993) Analysis of the brittle fracture in composites using singularities, in Mecamat'93- Int. seminar on micromechanics of materials, Eyrolles, Paris, 60-71.

[11] Leguillon D. (1999) Interface crack tip singularity with contact and friction, C.R. Acad. Sci. Paris : 327, série IIb, 437-442.

[12] Leguillon D., Bein S., Dupeux M. & Frelat J. (2000) Mixit modale pour une fissure d'interface – Application l'essai brsilien, to appear in Revue Europenne des Elments Finis.

[13] Leguillon D. & Sanchez-Palencia E. (1987) *Computation of singular solutions in elliptic problems and elasticity*, Masson, Paris, J. Wiley, New York.

[14] Rice J.R. (1988) Elastic fracture mechanics concepts for interfacial cracks, J. Appl. Mech. : 55, 98-103.

[15] Qian W. & Sun C.T. (1998) A frictional interfacial crack under combined shear and compression. Composite Science and Technology : 58, 1753-1761.

[16] Williams M.L. (1959) The stress around a fault or a crack in dissimilar media, Bulletin of the Seismology Soc. of America : 49, 199-204.

On the geometric and algebraic multiplicities for eigenvalue problems on graphs [1]

JOSE A. LUBARY Dept. Matemàtica Aplicada II, Universitat Politècnica de Catalunya, 08028 Barcelona. Spain.

1 INTRODUCTION

In this paper some properties of eigenvalue problems for second order elliptic differential operators on graphs are presented, concerning the geometric and algebraic multiplicities of the eigenvalues and the selfadjoint or nonselfadjoint nature of the operators.

First, an upper bound can be obtained for the geometric multiplicity, depending only on some combinatorial features of the graph. This bound is optimal in the sense that it can be achieved by an appropriate selection of the equations coefficients.

It will be shown that the operator is selfadjoint when the graph is a tree, but an additional property is required for graphs containing circuits. In the selfadjoint cases both multiplicities coincide and all the eigenvalues are real. In order to begin a study of nonselfadjoint cases, the simplest case of graphs containing circuits, the circle, is considered. Some properties involving the existence of nonreal eigenvalues and its multiplicities are obtained for this case.

Let us consider a connected and finite network with M edges and N nodes. We allow the possibility that there are multiple edges and loops, and there can be also terminal nodes, in other words, nodes receiving

[1]Work partially supported by the UPC grant 642.2 and the DGICYT project PB95-0629-C02-02, Spain

only one edge. We refer to [12] for graph terminology.

By means of a convenient $\mathcal{C}^2$–parametrization (see [2]) we can identify each edge as a real interval $[-a_i, a_i]$, $(i = 1, ..., M)$, and suppose that on each of them a second order differential equation is considered:

$$a^i(x)u_i''(x) + b^i(x)u_i'(x) + c^i(x)u_i(x) + \lambda u_i(x) = 0 \quad (i = 1, ..., M), \quad (1)$$

with continuous coefficients and $a^i(x) > 0$ for all x in $[-a_i, a_i]$. We suppose that in the interior nodes *continuity* conditions are imposed: in a node receiving the edges $[-a_1, a_1]$, ..., $[-a_k, a_k]$, $[-a_{k+1}, a_{k+1}]$, ..., $[-a_l, a_l]$, and such that the extremes $-a_1, ..., -a_k$, $a_{k+1}, ..., a_l$ coincide, the following conditions are set:

$$u_1(-a_1) = ... = u_k(-a_k) = u_{k+1}(a_{k+1}) = ... = u_l(a_l). \quad (2)$$

On the other hand, we impose at every node general *generalized Kirchhoff conditions*: for a node j like above,

$$-\sum_{i=1}^{k} \alpha_j^i u_i'(-a_i) + \sum_{i=k+1}^{l} \alpha_j^i u_i'(a_i) + r_j u_{0j} = 0 \quad (j = 1, ..., N), \quad (3)$$

where u_{0j} is the common value of the u_i at node j, $\alpha_j^i > 0$ for interior nodes, and $\alpha_j \geq 0$, $\alpha_j^2 + r_j^2 \neq 0$ for terminal nodes. For an interior node, the condition is of *Kirchhoff-law* type. If for a terminal node $r_j = 0$, the condition is of *Neumann* type; and when $\alpha_j = 0$ for a terminal node, it is of *Dirichlet* type.

Suppose that we fix the value of λ. If we ignore the coupling and boundary conditions, every edge carries a general solution depending on two parameters; then, the set of transition and boundary conditions is a linear algebraic system with $2M$ unknowns (the mentioned parameters) and $2M$ equations. Let $\mathcal{M}$ denote a matrix corresponding to these equations. The dimension of the kernel of $\mathcal{M}$ is the geometric multiplicity $m(\lambda)$ of λ (of course, $m(\lambda) = 0$ if λ is not an eigenvalue):

$$m(\lambda) = 2M - rank(\mathcal{M}).$$

If we eliminate all the bridges of a network, we obtain a set of subnetworks without bridges, together with isolated vertices (nodes). Reducing that subnetworks to vertices, and getting back the bridges, we will obtain a tree. If this *reduced tree* has T terminal vertices, we will say that the original network *has T terminals*. If there are no bridges, we define $T = 2$.

2 MAIN RESULTS ON GEOMETRIC MULTIPLICITY

The main results on the geometric multiplicity are the following theorems, that establish an accessible bound, depending only on M, N and T, for $m(\lambda)$. This results have been published in [5]. Previous results about the multiplicity of the eigenvalues in some particular cases can be found in [2]–[4], [9], [10] and [11].

THEOREM 1 *Given a network having M edges, N nodes and T terminals, the multiplicity $m(\lambda)$ described above for the problem* (1)–(2)–(3) *satisfies $m(\lambda) \leq M - N + T$.*

THEOREM 2 *Given a network having M edges, N nodes and T terminals and given λ, there exists a choice for the functions $a^i(x), b^i(x)$, $c^i(x)$ $(i = 1, ..., M)$ of* (1) *such that $m(\lambda) = M - N + T$.*

COROLLARY *For all functions $a^i(x), b^i(x)$, $c^i(x)$ $(i = 1, ..., M)$ of* (1) *the maximum multiplicity on a network is* 1 *if and only if it is a path (a two-terminals tree).*

3 SELFADJOINTNESS

It is well known that the geometric and algebraic multiplicities coincide for each eigenvalue of a selfadjoint operator. In this section we are going to investigate the selfadjointness of the operator described in Section 1. The results of this section are part of the author's thesis [8], Chapter 3.

So let us consider a finite connected network $\mathcal{G}$ having M edges and N nodes. Let I_i $(i = 1, \ldots, M)$ be the intervals identifying each edge of $\mathcal{G}$. Without loss of generality we can assume $I_i = [0, 1]$ $(i = 1, ..., M)$. We understand that a function u defined on $\mathcal{G}$ is a M–vector $(u_1, \ldots, u_M)$, where each u_i is a function defined on I_i.

Let $L^2(\mathcal{G})$ be the space of functions $u = (u_1, \ldots, u_M)$ defined on $\mathcal{G}$ such that u_i is in $L^2(I_i)$ for all $i \in \{1, \ldots, M\}$, and consider the operator L on $L^2(\mathcal{G})$ such that $Lu = (v_1, ..., v_M)$ with

$$v_i = a^i(x)u_i''(x) + b^i(x)u_i'(x) + c^i(x)u_i(x) \quad (i = 1, ..., M), \qquad (4)$$

and with domain $D(L) = H^2(\mathcal{G})$, the space of functions $u = (u_1, \ldots, u_M)$ defined on $\mathcal{G}$ such that u_i is in $H^2(I_i)$ and verify the continuity and generalized Kirchhoff conditions mentioned above.

3.1 The symmetry condition.

The right hand of (4) can be written (formally selfadjoint form)

$$\frac{1}{r_i(x)}(p_i(x)u_i'(x))' + q_i(x)u_i(x) \quad (i = 1, ..., M),$$

where

$$p_i(x) = k_i \exp\left(\int_0^x \frac{b^i(t)}{a^i(t)}\,dt\right), \quad r_i(x) = \frac{p_i(x)}{a^i(x)}, \tag{5}$$

with k_i an arbitrary (positive) constant for each edge. We define in the Hilbert space $L^2(\mathcal{G})$ a scalar product by

$$(u,v)_r = \int_{\mathcal{G}} r(x)u(x)\overline{v}(x) = \sum_{i=1}^{M} \int_0^1 r_i(x)u_i(x)\overline{v_i}(x)\,dx, \tag{6}$$

which induces a norm that is equivalent to the usual one on $L^2(\mathcal{G})$ (note that this scalar product depends on the choice of the k_i's in (5)).

The following lemma establishes a necessary and sufficient condition for the symmetry of L with respect to the metric induced by (6).

LEMMA *Given a choice of the k_i's in (5), L is symmetric with respect to the metric induced by (6) if and only if for every interior node j there exists a number μ_j such that $p_i(e_{ji}) = \mu_j \alpha_{ji}$ for all the endpoints e_{ji} of the edges incident to j.*

Proof: For all u, v in $D(L)$ we have

$$\begin{aligned}
(Lu,v)_r &= \int_G rLu\overline{v} \\
&= \sum_{i=1}^{M} \int_0^1 ((p_iu_i')'\overline{v_i} + r_iq_iu_i\overline{v_i}) \\
&= \sum_{j=1}^{N} \sum_{i \in B(j)} p_i(j)u_i'(j)\overline{v_i}(j) - \sum_{j=1}^{N} \sum_{i \in B(j)} p_i(j)u_i(j)\overline{v_i}'(j) \\
&+ \sum_{i=1}^{M} \int_0^1 ((p_iv_i')'\overline{u_i} \\
&= \sum_{j=1}^{N} \sum_{i \in B(j)} p_i(j)u_i'(j)\overline{v_i}(j) - \sum_{j=1}^{N} \sum_{i \in B(j)} p_i(j)u_i(j)\overline{v_i}'(j) \\
&+ (u, Lv)_r,
\end{aligned}$$

where $B(j)$ is the set of edges incident to j, the derivatives are exterior and we have integrated twice by parts.

So L is symmetric if and only if for all $u, v \in D(L)$ we have

$$E := \sum_{j=1}^{N} \sum_{i \in B(j)} p_i(j) u_i'(j) \overline{v_i}(j) - \sum_{j=1}^{N} \sum_{i \in B(j)} p_i(j) u_i(j) \overline{v_i}'(j) = 0.$$

For a terminal node j, it is easy to see that

$$p(j)u'(j)\overline{v}(j) - p(j)u(j)\overline{v}'(j) = 0.$$

For an interior node j we choose an appropriate $u \in D(L)$, substitute it in the expression E and take in account the generalized Kirchhoff condition at j, and we easily obtain that the values $p_i(j)$ have to be proportional to the values α_{ji}.

Conversely, if we suppose $p_i(j) = \mu_j \alpha_{ji}$ we have

$$\begin{aligned}
E &= \sum_{j=1}^{N} \sum_{i \in B(j)} p_i(j) u_i'(j) \overline{v_i}(j) - \sum_{j=1}^{N} \sum_{i \in B(j)} p_i(j) u_i(j) \overline{v_i'}(j) \\
&= \sum_{j \in \mathcal{N}_{int}} \sum_{i \in B(j)} p_i(j) (u_i'(j) \overline{v_i}(j) - u_i(j) \overline{v_i}'(j)) \\
&= \sum_{j \in \mathcal{N}_{int}} \sum_{i \in B(j)} \mu_j \alpha_{ji} (u_i'(j) \overline{v_i}(j) - u_i(j) \overline{v_i}'(j)) \\
&= \sum_{j \in \mathcal{N}_{int}} \overline{v}(j) \mu_j \sum_{i \in B(j)} \alpha_{ji} (u_i'(j) - \mu_j u(j) \sum_{i \in B(j)} \alpha_{ji} \overline{v_i}'(j)) \\
&= \sum_{j \in \mathcal{N}_{int}} \overline{v}(j) \mu_j \beta_j u(j) - \mu_j u(j) \beta_j \overline{v}(j)) = 0,
\end{aligned}$$

where $\mathcal{N}_{int}$ is the set of interior nodes. ■

3.2 The case of trees.

THEOREM 3 *If $\mathcal{G}$ is a tree, then there exists a choice of the k_i in* (5) *such that L is selfadjoint for the metric induced by* (6).

Proof: Putting $\delta_{ij} = p_i(j)/k_i$, the symmetry condition is equivalent to say that there exist k_i $(i = 1, ..., M)$ and μ_j $(j \in \mathcal{N}_{int})$ positive such that $k_i \delta_{ij} = \mu_j \alpha_{ji}$. These conditions give a linear homogeneous system with unknowns k_i, μ_j. We observe that the sign has to be the same for all them because the network is connected and the δ_{ij} and the α_{ji} are positive. It is clear that this system has non trivial solutions, because the number of equations is $2M - N + |\mathcal{N}_{int}|$, which is less than the number of unknowns $M + |\mathcal{N}_{int}|$, because $M < N$ for a tree.

In order to see that L is selfadjoint we need only to proof (see, for instance, [13]) that the equation

$$Lu + \lambda u = f \tag{7}$$

has a solution for every $f \in L^2(\mathcal{G})$ for some value of λ. Let X be the space of functions defined on $\mathcal{G}$ with each u_i in $H^1(I_i)$ and verify the continuity conditions and the Dirichlet conditions prescribed in some terminal nodes, if any. Then, by definition, u is a weak solution of (7) if for each $v \in X$:

$$-\sum_{i=1}^{M} \int_0^1 p_i(x)u_i'(x)v_i'(x)\,dx + \sum_{i=1}^{M} \int_0^1 \left(r_i(x)q_i(x) + \lambda\right) u_i(x)v_i(x)\,dx$$

$$-\sum_{j\in\mathcal{N}'} \beta_j u(j)v(j) = \sum_{i=1}^{M} \int_0^1 f_i(x)v_i(x)\,dx\,, \tag{8}$$

where $\mathcal{N}'$ is the set of nodes not having Dirichlet condition. We define the symmetric bilinear form

$$\begin{aligned} B(u,v) = &-\sum_{i=1}^{M} \int_0^1 p_i(x)u_i'(x)v_i'(x)\,dx \\ &+\sum_{i=1}^{M} \int_0^1 \left(r_i(x)q_i(x) + \lambda\right) u_i(x)v_i(x)\,dx \\ &-\sum_{j\in\mathcal{N}'} \beta_j u(j)v(j); \end{aligned}$$

by using the inequality

$$u^2(x) \le \int_0^1 u^2(y)dy + \frac{1}{\epsilon}\int_0^1 u^2(y)dy + \epsilon \int_0^1 (u')^2(y)dy$$

we obtain

$$B(u,u) \le -\alpha(u,u)_{H^1(\mathcal{G})}$$

for some positive α if $-\lambda$ is large enough. Now, by applying the Riesz Theorem, we obtain existence and uniqueness of weak solution for (8). Choosing an appropriate v, it is easy to see that the weak solution is in fact a strong solution. ■

3.3 The case of networks with circuits.

Let us suppose now that $\mathcal{G}$ contains at least one circuit. Let $\mathcal{C}$ be one of these circuits, having $M_{\mathcal{C}}$ edges and, obviously, $M_{\mathcal{C}}$ nodes. Let us order them by putting that the edge i connects the nodes i and $i+1$. We denote as $\alpha^{\mathcal{C}}_{ij}$ the coefficient of the generalized Kirchhoff condition for the edge i at the node j. By running along the circuit we find the coefficients $\alpha^{\mathcal{C}}_{11}$, $\alpha^{\mathcal{C}}_{12}$, $\alpha^{\mathcal{C}}_{22}$, $\alpha^{\mathcal{C}}_{23}$, ..., $\alpha^{\mathcal{C}}_{M_{\mathcal{C}},1}$.

THEOREM 4 *If $\mathcal{G}$ is not a tree, then the necessary and sufficient condition for the existence of a choice of the k_i in* (5) *such that L is selfadjoint for the metric induced by* (6) *is the following:*

$$\sum_{i=1}^{M_{\mathcal{C}}} \int_0^1 \frac{b(x)}{a(x)}\, dx = \sum_{i=1}^{M_{\mathcal{C}}} \ln \frac{\alpha_{i,i+1}}{\alpha_{ii}}$$

for every circuit $\mathcal{C}$ of $\mathcal{G}$.

Proof: Let us consider first the simplest case, in which $\mathcal{G}$ is simply a circuit having N edges and N nodes. With the same notation used in the proof of the precedent theorem, we can easily write the equations of the system mentioned there in such a way that the corresponding matrix is (the ordering of the columns is $k_1, \mu_2, k_2, \mu_3, ..., k_N, \mu_1$)

$$\begin{pmatrix} \delta_{12} & -\alpha_{12} & 0 & 0 & \dots & \dots & 0 \\ 0 & -\alpha_{22} & \delta_{22} & 0 & \dots & \dots & 0 \\ 0 & 0 & \delta_{23} & \alpha_{23} & \dots & \dots & 0 \\ \vdots & \ddots & \ddots & \ddots & \ddots & \ddots & \vdots \\ \delta_{11} & 0 & 0 & 0 & \dots & \dots & -\alpha_{11} \end{pmatrix},$$

which has determinant zero if and only if

$$\frac{\delta_{12}}{\alpha_{12}}\frac{\delta_{23}}{\alpha_{23}}\cdots\frac{\delta_{N-1,N}}{\alpha_{N-1,N}}\frac{\delta_{N1}}{\alpha_{N1}} = \frac{\delta_{11}}{\alpha_{11}}\frac{\delta_{22}}{\alpha_{22}}\cdots\frac{\delta_{NN}}{\alpha_{NN}}. \tag{9}$$

If $\mathcal{G}$ contains more than one circuit, then the condition (9) in all of them is necessary and sufficient for the existence of non trivial solutions of that system, because:

a) Suppose that one of the circuits does not verify (9); we order the rows and columns in such a way that the first ones correspond to the equations involving the nodes of this circuit; the remaining equations can be ordered in such a way that every new column has only one non-zero element. The resulting matrix corresponds obviously to a system without non trivial solutions.

b) Suppose that (9) is fulfilled in every circuit. Every minor containing all the columns involves all the N_{int} interior nodes and at least N_{int} edges connecting them, so some circuit is present, and by proceeding as before we deduce that the minor is zero.

Now recall the definition of the δ_{ij}, and write (9) in the form

$$\prod_{i=1}^{M_C} \exp\left(\int_0^1 \frac{b_i(x)}{a_i(x)}\,dx\right) = \prod_{i=1}^{M_C} \frac{\alpha_{i,i+1}}{\alpha_{ii}} ,$$

which is equivalent to

$$\sum_{i=1}^{M_C} \int_0^1 \frac{b(x)}{a(x)}\,dx = \sum_{i=1}^{M_C} \ln \frac{\alpha_{i,i+1}}{\alpha_{ii}}. \tag{10}$$

Assuming that L is symmetric, it becomes selfadjoint by reasoning as in the proof of the preceding theorem. ■

We have to point out that we have worked here with a metric that is induced in a quite natural way by the coefficients of the differential equations on each edge. The symmetry condition obtained, also called *consistency* condition, depends heavily on the metric (6).

There is the possibility of working with other metrics, and we refer to [3], where it is shown that for certain networks is possible to have real spectra even in the case that the consistency condition is not fulfilled.

Also we can find in [2] an example in which there are non real eigenvalues due to inconsistent Kirchhoff conditions.

Furthermore, for nonselfadjoint elliptic operators and for the operator matrix describing wave propagation on networks the properties of the spectrum and resolvent ensuring the generation of a strongly continuous semigroup have been proved in [1] (Chapter 2).

It will be shown in a forthcoming paper that for every network having at least one circuit we can find operators of this kind that have an infinite number of non real eigenvalues (see Chapter 5 in [8]).

4 THE CASE OF THE CIRCLE

The purpose of this section is to present the results of studying the problem defined in Section 1 for the case in which the network consists in one branch forming a loop, or in other words, one edge joining the initial node with itself. We will not assume that the corresponding operator is not symmetric in the sense explained in the preceding section.

These results have been published in a joint work with J. Solà–Morales [7], and they exhibit the typical phenomena that are associated to non-selfadjointness: existence of nonreal complex eigenvalues and existence of eigenvalues with algebraic multiplicity larger than the geometric multiplicity.

So let us consider the eigenvalue problem

$$a(x)u''(x) + b(x)u'(x) + c(x)u(x) + \lambda u(x) = 0 \quad (x_1 \leq x \leq x_2), \qquad (11)$$

where a, b, c are real functions on $[x_1, x_2]$ and $a(x) \geq \epsilon > 0$ for some ϵ, with the following conditions:

$$u(x_1) = u(x_2); \qquad \alpha_1 u'(x_1) + \beta u(x_1) = \alpha_2 u'(x_2), \qquad (12)$$

where α_1, α_2 and β are real numbers and α_1, $\alpha_2 > 0$ (the case $\alpha_1 = \alpha_2$, $\beta = 0$ corresponds to the *periodic* problem). Without loss of generality we can suppose $[x_1, x_2] = [0, 1]$.

It is well known that the equation (11) admits a *formally* selfadjoint form

$$(p(x)u'(x))' + q(x)u(x) + \lambda\gamma(x)u(x) = 0,$$

where $p(x) > 0$ and $\gamma(x) > 0$ for $0 \leq x \leq 1$, and, if

$$\int_0^1 \frac{b(x)}{a(x)}\,dx = \ln\frac{\alpha_2}{\alpha_1}\,, \qquad (13)$$

then the operator T of $L^2(0,1)$ defined by $Tu = \frac{1}{\gamma}((pu')' + qu)$ with

$$D(T) = \{u \in H^2(0,1) \mid u(0) = u(1);\ \alpha_1 u'(0) + \beta u(0) = \alpha_2 u'(1)\}$$

is selfadjoint for the metric of $L^2(0,1)$ defined by $(u,v) = \int_0^1 \gamma u\bar{v}\,dx$ and has an infinite number of real eigenvalues with geometric multiplicities equal to 1 or 2, and for every eigenvalue the algebraic and geometric multiplicities coincide.

We have investigated the case in which the condition (13) fails to hold. Assume

$$2k := \int_0^1 \frac{b(x)}{a(x)}\,dx - \ln\frac{\alpha_2}{\alpha_1} \neq 0;$$

then T is not symmetric (and consequently it is not selfadjoint) for the metric defined above.

In Theorems 5, 6 and 7 below we make the hypothesis that the coefficient functions $a(x)$, $b(x)$ and $c(x)$ that appear in (11) are functions of bounded variation, and in Theorems 8 and 9 that they are in $L^1(0,1)$. These are quite general assumptions and are motivated by the fact that discontinuous coefficients appear very naturally in applications as a consequence of changes of materials along the circle. For weak solutions one can see that one has the usual results of existence and uniqueness for the initial value problem, but details will not be presented here. It is also clear that any change in the coefficient functions on a set of zero measure does not change the solutions at all.

By performing some changes of variables in (11)–(12) we can write the problem in the form

$$v''(t) + 2kv'(t) + q(t)v(t) + \lambda r(t)v(t) = 0; \tag{14}$$

$$v(0) = v(1)\ ;\ v'(0) = v'(1)\ , \tag{15}$$

with $r(t) > 0$ for $0 \le t \le 1$.

It is known that if $r(t)$ is smooth and periodic, one can find a new change of variables (see [5]) such that the problem (14)–(15) keeps the same form except for changes in the function q and that the new funcion r becomes the function constantly equal to 1, but this seems not to be possible for a general $r(t)$.

Putting now $w = e^{kt}v$, we obtain what we call the *normalized form*

$$w''(t) + h(t)w(t) + \lambda r(t)w(t) = 0,\ \ (0 \le t \le 1); \tag{16}$$

$$w(1) = e^k w(0)\ ,\ w'(1) = e^k w'(0). \tag{17}$$

We call $BV(0,1)$ the class of functions of bounded variation in $(0,1)$.

THEOREM 5 *If $h, r \in BV(0,1)$, then there exists an infinite number of eigenvalues for the problem* (16)–(17).

THEOREM 6 *For $h, r \in BV(0,1)$, there exists $k_0 > 0$ such that, if $|k| > k_0$, then there is only one real eigenvalue for the problem* (16)–(17).

THEOREM 7 *For h in $BV(0,1)$ and r in $\mathcal{C}^2[0,1]$ with $r(0) = r(1)$, $r'(0) = r'(1)$, and $k \neq 0$, there exists only a finite number of real eigenvalues for the problem* (16)–(17).

Note that can replace $\mathcal{C}^2[0,1]$ in the precedent theorem by $\mathcal{W}^{2,1}(0,1)$, but we cannot relax too much this condition, as we can see in the following example. Suppose $h(t) = 0$ for all t and $r(t) = r_1$ for $0 \leq t \leq 1/2$, $r(t) = r_2$ for $1/2 \leq t \leq 1$, with r_1, r_2 positive constants, and $r_1 \neq r_2$. An elementary calculation shows that, in this case, we have an infinite number of real eigenvalues.

THEOREM 8 *For $h, r \in L^1(0,1)$, the eigenvalues for the problem* (16)–(17) (with $k \neq 0$) *have geometric multiplicities equal to 1; if any of them is not real, then its algebraic multiplicity is also equal to 1.*

THEOREM 9 *For $h, r \in L^1(0,1)$, the eigenvalues for the problem* (16)–(17) *have algebraic multiplicities less than or equal to 2.*

Note that for the operator L defined by

$$Lu = u'' - (\sin^2 x)u' + (\sin x \cos x)u,$$

with periodic boundary conditions on $[0, 2\pi]$, $\lambda = 1$ is an eigenvalue of L with geometric multiplicity 1 and algebraic multiplicity 2.

REFERENCES

1. F. Ali Mehmeti, Nonlinear Waves in Networks, Mathematical Research, Vol. 80, Akadenie Verlag, Berlin (1994).
2. J. von Below, A characteristic equation associated to an eigenvalue problem on C^2–networks, Linear Algebra Appl. 71: 309–325 (1985).
3. J. von Below, Kirchhoff Laws and Diffusion on Networks, Linear Algebra Appl. 121: 692–697 (1989).

4. J. von Below, Sturm–Liouville eigenvalue problems on networks, Math. Meth. Appl. Sciences 10: 383–395 (1988).

5. B. M. Levitan and I. S. Sargsjan, Sturm–Liouville and Dirac Operators, Kluwer Academic (1991).

6. J. A. Lubary, Multiplicity of solutions of second-order linear differential equations on networks, Linear Algebra Appl. 274: 301–315 (1998).

7. J. A. Lubary and J. Solà–Morales, Some nonselfadjoint problems on the circle, Z. angew. Math. Phys. 51: 318–331 (2000).

8. J. A. Lubary, Doctoral Thesis, Universitat Politècnica de Catalunya (1999).

9. S. Nicaise, Some results on spectral theory over networks, applied to nerve impulse transmission, Lec. Not. in Math. 1171: 532–541 (1985).

10. S. Nicaise, Approche spectrale des problemes de diffusion sur les réseaux, Lec. Not. in Math. 1235: 120–140 (1987).

11. J. P. Roth, Le spectre du laplacien sur un graphe, Lec. Not. in Math. 1096: 521–529 (1984).

12. R. J. Wilson, Introduction to Graph Theory, Oliver & Boyd, Edinburgh (1972).

13. K. Yosida, Functional Analysis, Springer Verlag Berlin (1978).

The Asymptotic Laplace Transform: New Results and Relation to Komatsu's Laplace Transform of Hyperfunctions

GÜNTER LUMER Institut de Mathématiques et d'Informatique, Université de Mons–Hainault, 6, Ave. du Champ de Mars, B–7000 Mons, Belgium.

FRANK NEUBRANDER Departement of Mathematics, Lousiana State University, Baton Rouge, La. 70803-4918 U.S.A.

0 Introduction

Based on the ground-breaking work of J. Vignaux [Vi] from 1939, a basic theory of asymptotic Laplace transforms applicable to locally Bochner integrable, Banach space valued functions on finite intervals $[0, T)$ or on the half line $[0, \infty)$ with arbitrary growth at infinity was recently developed by the present authors in [L-N]. However, several matters of considerable importance remained unclear, unsettled, or unexplored. In this paper we will focus on the following topics.

(I) We show why and how the basic definitions of [L-N] must and can be modified to get an optimal setting for certain theoretical results (see Section 3) and applications of asymptotic Laplace transforms Lf for $f : [0, \infty) \to X$, X Banach space, see Section 4. We do this in Section 2 after introducing notations, the basic context, and recalling known facts in Section 1. The main problem addressed in Section 2 is that the coclasses $Lf^{(1)}$ which define the asymptotic Laplace

(1) why we have coclasses is briefly explained in (II) below.

transforms in [L-N] (or in any of the previous works on the subject, see [Vi], [V-C], [Be], [Ly]) are too big. Specifically, the basic operational property $L((-t)^n f) = (Lf)^{(n)}$ $(n = 1, 2, \cdots)$ does not hold in the original set-ups (see also footnote 6 of [L-N], where a possible, yet not optimal, remedy is indicated). The necessary modifications in the definition of the coclasses Lf are treated in Section 2.

(II) The theory of generalized Laplace transforms for hyperfunctions was developed several years ago by H. Komatsu [K1], [K2], [K3]. On general grounds, and because of the implications for the concrete and effective computability of both transforms, the asymptotic L-transform[(2)] L and Komatsu's L-transform $\mathcal{L}$, it is essential to know the exact relation between these two extensions of the classical L-transform. Since both extended L-transforms, L and $\mathcal{L}$, must be able to "sum" divergent Laplace integrals, it is not surprising that these extended summations (integrations), like for divergent series, lead not to a single function but to an equivalence class (coclass) Lf, $\mathcal{L}f$ for any $f \in L^1_{loc}$[(3)]

In Section 3, we prove that already in the context of [L-N], the coclass constituting $\mathcal{L}f$ is contained in the one constituting Lf. Moreover, once the modification of (I) is made, one gets $Lf = \mathcal{L}f$ for all $f \in L^1_{loc}$; i.e., Komatsu's L-transform $\mathcal{L}$ coincides with the asymptotic L-transform L on locally Bochner integrable functions.

(III) The results obtained treating (I) and (II) yield several useful consequences on asymptotic L-transforms and asymptotic approximation. This we take up in Section 4. There we show that if the coclass Lf is restricted to those analytic functions $\hat{f}$ in a post-sectorial region (see below) which are of minimal exponential type, then the asymptotic approximation

$$\limsup_{\lambda \to \infty} \frac{1}{\lambda} \ln \| \int_0^T e^{-\lambda t} f(t)\, dt - \hat{f}(\lambda) \| \leq -T$$

required in the setup of [L-N] for positive reals λ extends to complex

(2) Hereafter, "Laplace transform" will usually be abbreviated "L-transform".

(3) See references above and more details in Section 1 below.

$\lambda = re^{i\phi}$, $|\phi| < \pi/2$, $r \to \infty$, in the analogous form

$$\limsup_{r\to\infty} \frac{1}{r} \ln \| \int_0^T e^{-tre^{i\phi}} f(t)\, dt - \hat{f}(re^{i\phi}) \| \leq -T\cos\phi. \qquad (1)$$

We show that one can extend the asymptotic theory to general hyperfunctions(2) $f \in B([0,\infty); X)$, where $\int_0^T e^{-\lambda t} f(t)\, dt$ is defined in terms of a compactly supported restriction of f to $(-\infty, T]$ considered as an analytic functional.

Also, as a consequence of what we obtain, one sees that any hyperfunction $b \in B([0,\infty); X)$, in general not of compact support, can be approximated in an asymptotic sense by its restrictions to $[0, T)$ as $0 < T \to \infty$, and hence by δ-expansions; i.e.,

$$b \sim \sum_{j=0}^{\infty} c_j^T \delta^{(j)}. \qquad (2)$$

Finally, let us mention that conceptually the result $L = \mathcal{L}$ (i.e., the fact that both approaches eventually yield the same coclasses for the generalized L-transform) is reassuring for those who might have felt uneasy about the multiple-valuedness of the hyperfunction construction of $\mathcal{L}f$. The multiple-valued feature is a priori a natural ingredient of any asymptotic approach and is recognized nowadays as being often an advantage rather than a drawback(3) As for some numerous applications to evolution equations and singular PDE of the extended L-transform, see [L-N], in particular pp.55-56 and pertinent references given there.

1 Notations, Basic Context, Known Facts

---

(2) $B([a,\infty); X) = \dfrac{\mathcal{O}(\mathbb{C}\setminus[a,\infty); X)}{\mathcal{O}(\mathbb{C}; X)}$, where $\mathcal{O}(V; X)$ denotes the space of analytic (holomorphic) functions on the open set $V \subset \mathbb{C}$ with values in a Banach space X. See [K1-3] and [Lu] for details.

(3) For similar phenomena in summation methods and related earlier ideas by Borel, Hardy, Galois, and Picard, see J.-P. Ramis [Ra] (Section 3.5).

We denote by Σ a postsectorial region in $\mathbb{C}$; i.e., an open subset of the right half plane such that for all angles $0 < \phi_0 < \pi/2$ there exists $r_0 > 0$ such that $\lambda = re^{i\phi} \in \Sigma$ for all $r > r_0$ and $|\phi| < \phi_0$. Notice that these postsectorial regions Σ constitute a basis of neighborhoods of the open half circle $\Sigma_\infty = \{\infty e^{i\phi} : |\phi| < \pi/2\}$ at ∞ in the radial compactification $\mathbb{C}^*$ of $\mathbb{C}$. Below, functions are of course treated in "sheaf theoretic fashion"; i.e., for f_1 on Σ_1, f_2 on Σ_2, $f_1 + f_2$ is considered on some $\Sigma_3 \subset \Sigma_1 \cap \Sigma_2$, etc... .

Let $\hat{a} : (\omega, \infty) \to X$ and $T > 0$. To measure asymptotic error (or equivalence) at infinity, we recall from [L-N] the notation $\hat{a} \approx_T 0$, meaning

$$\limsup_{\lambda\to\infty} \frac{1}{\lambda} \ln \|\hat{a}(\lambda)\| \leq -T. \tag{3}$$

Unless otherwise stated, let f denote an element of $L^1_{loc}([0,\infty); X) = L^1_{loc}$. In [L-N] we defined for $0 < T < \infty$ the T-asymptotic L-transform of f as

$$\{f\}_T = \{\hat{f} \in \mathcal{O}(\Sigma; X) \text{ for some } \Sigma \text{ and } \int_0^T e^{-\lambda t} f(t)\, dt - \hat{f}(\lambda) \approx_T 0\}, \tag{4}$$

and the asymptotic L-transform of f by

$$Lf = \{f\} = \cap_{T>0} \{f\}_T. \tag{5}$$

For the fact that L is well-defined and injective on L^1_{loc}, i.e.,

$\{f\} \neq \emptyset$ and $\{f\} \cap \{g\} \neq \emptyset$ if and only if $f = g$ a.e. for all $f, g \in L^1_{loc}$,

and other known properties of the asymptotic L-transform, see [L-N]. We say that two functions $\hat{f}, \hat{g} \in \mathcal{O}(\Sigma; X)$ are asymptotically equivalent if $\hat{f} \approx \hat{g}$, meaning that $\hat{f} - \hat{g} \approx_T 0$ for all $T > 0$. We say that $\hat{f}, \hat{g}$ are fully asymptotically equivalent and write $\hat{f} \approx_\Sigma \hat{g}$, if $\approx_T$ is replaced above by $\approx_{\Sigma_T}$, where $\hat{a} \approx_{\Sigma_T} 0$ means

$$\limsup_{r\to\infty} \frac{1}{r} \ln \|\hat{a}(re^{i\phi})\| \leq -T \cos\phi \quad \text{for all } |\phi| < \pi/2. \tag{6}$$

For the Komatsu L-transform $\mathcal{L}$ we refer the reader to [K1], [K2], and [K3]. For our use here let us explicitely recall that given any hyperfunction $b = [F] \in B([0,\infty);X)$ there exists $F_0 \in [F]$ of exponential type on $\mathbb{C}^* \setminus [0,\infty]$ such that the L-transform

$$\tilde{F}_0(\lambda) = \int_\Gamma e^{-\lambda\tau} F_0(\tau)\, d\tau \tag{7}$$

exists for all λ in some translated postsector Σ with an integration path Γ around a real half line $[\omega,\infty)$ as detailed in [K1]. Then, and in particular for $b = f$, the generalized L-transform $\mathcal{L}f$ constructed in the Komatsu theory is simply

$$\mathcal{L}f = \{\tilde{F}_0 + \hat{h} : \hat{h} \approx_\Sigma 0\} = \{\hat{f} : \hat{f} \approx_\Sigma \tilde{F}_0\} = \tilde{F}_0 + \mathcal{L}B^{exp}_{[\infty]}. \tag{8}$$

We will need the following result from [K3] which is a bit stronger than the important surjectivity theorem in [K1].

1.1. Lemma (Komatsu, Kaneko). *Let $[F] \in B([a,\infty);X)$ and V be an open neighborhood of $[a,\infty)$ in $\mathbb{C}$. Then, for every $\epsilon > 0$, there exists $F_0 \in [F]$ such that*

$$\|F_0(\cdot)\| \le \epsilon \quad on\ \mathbb{C} \setminus V. \tag{9}$$

In what follows, an important issue is to see under what conditions we can pass from an approximation governed by (3) to a stronger one governed by (6). To this end, several things are discussed next. Let us recall first that $\hat{f} \in \mathcal{O}(\Sigma;X)$ is said to be of exponential type on a postsector Σ if for every sector $\Sigma_\phi = \{\lambda : |arg(\lambda)| \le \phi < \pi/2$ and $Re\lambda > 0\}$ there exist $M,\omega \ge 0$ such that $\|\hat{f}(\lambda)\| \le Me^{\omega|\lambda|}$ in $\Sigma \cap \Sigma_\phi$. We say that $\hat{f}$ is of bounded type on Σ if we can always take $\omega = 0$ and $\hat{f}$ is said to be of minimal exponential type if we can take $\omega > 0$ arbitrarily small. Further recall that for a function $\hat{f}$ of bounded or exponential type, we define the Phragmen-Lindelöf function h by[4]

$$h(\phi) = h(\phi,\hat{f}) = \limsup_{r\to\infty} \frac{1}{r} \ln \|\hat{f}(re^{i\phi})\| < \infty. \tag{10}$$

---

[4] see [Ti], p. 180-184

Now, using 5.71 of [Ti][(5)] with $\phi_1 = 0$, $\phi_2 = \pi/2 - \mu$, $0 < \mu$ small and eventually going to zero, $h(\phi_1) = h(0) \le -T$, and $h(\phi_2) \le 0$ yields $h(\phi) \le \frac{1}{\sin(\pi/2-\mu)}[-T\sin(\pi/2-\mu-\phi)] = \frac{1}{\cos\mu}[-T\cos(\mu+\phi)]$ for all $0 < \phi < \pi/2 - \mu$. Thus, if $\hat{f}$ is of minimal exponential type and $h(0) \le -T$, then

$$h(\phi) \le -T cos \phi \text{ for } |\phi| < \pi/2.$$

Moreover, we can use the uniform estimate $\|\hat{f}(re^{i\phi})\| \le e^{r(h(\phi)+\epsilon)}$ ($r > r_0(\epsilon)$, $|\phi| \le \phi_0 < \pi$) described in [Ti] (p. 184, last two lines before 5.72). This proves the following lemma.

1.2. Lemma. *Let $\hat{a} \in \mathcal{O}(\Sigma; X)$ be of minimal exponential type. If*

$$\limsup_{\lambda\to\infty} \frac{1}{\lambda} \ln \|\hat{a}(\lambda)\| \le -T,$$

then $\limsup_{r\to\infty} \frac{1}{r} \ln \|\hat{a}(re^{i\phi})\| \le -T\cos\phi$ for all $|\phi| < \pi/2$; i.e., (3) implies (6). Moreover, for any $\phi_0 < \pi/2$ and any $\epsilon > 0$ there exists $r_0 > 0$ such that

$$\|\hat{a}(re^{i\phi})\| \le e^{r(-T\cos\phi+\epsilon)} \quad \text{for all } r \ge r_0 \text{ and } |\phi| \le \phi_0. \tag{11}$$

Finally, we need a quite elementary, but often (implicitly or explicitely) used fact.

1.3. Lemma. *Let $\hat{f}_1, \hat{f}_2 \in \mathcal{O}(\Sigma; X)$ be of exponential type. Then*

$$h(\phi, \hat{f}_1 + \hat{f}_2) \le \max(h(\phi, \hat{f}_1), h(\phi, \hat{f}_2)) \text{ for } |\phi| < \pi/2. \tag{12}$$

Proof. Set $h(\phi) = \max(h(\phi, \hat{f}_1), h(\phi, \hat{f}_2))$ and assume that $h(\phi) = h(\phi, \hat{f}_1)$. For $j = 1, 2$, $2\epsilon_2 = \epsilon_1 > 0$, and r large, $\|\hat{f}_j(re^{i\phi})\| \le e^{(h(\phi,\hat{f}_j)+\epsilon_j)r}$. It follows from

$$\|\hat{f}_1(re^{i\phi}) + \hat{f}_2(re^{i\phi})\| \le e^{(h(\phi)+\epsilon_1)r}\left(1 + e^{(h(\phi,\hat{f}_2)-h(\phi)+\epsilon_2-\epsilon_1)r}\right)$$

that $\ln\|(\hat{f}_1 + \hat{f}_2)(re^{i\phi})\| \le (h(\phi) + \epsilon_1)r + e^{(h(\phi,\hat{f}_2)-h(\phi)-\epsilon_1/2)r}$, where we use that $\ln(1+x) \le x$ for $x \ge 0$. Since the exponent in the last term is negative, $h(\phi, \hat{f}_1 + \hat{f}_2) \le h(\phi) + \epsilon_1$ for $\epsilon_1 > 0$ arbitrary. ◊

---

(5) If (10) holds, then 5.71 of [Ti] states that if $-\pi/2 < \phi_1 < \phi_2 < \pi/2$, $\phi_2 - \phi_1 < \pi$, $h(\phi_1) \le h_1$, and $h(\phi_2) \le h_2$, then $h(\phi) \le \frac{1}{\sin(\phi_2-\phi_1)}[h_1\sin(\phi_2 - \phi) + h_2\sin(\phi - \phi_1)]$ for all $\phi_1 < \phi < \phi_2$.

In particular, since $\hat{a} \approx_{\Sigma_T} 0$ means $h(\phi, \hat{a}) \leq -T\cos\phi$ (see (6)), (12) implies that $\mu_1\hat{a}_1 + \mu_2\hat{a}_2 \approx_{\Sigma_T} 0$ for $\mu_1, \mu_2 \in \mathbb{C}$ if $\hat{a}_1 \approx_{\Sigma_T} 0$ and $\hat{a}_2 \approx_{\Sigma_T} 0$.

2. The Optimal Definition of Asymptotic L-Transforms

The following example shows well why the coclasses

$$Lf = \{\hat{f} \in \mathcal{O}(\Sigma; X) \text{ for some } \Sigma$$

and

$$\int_0^T e^{-\lambda t} f(t)\,dt - \hat{f}(\lambda) \approx_T 0 \text{ for all } T > 0\}$$

as defined in [L-N] (see also (4) and (5)) and other earlier works (see [Be], [Ly], [Vi], [V-C]) are too large to obtain an optimal theory for asymptotic L-transforms.

2.1. Example *Let X be a Banach space and $f \in L^1_{loc}(\mathbb{R}_+; X)$. If the asymptotic L-transform is defined as in (5), then the operational property*

$$L(-tf) = (Lf)' \tag{13}$$

does not hold. In fact, not even a weaker version of (13), with $\supset$ instead of $=$, holds.

First take $X = \mathbb{C}$, $1 < \alpha < 2$, and $\hat{h}_0(\lambda) := e^{-\lambda^\alpha}\sin(e^{\lambda^2})$. Since $\hat{h}_0 \approx 0$, it follows that $\hat{h}_0 \in L0$. An easy estimate shows that $\limsup_{\lambda\to\infty} \frac{1}{\lambda}\ln|\hat{h}_0'(\lambda)| = +\infty$. Thus $\hat{h}_0' \notin L0 = L(-t0)$, or $(L0)'$ is not contained in $L(-t0)$. Now let X be a Banach space, $0 \neq x_0 \in X$, and $\hat{f}_0(\lambda) := \hat{h}_0(\lambda)x_0$. Assume that (13) holds for some $f \in L^1_{loc}(\mathbb{R}_+; X)$. Let $\hat{g} \in L(-tf)$. Then $\hat{g} \approx \hat{f}'$. Since $\hat{f}_0 \approx 0$, it follows that $\hat{f} + \hat{f}_0 \in Lf$. Suppose $(\hat{f} + \hat{f}_0)' \in (Lf)' = L(-tf)$. Then $\hat{f}' + \hat{f}_0' - \hat{g} \approx 0$ and $\hat{g} - \hat{f}' \approx 0$. By Lemma 1.3, $\hat{f}_0' \approx 0$ which is a contradiction. This also proves the last sentence of 2.1. $\diamondsuit$

One should notice that the function $\hat{h}_0(\lambda) = e^{-\lambda^\alpha}\sin(e^{\lambda^2})$ is not of exponential type on any postsector Σ and that when the latter is not the case, the behavior of $\hat{h}_0$ on $\mathbb{R}_+$ yields little information on the growth of $\hat{h}_0(\lambda)$ where $\lambda = re^{i\phi}$ for $r > r_0$ and $|\phi| < \phi_0$. This example shows that the coclasses $Lf = \{f\}$ defining the asymptotic L-transform of f are too large and an adjustment in the definition has to be made. We denote by $\mathcal{O}_{exp}(X)$ the set of analytic functions $\hat{f}$ defined on some postsector Σ with values in X which are of minimal exponential type on Σ; i.e., for every sector $\Sigma_\phi = \{\lambda : |arg(\lambda)| \leq \phi < \pi/2 \text{ and } Re\lambda > 0\}$ and all $\omega > 0$ there exists $M > 0$ such that $\|\hat{f}(\lambda)\| \leq Me^{\omega|\lambda|}$ in $\Sigma \cap \Sigma_\phi$. In view of the example above, we consider the following subset of Lf:

$$L_{exp}f := Lf \cap \mathcal{O}_{exp}(X).$$

In the next section we shall prove that the Komatsu L-transform $\mathcal{L}f$ is contained in the asymptotic L-transform Lf for all $f \in L^1_{loc}$. For the remainder of this section we will consider this to be a known fact. For $\hat{f} \in \mathcal{L}f$, we know from [K1] that $\hat{f} \in \mathcal{O}_{exp}(X)$. Thus, $\mathcal{L}f \subset L_{exp}f$. Notice that if $\hat{f} \in L_{exp}f$, then it follows from Lemma 1.2 that the functions $\tilde{f}_T$ yield a full asymptotic approximation of $\hat{f}$; i.e.,

$$\tilde{f}_T \approx_{\Sigma_T} \hat{f} \text{ for all } T > 0,$$

where $\tilde{f}_T(\lambda) = \int_0^T e^{-\lambda t} f(t)\,dt = \int_0^\infty e^{-\lambda t} f_T(t)\,dt$, $f_T := f\chi_T$, and χ_T denotes the characteristic function of $[0, T]$. In fact, it is easy to prove the following statement.

2.2. Proposition *Let $f \in L^1_{loc}(\mathbb{R}_+; X)$ and $\hat{f} \in Lf$. The following are equivalent.*

(i) $\hat{f}$ is of minimal exponential type; i.e., $\hat{f} \in L_{exp}f$.
(ii) $\tilde{f}_T \approx_{\Sigma_T} \hat{f}$ for all $T > 0$.

The next result shows that if the range of the asymptotic L-transform is restricted to analytic functions of minimal exponential type, then the important operational property (13) holds.

2.3. Theorem *Let X be a Banach space and $f \in L^1_{loc}(\mathbb{R}_+; X)$. Then*

$$L_{exp}(-tf) = (L_{exp}f)' = \{\hat{f}' : \hat{f} \in L_{exp}f\}. \tag{14}$$

Proof. Let $\hat{f} \in L_{exp}f$ be analytic on a postsector Σ and consider $\hat{g} = \hat{f} - \tilde{f_T}$. Then

$$\hat{g}'(\lambda) = \hat{f}'(\lambda) - \tilde{f_T}'(\lambda) = \hat{f}'(\lambda) - \widetilde{(-tf)_T}(\lambda).$$

Take a large real $\lambda \in \Sigma$ and choose $r > 0$ such that the disc $\Omega_r = \{\xi : |\xi - \lambda| \le r\}$ is contained in Σ. By Cauchy's Theorem,

$$\|\hat{g}'(\lambda)\| = \|\frac{1}{2\pi i}\int_{\partial\Omega_r} \frac{\hat{g}(\xi)}{(\xi-\lambda)^2}\, d\xi\| \le \frac{1}{r}\max_{\xi\in\partial\Omega_r} \|\hat{g}(\xi)\|.$$

Let $T, \epsilon > 0$. By the uniform estimate provided in Lemma 1.2 and Proposition 2.2, $\|\hat{g}(\xi)\| \le e^{(1-\epsilon)(-T+\epsilon)|\xi|}$. Thus

$$\limsup_{\lambda\to\infty} \frac{1}{\lambda} \ln \|\hat{g}'(\lambda)\|$$

$$\le \limsup_{\lambda\to\infty} (1-\epsilon)(-T+\epsilon)(1-\frac{r}{\lambda}) - \frac{\ln r}{\lambda} \le (1-\epsilon)(-T+\epsilon).$$

Since $\hat{g}$ is of minimal exponential type, the same holds for $\hat{g}'$. Therefore, by Lemma 1.2, $\widetilde{(-tf)_T} \approx_{\Sigma_T} \hat{f}'$ for all $T > 0$. This proves that $(L_{exp}f)' \subset L_{exp}(-tf)$.

To prove the other inclusion, we observe that $L_{exp}f = \hat{h}_f + \{0\}$ and $L_{exp}(-tf) = \hat{h}_{-tf} + \{0\}$, where $\{0\} = \{\hat{h} : \hat{h} \approx_\Sigma 0\}$. Fix $\hat{h} \in (L_{exp}f)' \subset L_{exp}(-tf)$. Then there exist $\hat{h}_0, \hat{h}_1 \in \{0\}$ such that $\hat{h} = (\hat{h}_f + \hat{h}_0)' = \hat{h}_{-tf} + \hat{h}_1$. Now let $\hat{g} \in L_{exp}(-tf)$. Then there exists $\hat{h}_2 \in \{0\}$ such that $\hat{g} = \hat{h}_{-tf} + \hat{h}_2 = \hat{h}_{-tf} + \hat{h}_1 + (\hat{h}_2 - \hat{h}_1) = (\hat{h}_f + \hat{h}_0)' + \hat{h}_3$, where $\hat{h}_3 = \hat{h}_2 - \hat{h}_1 \in \{0\}$. Thus, in order to go from the inclusion just obtained to the equality (14), it suffices to show that for all $\hat{h}_3 \approx_\Sigma 0$ exists $\hat{h}_4 \approx_\Sigma 0$ such that $\hat{h}_4' = \hat{h}_3$. The required $\hat{h}_4$ is given by $\hat{h}_4(\lambda) = -\int_\lambda^\infty \hat{h}_3(\mu)\, d\mu$, where we integrate along the ray joining λ to ∞ and using (6) to check that $\hat{h}_4 \approx_\Sigma 0$. (Of course the postsectors involved are adjusted as needed.) ◇

A subset $L^{\#}f$ of Lf is called asymptotically closed if $\hat{f} \in L^{\#}f$ and $\hat{g} \approx_{\Sigma} \hat{f}$ implies that $\hat{g} \in L^{\#}f$. The operator $L^{\#}$ is called asymptotically closed if $L^{\#}f$ is asymptotically closed for all f in the domain of $L^{\#}$. We consider the following properties:

(i) Any $\hat{f} \in L^{\#}f$ is of minimal exponential type.
(ii) $\tilde{f_T} \approx_{\Sigma_T} \hat{f}$ for all $T > 0$ and $\hat{f} \in L^{\#}f$.
(iii) $L_{exp}(-tf) = (L_{exp}f)' = \{\hat{f}' : \hat{f} \in L_{exp}f\}$ for all $\hat{f} \in L^{\#}f$.
(iv) $L^{\#}$ is asymptotically closed.

Since the Komatsu L-transform $\mathcal{L}$ satisfies (i), (ii), and (iv), it also satisfies (iii) which can be easily seen from the proof of Theorem 2.4. So by "cutting" Lf down to $\mathcal{L}f$, where by (8) and Lemma 1.2 we simply have

$$\mathcal{L}f = \{\hat{f} \in Lf : \hat{f} \text{ is of minimal exponential type }\},$$

we gain the properties (i), (ii), and (iii) [(iv) we already had]. Since there is no reason of giving up any of these gained properties by using another nonempty restriction $L^{\#}$ of L, we demand these properties to hold for any acceptable $L^{\#}$. We then have immediately:

2.4. Proposition *Any acceptable $L^{\#}$, satisfying the properties (i) to (iv), necessarily satisfies $\mathcal{L} \subset L^{\#}$. (So, $\mathcal{L}$ is the smallest acceptable one.)*

Proof. Let $\hat{f}_0 \in L^{\#}f$. By (ii), $\tilde{f_T} \approx_{\Sigma} \hat{f}_0$. For all $\hat{f} \in \mathcal{L}f \subset L_{exp}f$, we have that $\tilde{f_T} \approx_{\Sigma} \hat{f}$. So $\hat{f}_0 = \hat{f}$ by (iv). ◇

It follows from the last proposition that $\mathcal{L}$ is the minimum nonempty acceptable restriction of L, and in this sense we get the following optimal definition for the asymptotic L-transform which will again be denoted by L.

2.5. Definition *The asymptotic L-transform of $f \in L^1_{loc}(\mathbb{R}_+; X)$ is given by*

$$Lf := \{\hat{f} \in \mathcal{O}(\Sigma; X) : \hat{f} \textit{ is of minimal exponential type on some postsector } \Sigma, \textit{ and } \tilde{f_T} \approx_{\Sigma_T} \hat{f} \textit{ for all } T > 0\}.$$

Clearly, by what we established above, we have now that the asymptotic L-transform L coincides with the Komatsu L-transform $\mathcal{L}$; i.e.,

$$Lf = \mathcal{L}f \text{ for all } f \in L^1_{loc}(\mathbb{R}_+; X). \tag{15}$$

3. Asymptotic Behavior of the Komatsu L-Transform

We use L_0 for the letter L used above to denote the asymptotic L-transform before the new Definition 2.5 was made.

3.1 Theorem $\mathcal{L}f \subset L_\circ f$ *for all* $f \in L^1_{loc}(\mathbb{R}_+; X)$.

Proof. For $0 < T < T'$ consider the characteristic functions χ_T, $\chi_{TT'}$, $\chi_{T'\infty}$ of $[0, T]$, $[T', T]$, $[T', \infty)$, respectively. Define $f_T = f\chi_T$, $f_{TT'} = f\chi_{TT'}$, $f_{T'\infty} = f\chi_{T'\infty}$ and set, respectively, $F, F_T, F_{TT'}, F_{T'\infty}$ for the standard ("Cauchy integral") representatives. Then we have

$$f = f_T + f_{TT'} + f_{T'\infty}, \quad F = F_T + F_{TT'} + F_{T'\infty},$$

$$[F] = [F_T] + [F_{TT'}] + [F_{T'\infty}]. \tag{16}$$

Now we take some $\mu > 0$ such that $T < T' - \mu$, and construct the following open complex neighborhood V of $[T', \infty)$:

$$V = \{z = x + iy : |y| < \mu \text{ if } x \geq T' \text{ and } |z - T'| < \mu \text{ if } x \leq T'\}, \tag{17}$$

so that ∂V is the union of a half circle Γ and two straight lines parallel to the x-axis. Moreover, consider the rays $\{re^{\pm i\alpha}; r \geq 0\}$ tangent to Γ at the points $z_1 = x_0 + iy, z_2 = x_0 - iy$, where $T' - \mu < x_0 < T'$. Finally, let us define what will be later a path of integration C^*, formed by the arc of Γ between (and to the left of) z_1 and z_2, and the two rays of angle $\pm\alpha$ from z_1 to ∞, z_2 to ∞, oriented from ∞ to z_2, z_2 to z_1, and z_1 to ∞. Now, to construct $\mathcal{L}f$ (for details see [K1]), we need a lifting of $[F]$ to $[F^*] \in B^{exp}_{[0,\infty]}$. We do this explicitely by fixing first $\epsilon > 0$, then using Lemma 1.1 to find $F_0 = F^*_{T'\infty}$ such that $\|F^*_{T'\infty}(\cdot)\| \leq \epsilon$ on $\mathbb{C} \setminus V$ (see (9), where of course the role of a, F in Lemma 1.1 is now played by $T', F_{T'\infty}$), and finally setting

$$F^* = F_T + F_{TT'} + F^*_{T'\infty}. \tag{18}$$

Then the coclass $\mathcal{L}f$ will be given explicitely as $\mathcal{L}f = \tilde{F}^* + \mathcal{L}B^{exp}_{[\infty]}$, where

$$\tilde{F}^*(\lambda) = \int_{\mathcal{C}^*} e^{-\lambda\xi} F^*(\xi)\, d\xi \qquad \text{(similarly for } \tilde{F}^*_{T'\infty}(\lambda)). \tag{19}$$

Then, from (18) and standard use of Fubini in the first two summands of (19), we have

$$\tilde{F}^*(\lambda) = \int_0^T e^{-\lambda t} f(t)\, dt + \int_T^{T'} e^{-\lambda t} f(t)\, dt + \tilde{F}^*_{T'\infty}(\lambda). \tag{20}$$

It is immediate that the Phragmen-Lindelöf functions for the first two terms of (20) (as functions of $\lambda > 0$; see (10)) satisfy $h(0) \leq -T$. Next we show that

$$\limsup_{\lambda\to\infty} \frac{1}{\lambda} \ln \|\tilde{F}^*_{T'\infty}(\lambda)\| \leq -T. \tag{21}$$

If (21) holds, then it follows using Lemma 1.3 and

$$\limsup_{\lambda\to\infty} \frac{1}{\lambda} \ln \|\tilde{G}(\lambda)\| = -\infty \qquad \text{for all } \tilde{G} \in \mathcal{L}B^{exp}_{[\infty]}$$

(see [K1]), that $\tilde{F}^* \approx_T 0$ for our $T > 0$ which was arbitrarily chosen, and finally that

$$\mathcal{L}f \subset \{f\}_T \text{ for all } T > 0, \text{ hence } \mathcal{L}f \subset \{f\} = L_0 f. \tag{22}$$

To obtain (21) we estimate

$$\tilde{F}^*_{T'\infty}(\lambda) = \int_\Gamma e^{-\lambda\xi} F^*_{T'\infty}(\xi)\, d\xi + \int_{\mathcal{C}^*\setminus\Gamma} e^{-\lambda\xi} F^*_{T'\infty}(\xi)\, d\xi.$$

Recall from above that $\|F^*_{T'\infty}(\cdot)\| \leq \epsilon$ along the path $\mathcal{C}^*$. Therefore, for $\lambda > 0$,

$$\|\tilde{F}^*_{T'\infty}(\lambda)\| \leq \pi\mu\epsilon e^{-\lambda(T'-\mu)} + \frac{\epsilon}{\lambda\cos\alpha} e^{-\lambda x_0}.$$

Since $T < T' - \mu < x_0$, the statement (21) follows at once. ◇

4. Asymptotics and Asymptotic L-Transform for General Hyperfunctions

The idea of extending the asymptotic L-transform to general hyperfunctions f stems from the fact that within our asymptotically permissible error we can give, for any general $f \in B([0,\infty);X)$, a well defined meaning to f_T (for any $T > 0$) and to

$$\tilde{f_T}(\lambda) = \int_0^T e^{-\lambda t} f(t)\, dt \tag{23}$$

and then proceed similarly to the L^1_{loc}-case treated above. We need the following simple lemma.

4.1. Lemma *Let b be a hyperfunction on $\mathbb{R}$ with support in $[T, T']$, $T \leq T'$, let μ_b be the analytic functional associated to b, and let $\tilde{\mu}_b(\lambda) = \langle \mu_b, e^{-\lambda \cdot}\rangle$. Then the Phragmen-Lindelöf function for $\tilde{\mu}_b$ satisfies $h(0) \leq -T$.*

Proof. Let F_b be the standard representative of b and let Γ denote a contour around $[T, T']$. Then

$$\tilde{\mu}_b(\lambda) = -\int_\Gamma e^{-\lambda\xi} F_b(\xi)\, d\xi \tag{24}$$

(see [Lu]). For $\epsilon > 0$ we can take Γ close enough to $[T, T']$ so as to get from (24) the estimate $\|\tilde{\mu}_b(\lambda)\| \leq const \cdot e^{-\lambda(T-\epsilon)}$. ◊

To reach a unique well defined meaning of (23), consider first $b \in B_{cp}$. For such b there is a unique "principal determination" in $\mathcal{L}b$, $(\mathcal{L}b)_p(\lambda) = \tilde{\mu}_b(\lambda)$ as defined in (24) because there is a distinguished lifting $F_b(\xi)$ (the so-called standard representative of b; see also [Lu]). Of course any other choice of $\hat{b}(\lambda) \in \mathcal{L}b$ satisfies $\hat{b} \approx \tilde{\mu}_b$ (even $\approx_\Sigma$) and remains thus within the equivalence in terms of the admissible asymptotic error. So, in the case of compact support, $supp(b) \subset [0, T]$, we have a unique well defined meaning for what we can write symbolically as

$$\int_0^T e^{-\lambda\xi} b(\xi)\, d\xi \tag{25}$$

instead of $\langle \mu_b, e^{-\lambda \cdot} \rangle$. Given a general $f = b \in B([0,\infty);X)$ and any $T > 0$ we can "restrict f to $(-\infty, T]$ ", meaning that we can find a (not unique) f_T with support in $[0,T]$ which coincides with f on $(-\infty, T)$ and where any other such $f_T^{\#}$ satisfies $(\mathcal{L}f_T)_p \approx_T (\mathcal{L}f_T^{\#})_p$ by Lemma 4.1. This argument uses flabbiness, indeed "decomposition of supports", see [Lu] 5.9, p.23).[(6)] So (23) is interpreted as (25) with $b = f_T$, and the theory of asymptotic L-transforms for general hyperfunctions can now be developed as for L^1_{loc}. In particular, as in the proof of Theorem 3.1 one can show that

$$\tilde{f}_T \approx_{\Sigma_T} \mathcal{L}f \text{ for all } T > 0. \tag{26}$$

If (26) holds we say that f_T gives an asymptotic L-transform approximation to f and write $f_T \sim_{Lta} f$ (as $T \to \infty$). Using δ-expansions for f_T (see [Lu]) this takes the form

$$\sum_{j=0}^{\infty} c_j^T \delta^{(j)} \sim_{Lta} f \quad \text{as } T \to \infty \text{ for all } f \in B([0,\infty);X), \tag{26}$$

where $c_j^T = c_j(f_T)$.

5. Remarks

In Section 2 we proved via asymptotic L-transforms that in particular $(*)$ $\mathcal{L}(-tf) = (\mathcal{L}f)'$ for all $f \in L^1_{loc}$. Strangely enough, when looking for a direct proof of this fact within the $\mathcal{L}$ methods, we seemed to encounter a bit more difficulties than expected (but maybe we did not look at it the right way). The point seems to be that if the distribution

---

(6) Notice that it makes no difference for the definitions and proofs concerning the general asymptotic L-transform whether we talk about $[0,T)$ or $[0,T]$ (this is important when considering partitions of the type $f_{[0,T]} + f_{[T,T']} + \cdots$) because given f_T as defined above and any other $f_T^{\#}$ with support in $[0,T]$ that coincides with f_T on $(-\infty, T)$, the support of the difference $f_T - f_T^{\#}$ is contained in $\{T\}$. Since by Lemma 4.1, $(\mathcal{L}f_T)_p - (\mathcal{L}f_T^{\#})_p \approx_T 0$, f_T and $f_T^{\#}$ give the same $\{f\}_T$, and eventually the same $\{f\}$.

f is equal to $[F]$, where F is lifted (i.e., of exponential type in $\mathbb{C}^* \setminus [0, \infty]$), then it is clear that $(\int_C e^{-\lambda\xi} F(\xi)\, d\xi)' = -\int_C e^{-\lambda\xi} \xi F(\xi)\, d\xi$ and we must simply use $[\xi F(\xi)] = tf(t)$ to prove $(*)$. However, it takes some effort to prove "from scratch" that

$$(\xi F(\xi))(t + i\epsilon) - (\xi F(\xi))(t - i\epsilon) \to tf(t)$$

for $f \in L^1_{loc}$, where the limit is taken in the $\mathcal{D}'$ sense. Anyway, all this can be proved, even with $\Psi(\xi)F(\xi)$, Ψ entire, instead of $\xi F(\xi)$, but the arguments we use are not totally immediate.

References

[Be] L. Berg, Asymptotische Auffassung der Operatorenrechnung, *Studia Math.* 21 (1962), 215-229.

[K1] H. Komatsu, Operational calculus and semi-groups of operators. *Functional Analysis and Related Topics*, 1991 (Kyoto), Lecture Notes Math. 1540, Springer 1993, 213-234.

[K2] H. Komatsu, Laplace transforms of Hyperfunctions - a new foundation of the Heaviside calculus. *J. Fac. Sci. Tokyo, Sect. IA. Math.*, 34 (1987), 805-820.

[K3] H. Komatsu, Operational calculus, hyperfunctions and ultradistributions, *Algebraic Analysis*, Vol.1, Academic Press 1988, 357-372.

[L-N] G. Lumer and F. Neubrander, Asymptotic Laplace transforms and evolution equations. *Advances in Partial Differential Equations*, Math. Topics Vol. 16, Wiley-VCH 1999, 37-57.

[Lu] G. Lumer, An introduction to hyperfunctions and δ-expansions, *Generalized Functions, Operator Theory and Dynamical Systems*, Chapman & Hall/CRC Res. Notes in Math, Vol. 399, 1999, 1-25.

[Ly] Y. I. Lyubich, The classical and local Laplace transformation in an abstract Cauchy problem. *Usepi Math. Nauk* 21 (1966), 3-51.

[Ra] J.-P. Ramis, *Séries Divergentes et Théories Asymptotiques*, Soc. Math. de France, Panoramas et Synthèses, Vol. 121, 1993.

[Ti] E.C. Titchmarsh, *The Theory of Functions*, 2nd ed., Oxford Univ. Press 1939 (1975 printing).

[Vi] J. C. Vignaux, Sugli integrali di Laplace asintotici, *Atti Accad. naz. Lincei, Rend. Cl. Sci. fis. mat.*6 (29) (1939), 396-402.

[V-C] J. C. Vignaux and M. Cotlar, Asymptotic Laplace-Stieltjes integrals. *Univ. Nac. La Plata. Publ. Fac. Ci. Fisicomat.(2),* 3(14) (1944), 345-400.

Some systems of PDE on polygonal networks

D. MERCIER, University of Valenciennes, MACS, B.P.311 59313 Valenciennes Cedex 9 France. e-mail dmercier@univ-valenciennes.fr

1 INTRODUCTION

Various models of systems of partial differential equations on multiples structures made of a finite number of interconnected flexible elements like strings, beams, shells, plates or combination of them have been recently proposed. Up to now, existence results of such problems have been given for each model and have clearly some similarities. Therefore the main goal of this work is to present a general framework as large as possible to include the former two-dimensional models. More precisely, we study transmission problems for elliptic systems of P.D.E.'s on two-dimensional polygonal networks with general boundary and interface conditions.

We introduce covering conditions along the boundary and the interfaces similar to the Agmon-Douglis-Nirenberg conditions (see [1]).

We consider a class of boundary and transmission operators as large as possible so that a weak formulation of these problems is possible with the help of Green's formula.

Then we study the regularity of the variational solution. This study is made in a more or less classical way (see [2, 3, 4] for example). We show that the weak solution may be split up into a regular part and a singular part. The regular part has the optimal regularity while the singular part is a finite linear combination of singular functions depending only on the domain and the operators. The coefficients depend continuously on the data.

For some mechanical examples, we finally give some numerical results about the singular exponents giving informations on the minimal regularity of the solutions.

2 SOME EXAMPLES

We start with some examples that are in our general framework.

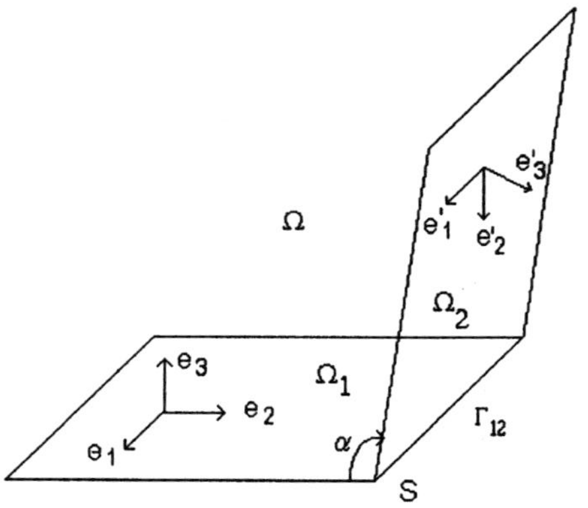

Figure 1:

2.1 Interconnected elastic membranes

This first example comes from [5], where the authors study the movement of a body constituted of several elastic membranes rigidly interconnected. Here we study the equations of the movement of two rectangular plates making an angle α between them (see Figure 1). Each body $\Omega_k, k = 1, 2$ is equipped with a system of coordinates such that u^k represents the displacement in the plane of the plate Ω_k and u_3^k represents the displacement in the orthogonal direction of the plate.

The system of equations after linearization is:

$$\sum_{j=1}^{2} \frac{\partial \sigma_{ij}(\mathbf{u}^k)}{\partial x_j} = f_i^k \ in \ \Omega_k, \ i, k = 1, 2, \qquad (2.\ 1)$$

$$\mu_k \Delta u_3^k = f_3^k \ in \ \Omega_k, k = 1, 2, \qquad (2.\ 2)$$

$$u_i^k = 0 \ on \ \Gamma_k, i = 1, 2, 3, \ k = 1, 2, \qquad (2.\ 3)$$

$$\begin{cases} u_1^1 = u_1^2 \\ u_2^1 = \cos\alpha\, u_2^2 + \sin\alpha\, u_3^2 \\ u_3^1 = -\sin\alpha\, u_2^2 + \cos\alpha\, u_3^2 \end{cases} \quad on \ \Gamma_{12}, \qquad (2.\ 4)$$

$$\begin{cases} \sigma_{12}(\mathbf{u}^1) = -\sigma_{12}(\mathbf{u}^2) \\ \sigma_{22}(\mathbf{u}^1) = -\cos\alpha\, \sigma_{22}(\mathbf{u}^2) - \mu_2 \sin\alpha\, \frac{\partial u_3^2}{\partial \nu_2} \\ \mu_1 \frac{\partial u_3^1}{\partial \nu_1} = \sin\alpha\, \sigma_{22}(\mathbf{u}^2) - \mu_2 \cos\alpha \frac{\partial u_3^2}{\partial \nu_2} \end{cases} \quad on \ \Gamma_{12}, \qquad (2.\ 5)$$

where Γ_k is the external boundary of Ω_k ($k = 1, 2$), Γ_{12} is the common boundary between Ω_1 and Ω_2, λ_k and μ_k are the Lamé constants in Ω_k, $k = 1, 2$ and

$$\sigma_{ij}(\mathbf{u}^k) = \mu_k\left(\frac{\partial u_i^k}{\partial x_j} + \frac{\partial u_j^k}{\partial x_i}\right) + \lambda_k \mathrm{div}(\mathbf{u}^k)\delta_{ij}, \forall i, j, k = 1, 2.$$

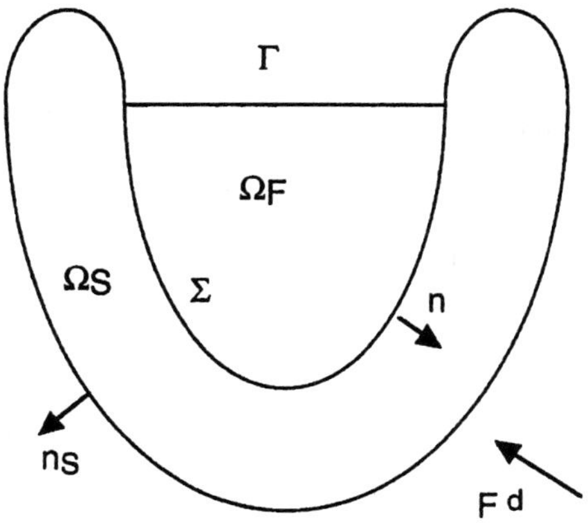

Figure 2:

2.2 Hydroelastic vibration (fluid-structure interaction)

We study the vibrations in harmonic mode of an elastic structure containing an incompressible liquid and presenting a smooth surface (see Figure 2 and [6]).

The function φ represents a potential displacement for the liquid Ω_F and u represents the displacement field of the elastic structure Ω_S.

$$\Delta\varphi = 0 \quad in\ \Omega_F, \tag{2. 6}$$

$$\varphi = 0 \quad on\ \Gamma, \tag{2. 7}$$

$$\frac{\partial\varphi}{\partial n} = \mathbf{u}.n \quad on\ \Sigma, \tag{2. 8}$$

$$\sigma_{ij}(\mathbf{u})n_j^S = \rho_F\omega^2\varphi n_i \quad on\ \Sigma, \tag{2. 9}$$

$$\sigma_{ij},_j(\mathbf{u}) + \rho_S\omega^2 u_i = 0 \quad in\ \Omega_S, \tag{2. 10}$$

$$\sigma_{ij}(\mathbf{u})n_j^S = F_i^d \quad on\ \partial\Omega_S\backslash\Sigma. \tag{2. 11}$$

The constants ρ_F and ρ_S are positive and represent respectively the volumic masses of the solid and fluid structures, F^d represents a force field acting on the solid structure.

2.3 Interface problem between the Laplace and the biharmonic operators

This example is a coupling between a plate and a membrane (see [7] and Figure 3). The system of operators are the Laplace operator and the biharmonic one. The main interest of this example is that the operators are of different orders.

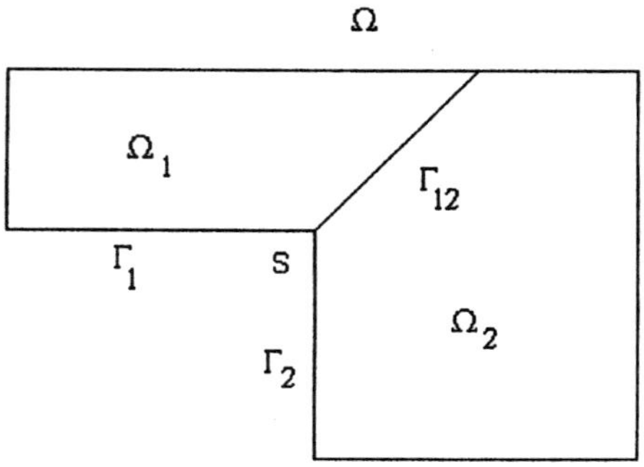

Figure 3:

The equations are:

$$\Delta u_1 = f_1 \ in \ \Omega_1, \quad (2.\ 12)$$

$$\Delta^2 u_2 = f_2 \ in \ \Omega_2, \quad (2.\ 13)$$

$$u_1 = 0 \ on \ \Gamma_1, \quad (2.\ 14)$$

$$u_2 = \frac{\partial u_2}{\partial \nu_2} = 0 \ on \ \Gamma_2, \quad (2.\ 15)$$

$$u_1 = u_2 \ on \ \Gamma_{12}, \quad (2.\ 16)$$

$$Mu_2 = h_1 \ on \ \Gamma_{12}, \quad (2.\ 17)$$

$$Nu_2 + \frac{\partial u_1}{\partial \nu_2} = h_2 \ on \ \Gamma_{12}. \quad (2.\ 18)$$

where

$$Mu_2 = \frac{E}{1-\sigma^2}(\sigma \Delta u_2 + (1-\sigma)\frac{\partial^2 u_2}{\partial \nu_2^2}),$$

$$and \ Nu_2 = \frac{E}{1-\sigma^2}(\frac{\partial \Delta u_2}{\partial \nu_2} + (1-\sigma)\frac{\partial^3 u_2}{\partial \nu_2 \partial \tau_2^2}).$$

$E > 0$, $\sigma \in]0, 1[$ are respectively the Young modulus and the Poisson coefficient of the constitutive material of the plate Ω_2.

3 FORMULATION OF THE PROBLEM

3.1 Domains and notations

The considered domains are two-dimensional polygonal networks in $I\!R^3$ denoted by Ω (they are ramified space in the sense of [8]). More precisely Ω is a union of polygons Ω_i (see for instance Figure 4) :

$$\Omega = \bigcup_{i=1}^{K} \Omega_i,$$

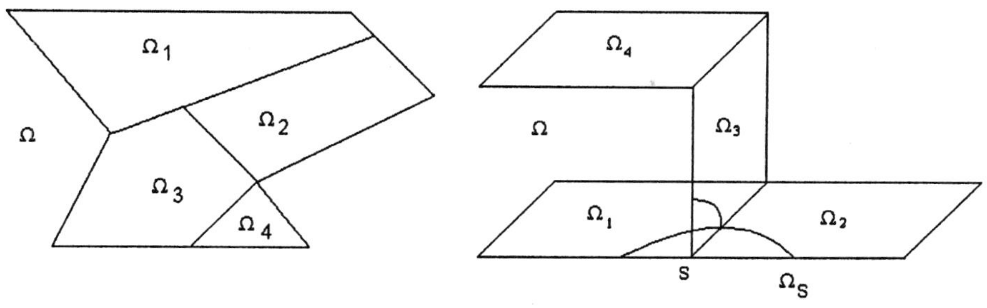

Figure 4:

with

$$\partial\Omega_i = \bigcup_{q=1}^{Q_i} \gamma_{iq}.$$

For convenience, we denote by $\mathcal{A}$ the set of all the edges γ_{iq}, by $\mathcal{S}$ the set of all the vertices of Ω and $I_\Gamma = \{i : \Gamma \text{ is a edge of } \Omega_i\}$.

3.2 The operators

On each face Ω_i, let us consider a system $l_i = (l_i^{ll'}(x, D))$ of $N_i \times N_i$ operators. We suppose that each system l_i is properly elliptic in the sense of Agmon-Douglis and Nirenberg (see for instance [9] for the exact definition), called **Condition I.** This implies that there exist integers s_{il} and t_{il} such that $\deg l_i^{ll'} = s_{il} + t_{il'}$ (with the assumption $s_{il} \leq 0$ and $l_i^{ll'} = 0$ if $s_{il} + t_{il'} < 0$). Moreover we suppose that the operators $l_i^{ll'}$ can be written in the divergence form:

$$l_i^{ll'} = \sum_{|\alpha| \leq m_{il'}, |\beta| \leq m_{il}} (-1)^\beta D^\beta (a_{ill'}^{\alpha\beta} D^\alpha), l, l' \in \{1, ..., N_i\},$$

with $a_{ill'}^{\alpha\beta} \in C^\infty(\overline{\Omega_i})$.

Now, for each edge Γ we consider a system of $m_\Gamma \times \sum_{i \in I_\Gamma} N_i$ boundary and/or transmission operators:

$$B_\Gamma = B_{\Gamma h}^{il}, \ l \in \{1, ..., N_i\}, h \in \{1, ..., m_\Gamma\},$$

where $m_\Gamma = \sum_{i \in I_\Gamma} \sum_{l=1}^{N_i} m_{il}$ and $\deg B_{\Gamma h}^{il} = r_{\Gamma h} + t_{il}$.

We suppose that B_Γ covers $(l_i)_{i \in I_\Gamma}$ (see also [9]), called **Condition II.**

In the sequel we write for simplicity $[L, B] = [(l_i)_i, (B_\Gamma)_\Gamma]$.

Consider the multi-indices $\vec{s} = (s_{il})$, $\vec{t} = (t_{il})$ and $\vec{r_\Gamma} = (r_{\Gamma h})$ then the transformation by $[L, B]$ in the Sobolev spaces gives :

$$u \in W^{k+\vec{t},\,p}(\Omega) \longmapsto Lu \in W^{k-\vec{s},\,p}(\Omega)$$

$$u \in W^{k+\vec{t},\,p}(\Omega) \longmapsto B_\Gamma u \in W^{k-\frac{1}{p}-\vec{r_\Gamma},\,p}(\Gamma), \forall \Gamma \in \mathcal{A},$$

where $u \in W^{k+\vec{t},\,p}(\Omega)$ means

$$u_{|\Omega_i} = u_i = (u_{i1}, ..., u_{il}, ..., u_{iN_i}),\ u_{il} \in W^{k+t_{il},\,p}(\Omega_i),$$

$W^{k-\vec{s},\,p}(\Omega)$ and $W^{k-\frac{1}{p}-\vec{r_\Gamma},\,p}(\Gamma)$ are similarly defined.

We further introduce the space $H_v^{\vec{m}}(\Omega)$ as follows: $u = (u_{il}) \in H_v^{\vec{m}}(\Omega)$ if each $u_{il} \in H_v^{m_{il}}(\Omega_i)$, where $H_v^{m_{il}}(\Omega_i)$ is the closure in $H^{m_{il}}(\Omega_i)$ of the set of functions which are in $C^\infty(\Omega_i)$ and equal to 0 near each vertex of Ω_i.

3.3 Formulation of the problem

We are now in position to formulate our problem: Let $k \in I\!N$, given $f \in W^{k-\vec{s},\,p}(\Omega)$, $\varphi_\Gamma \in W^{k-\frac{1}{p}-\vec{r_\Gamma},\,p}(\Gamma)$, for all $\Gamma \in \mathcal{A}$, find $u \in W^{k+\vec{t},\,p}(\Omega)$ satisfying

$$Lu = f \text{ in } \Omega, \tag{3. 1}$$

$$B_\Gamma u = \varphi_\Gamma \text{ on } \Gamma, \forall \Gamma \in \mathcal{A}. \tag{3. 2}$$

4 THE WEAK FORMULATION

The main question is to know which class of operators B_Γ may be chosen in order to get a weak formulation of problem (3. 1)-(3. 2)? The answer will follow from Green's formula.

First we consider the natural bilinear form associated with the operator L :

$$a(u, v) = \sum_{i=1}^{K} \sum_{l=1}^{N_i} \int_{\Omega_i} \sum_{l'=1}^{N_i} \sum_{|\alpha| \le m_{il'}, |\beta| \le m_{il}} a_{ill'}^{\alpha\beta} D^\alpha u_{il'} D^\beta \overline{v_{il}} dx, \forall u, v \in H^{\vec{m}}(\Omega). \tag{4. 1}$$

In order to obtain a Green formula, on each side γ_{iq} of Ω_i and for each $l \in \{1, 2, \cdots, N_i\}$ we fix a Dirichlet system of order m_{il} : $\{F_{iqlj}\}_{j=0}^{m_{il}-1}$ (see [10], Definition 2.2.1) with coefficient in $C^\infty(\overline{\gamma_{iq}})$ and, without loss of generality, we may suppose that the order of F_{iqlj} is j.

LEMMA 1 *For each l and $l' \in \{1, 2, ..., N_i\}$, and each side γ_{iq} of Ω_i, there exists a system of operators $\{\Phi_{iqll'j}\}_{j=0}^{m_{il}-1}$ on γ_{iq} with coefficients in $C^\infty(\overline{\gamma_{iq}})$, the order of $\Phi_{iqll'j}$ being less than $m_{il} + m_{il'} - 1 - j$, such that:*

$$\begin{aligned}(l(u), v)_{0,2,\Omega} &= \sum_{i=1}^{K} \sum_{l=1}^{N_i} \int_{\Omega_i} (\sum_{l'=1}^{N_i} l_i^{ll'} u_{il'}) \overline{v_{il}} dx \\ &= a(u, v) \\ &+ \sum_{i=1}^{K} \sum_{q=1}^{Q_i} \sum_{l=1}^{N_i} \sum_{j=0}^{m_{il}-1} \int_{\gamma_{iq}} (\sum_{l'=1}^{N_i} \Phi_{iqll'j} u_{il'}) F_{iqlj} \overline{v_{il}} d\sigma.\end{aligned} \tag{4. 2}$$

for all $u \in H^{\vec{t}}(\Omega)$ *and* $v \in H_v^{\vec{m}}(\Omega)$.

PROOF Similar arguments than in Theorem 22.6 of [10].■

We are now able to precise the class of boundary operators which allows to obtain a weak formulation of problem (3. 1)-(3. 2).

First, for each $\gamma_{iq}, l \in \{1, 2, ..., N_i\}, j \in \{0, 1, ..., m_{il} - 1\}$, we write

$$\Phi_{iqlj}(u_i) = \sum_{l'=1}^{N_i} \Phi_{iqll'j} u_{il'},$$

and introduce the column vectors F_Γ, Φ_Γ with m_Γ components as:

$$F_\Gamma(v) = (F_{iqlj}(v))_{i \in I_\Gamma, \gamma_{iq}=\Gamma, l \in \{1,\cdots,N_i\}, j \in \{0,\cdots,m_{il}-1\}}, \tag{4. 3}$$

$$\Phi_\Gamma(u) = (\Phi_{iqlj}(u_i))_{i \in I_\Gamma, \gamma_{iq}=\Gamma, l \in \{1,\cdots,N_i\}, j \in \{0,\cdots,m_{il}-1\}}. \tag{4. 4}$$

We then consider the boundary conditions in the following form

$$B_\Gamma u = \begin{pmatrix} 0 \\ B_{1,\Gamma}^2 \end{pmatrix} \Phi_\Gamma(u) - \begin{pmatrix} B_{2,\Gamma}^1 \\ 0 \end{pmatrix} F_\Gamma(u) = \begin{pmatrix} \varphi_\Gamma^1 \\ \varphi_\Gamma^2 \end{pmatrix}, \tag{4. 5}$$

where $\varphi_\Gamma = \begin{pmatrix} \varphi_\Gamma^1 \\ \varphi_\Gamma^2 \end{pmatrix} = (\varphi_{\Gamma h})_{h=1,\dots,m_\Gamma}$ and $B_{2,\Gamma}^1$ is a $m'_\Gamma \times m_\Gamma$ matrix with a rank equal to m'_Γ, $B_{1,\Gamma}^2$ is a $(m_\Gamma - m'_\Gamma) \times m_\Gamma$ matrix with a rank equal to $m_\Gamma - m'_\Gamma$ (conditions similar to the condition (19) of [11] in the one dimensional case).

Notice that the first m'_Γ conditions of (4. 5) have a meaning for u in $H^{\vec{m}}(\Omega)$ (stable conditions) while the other conditions have no sense for u in $H^{\vec{m}}(\Omega)$ (transversal conditions).

We now consider the variational spaces:

$$V = \{u \in H_v^{\vec{m}}(\Omega) / \forall \Gamma \in \mathcal{A}, B_{2,\Gamma}^1 . F_\Gamma(u) = 0\}, \tag{4. 6}$$

$$\widetilde{V} = \{u \in H^{\vec{m}}(\Omega) / \forall \Gamma \in \mathcal{A}, B_{2,\Gamma}^1 . F_\Gamma(u) = 0\}. \tag{4. 7}$$

THEOREM 2 *Assume that the conditions **I-II** hold and let* $f \in H^{-\vec{s}}(\Omega)$, $\varphi_{\Gamma h} \in H^{-r_{\Gamma h}-\frac{1}{2}}(\Gamma)$, *for all* $\Gamma \in \mathcal{A}, h = 1, ..., m_\Gamma$. *If* $u \in H^{\vec{t}}(\Omega) \cap \widetilde{V}$ *is solution of*

$$a(u, v) = (f, v)_{0,2,\Omega}, \forall v \in V, \tag{4. 8}$$

then u satisfies

$$\begin{cases} Lu = f \text{ in } \Omega, \\ B_\Gamma u = 0 \text{ on } \Gamma, \forall \Gamma \in \mathcal{A}. \end{cases} \tag{4. 9}$$

PROOF Applying (4. 8) with $v \in \mathcal{D}(\Omega)$ we see that $Lu = f$. The remainder follows from Green's formula (4. 2).■

REMARK 3 It is possible to consider a class of boundary conditions larger than (4. 5). In such a situation, boundary terms may appear in the bilinear form a. In the previous theorem we considered only homogeneous boundary conditions but owing to the properties of the operators F_Γ, and with the help of a lifting trace theorem we can consider non homogeneous boundary and transmission conditions.

REMARK 4 In the sequel we only study u in V solution of (4. 8). This solution will be called the weak solution of (4. 9). This is not restrictive because if $\widetilde{u} \in \widetilde{V}$ is solution of (4. 8), for all $\widetilde{v} \in \widetilde{V}$, setting $u = \widetilde{u} - \sum_{S \in \mathcal{S}} \Phi_S T_S \widetilde{u}$, we can show that u fullfills a similar problem, where the data f will be replaced by $f - L(\sum_{S \in \mathcal{S}} \Phi_S T_S \widetilde{u})$ and boundary conditions will be non homogeneous (one applies Remark 3) with data $-B_\Gamma(\sum_{S \in \mathcal{S}} \Phi_S T_S \widetilde{u})$, for all Γ in A . Here $(T_S \widetilde{u})_{il}$ is the limited Taylor expansion of $\widetilde{u}_{il}$ at S of order $m_{il} - 2$ and Φ_S a regular cut-off function fullfilling $\Phi_S = 1$ near S and $\Phi_S = 0$ near the other vertices.

5 REGULARITY IN WEIGHTED SOBOLEV SPACES

5.1 The operator$\mathcal{A}_S(\lambda)$

The weighted Sobolev spaces are well adapted to the study of our problem (as initiated in [2]). As usual we come back to ordinary spaces via the polynomial resolution.

First we define the corresponding weighted Sobolev spaces for our boundary transmission problem.

DEFINITION 5 *If $k \in \mathbb{N}$, $V_\beta^{k,p}(\Omega_i)$ is the closure of $C_v^\infty(\overline{\Omega_i})$ with respect to the norm*

$$\|f\|_{V_\beta^{k,p}(\Omega_i)} = \sum_{|\alpha\ | \leq m} \left\| r_i^{\beta-k+|\alpha|} D^\alpha f \right\|_{L_p(\Omega_i)}, \tag{5. 1}$$

where $r_i = r_i(x)$ is the distance of x_i to the vertices of Ω_i.

$V_\beta^{k+\overrightarrow{t},p}(\Omega) = \{u : u_{il} \in V_\beta^{k+t_{il},p}(\Omega_i)\}$ is equipped with the norm

$$\|u\|_{V_\beta^{k+\overrightarrow{t},p}(\Omega)} = \sum_i \sum_l \omega_{il} \|u_{il}\|_{V_\beta^{k+t_{il},p}(\Omega_i)}.$$

The space of traces on each edge γ_{iq} is the quotient space

$$V_\beta^{(k-\frac{1}{p}),p}(\gamma_{iq}) = V_\beta^{k,p}(\Omega_i) / \overset{\circ}{V}_\beta^{k,p}(\Omega_i, \gamma_{iq})$$

where $\overset{\circ}{V}_\beta^{k,p}(\Omega_i, \gamma_{iq})$ is the closure of

$$C_{\gamma_{iq}}^\infty(\Omega_i) = \{u \in C_v^\infty(\overline{\Omega_i}),\ supp\ u \cap \gamma_{iq} = \varnothing\}$$

for the norm (5. 1).

We now remark that $[L(x), B(x)]$ transforms

$$V_\beta^{k+\vec{t},p}(\Omega) \text{ in } V_\beta^{k-\vec{s},p}(\Omega) \times \prod_{\Gamma\in\mathcal{A},h=1,\cdots,m_\Gamma} V_\beta^{k-r_{\Gamma,h}-1/p,p}(\Gamma) = Y_\beta^{k,p}(\Omega). \quad (5.\ 2)$$

For $S \in \mathcal{S}$ and each $i \in I_S, C_i$ is the cone which coincides with Ω_i in a neighbourhood of S and we set $C_S = \bigcup_{i\in I_S} C_i$. The spaces defined in Ω are analogously defined in C_S.

We first consider a particular problem in C_S which is generated by the problem (3. 1)-(3. 2) taking the principal part of $[L, B]$ with frozen coefficients at S.

$$\begin{aligned} \sum_{l'=1}^{N_i} l_{i,0}^{ll',S}(D)u_{il} &= f_{il} \text{ in } C_i, \forall i = 1,\cdots,K, \forall l = 1,\cdots,N_i, && (5.\ 3)\\ B_{\Gamma h,0}^{S}(D)u &= \varphi_{\Gamma h} \text{ on } \Gamma, \forall\Gamma \in \mathcal{A}, \forall h = 1,\cdots,m_\Gamma. && (5.\ 4)\end{aligned}$$

We denote by $[L_0(S), B_0(S)]$ the operator associated with problem (5. 3)-(5. 4).

Introducing polar co-ordinates $(r, (\theta)_i)$ centered at S and using the Mellin transform

$$\frac{1}{\sqrt{2\pi}}\int_0^{+\infty} r^{-\lambda-1}u(r,\theta)dr = \hat{u}(\lambda,\theta),$$

the equations (5. 3)-(5. 4) become a boundary transmission problem with parameter λ in $\Omega_S = C_S \cap S^2$ (where S^2 is the unit sphere) :

$$\sum_{l'=1}^{N_i} \mathcal{L}_i^{ll',S}(\theta_i, D_\theta, \lambda + t_{il'})\widehat{u}_{il'}(\lambda + t_{il'}, \theta_i) = \widehat{f_{il}}(\lambda - s_{il}, \theta_i), \quad (5.\ 5)$$

for all $\theta_i \in]0;\omega_i[$, $i = 1,\cdots,K$, $l = 1,\cdots,N_i$ and

$$\sum_{i\in I_\Gamma}\sum_{l=1}^{N_i} \mathcal{B}_{\Gamma,h}^{il,S}(\omega_{iq}, D_\theta, \lambda + t_{il})\widehat{u}_{il}(\lambda + t_{il}, \omega_i) = \widehat{\varphi_{\Gamma h}}(\lambda - r_{\Gamma h}, \omega_{iq}), \quad (5.\ 6)$$

for all $\Gamma \in \mathcal{A}, h = 1,\cdots,m_\Gamma$, and where $\omega_{iq} = 0$ or ω_i depending on $(r, 0)$ or $(r, \omega_i) \in \gamma_{iq} = \Gamma$ and where we have set

$$\begin{aligned} \mathcal{L}_i^{ll',S}(\theta_i, D_\theta, r\partial r) &= r^{s_{il}+t_{il'}} l_{i,0}^{ll',S}(D),\\ \mathcal{B}_{\Gamma h}^{il,S}(\omega_{iq}, D_\theta, r\partial r) &= r^{r_{\Gamma h}+t_{il}} B_{\Gamma h,0}^{il,S}(D).\end{aligned}$$

We write $\mathcal{A}_S(\lambda) = [\mathcal{L}_S(\lambda), \mathcal{B}_S(\lambda)]$ the operator generated by (5. 5)-(5. 6), defined in Ω_S and which maps continuously

$$W^{k+\vec{t},p}(\Omega_S) \text{ into } W^{k-\vec{s},p}(\Omega_S) \times \prod_{\Gamma,h} \mathbb{C} = Y^{k,p}(\Omega_S) \quad (5.\ 7)$$

It follows from the ellipticity of the operator $[L_0(S), B_0(S)]$ in C_S, the ellipticity of $\mathcal{A}_S(\lambda)$ in the sense of [12]. Therefore, for all $\lambda \in \mathbb{C}$, except for a set of isolated points, $\mathcal{A}_S(\lambda)$ is an isomorphism between the spaces in (5. 7). If $\mathcal{A}_S(\lambda_0)$ is not invertible, we shall say that λ_0 is an eigenvalue of $\mathcal{A}_S(\lambda)$. Since $\mathcal{A}_S(\lambda) - \mathcal{A}_S(\lambda_0)$ is a compact operator, dim ker $\mathcal{A}_S(\lambda)$ is finite, for all $\lambda \in \mathbb{C}$. Finally we establish that the number of eigenvalues is finite in any strip $h \leq Re(\lambda) \leq h_1$ thanks to the following Lemma:

LEMMA 6 *Under the conditions* ***I*** *and* ***II****, there exist* $\delta > 0, N > 0$ *and* $C > 0$, *such that for all* $\lambda \in \sum_{\delta,N} = \{\xi \in \mathbb{C} : |\arg\xi \pm \frac{\pi}{2}| \leq \delta$ *and* $|\xi| \geq N\}$, *and for all* $u \in H^{k+\vec{t}}(\Omega_S) = W^{k+\vec{t},2}(\Omega_S)$, *we have:*

$$\begin{aligned} \|u\|_{k+\vec{t},2,\Omega_S,|\lambda|} &\leq C\{\|\mathcal{L}_S(\lambda)(u)\|_{k-\vec{s},2,\Omega_S,|\lambda|} \\ &+\sum_{\Gamma,h} \left\|\mathcal{B}^S_{\Gamma,h}(\lambda)(u)\right\|_{k-r_{\Gamma h}-\frac{1}{2},\omega_\Gamma,|\lambda|}\} \end{aligned} \tag{5. 8}$$

where we have set:

$$\begin{aligned} \|u\|_{k+\vec{t},\Omega_S,|\lambda|} &= \sum_{i,l} \|u_{il}\|_{k+t_{il},]0,\omega_i[,|\lambda|} \quad \textit{with} \\ \|u_{il}\|_{k,]0,\omega_i[,|\lambda|} &= (|\lambda|^{2k} \|u_{il}\|^2_{0,]0,\omega_i[} + |\, u_{il}\,|^2_{k,]0,\omega_i[})^{\frac{1}{2}} \end{aligned}$$

(see [3], Definition AA.17) and

$$\left\|\mathcal{B}^S_{\Gamma,h}(u)\right\|_{k-r_{\Gamma h}-1/2,\omega_\Gamma,|\lambda|} = \sum_{i\in I_\Gamma}\sum_{l=1}^{N_i} |\lambda|^{k-r_{\Gamma h}-1/2} \left|\mathcal{B}^{il,S}_{\Gamma h} u_{il}(\omega_{iq})\right|.$$

PROOF Proved as in Lemma 5.11 of [9].■

Now we recall the definition of eigenvalues, eigensolutions and Jordan chains.

DEFINITION 7 *The complex number* $\lambda = \lambda_0$ *is an eigenvalue of the operator* $\mathcal{A}_S(\lambda)$, *if there is a non-trivial function* $\phi^{S,\lambda_0,0} \in W^{k+\vec{t},p}(\Omega_S)$ *with* $\mathcal{A}_S(\lambda_0)\phi^{S,\lambda_0,0} = 0$. *The function* $\phi^{S,\lambda_0,0}$ *is called an eigensolution of* $\mathcal{A}_S(\lambda)$ *for* $\lambda = \lambda_0$.

DEFINITION 8 *The system* $\{\phi^{S,\lambda_0,\mu,k}\}_{\substack{k=0,\dots,N^{S,\lambda_0,\mu}\\ \mu=1,\dots,I^{S,\lambda_0}}}$ *is a system of Jordan chains if*

$$\sum_{q=0}^{k} \frac{1}{q!}(\partial/\partial\lambda)^q \mathcal{A}_S(\lambda)\phi^{S,\lambda_0,\mu,k-q}|_{\lambda=\lambda_0} = 0$$

for $k = 0, ..., N^{S,\lambda_0,\mu}$, $N^{S,\lambda_0,\mu}$ *being decreasing with respect to* μ.

The following theorem can be proved analogously as in [2] for $p = 2$, and [13] for $p \neq 2$.

THEOREM 9 *Under the conditions* ***I*** *and* ***II****, the operator* $[L, B]$ *defined by (5. 2) is a Fredholm operator if and only if no eigenvalue of* $\mathcal{A}_S(\lambda)$ *lies on the line* $Re\lambda = -\beta - 2/p + k$, *for all* $S \in \mathcal{S}$.

THEOREM 10 *Under the conditions* ***I*** *and* ***II****, suppose (for each* $S \in \mathcal{S}$*) that* $[L_0(S), B_0(S)] = [L(x), B(x)]$ *in a neighbourhood of S and that no eigenvalue of* $\mathcal{A}_S(\lambda)$ *lies on the lines* $Re\lambda = -\beta_1 - 2/p_1 + k_1 = h_1$ *and* $Re\lambda = -\beta_2 - 2/p_2 + k_2 = h_2$ *with* $h_1 < h_2$; *then for* $u \in V^{k_1+\vec{t},p}_\beta(\Omega)$ *solution of the (3. 1)-(3. 2) such that* $[f, \varphi] \in Y^{k_1,p_1}_{\beta_1}(\Omega) \cap Y^{k_2,p_2}_{\beta_2}(\Omega)$ *we have the following expansion in a neighbourhood of S:*

$$u = w + \Phi_S \sum_{\lambda,\mu,k} c_{S,\lambda,\mu,k}\sigma^{S,\lambda,\mu,k},$$

where $w \in V_{\beta_2}^{k_2+\vec{t},p_2}(\Omega), c_{S,\lambda,\mu,k} \in \mathbb{C}, \Phi_S$ *is a cut-off function equal to 1 in a neighbourhood of* S*, the sum concerns only the eigenvalues* λ *of* $\mathcal{A}_S(\xi)$ *situated in the strip* $Re\lambda \in]h_1, h_2[, \mu = 1, ..., I^{S,\lambda}, k = 0, ..., N^{S,\lambda_0,\mu}$ *and the singular functions are given by*

$$\sigma^{S,\lambda,\mu,k} = r^{\lambda+\vec{t}} \sum_{q=0}^{k} \left(\frac{(\ln r)^q}{q!}\right) \phi^{S,\lambda,\mu,k-q}. \tag{5. 9}$$

The coefficients $c_{S,\lambda,\mu,k}$ *depend continuously on* $[f, \varphi]$.

5.2 Decomposition in weighted Sobolev spaces

To study the regularity of the weak solution of problem (4. 8), we need to show that this weak solution belongs to a weighted space. For simplicity we suppose here that the operator $[L, B]$ is homogeneous with constant coefficient.

LEMMA 11 *Under the conditions* ***I-II****, let* $u \in V$ *be a solution of problem (4. 8) with right-hand sides* $f \in W^{-\vec{s},p}(\Omega)$ *and* $\varphi_{\Gamma,h} \in W^{-r_{\Gamma,h}-\frac{1}{p},p}(\Gamma)$ *where* $1 < p \leq 2$. *Then*

$$u \in V_\alpha^{\vec{t},p}(\Omega) \tag{5. 10}$$

with $\vec{t} = (t_{il}), \alpha = \gamma - \frac{2}{p} + \eta + 1, \forall \gamma \in]0; 1[$*, where* $\eta = \sup(t_{il} - m_{il})$.

PROOF Use a diadic covering and Agmon-Douglis-Nirenberg estimates (see Lemma 5.21 of [9], and also [3]).■

We are now able to give a regularity result when the data are in weighted Sobolev spaces.

THEOREM 12 *Under the conditions* ***I*** *and* ***II****, let* $k_1 \in \mathbb{N}$ *and suppose that* $k_1 - \beta \geq 0$ *and that no eigenvalue of* $\mathcal{A}_S(\lambda)$ *(for all* $S \in \mathcal{S}$*) lies on the line* $Re\lambda = -\beta - \frac{2}{p} + k_1$. *Then for all* $[f, \varphi] \in Y_\beta^{k_1,p}(\Omega)$ *with* φ *satisfying* $\varphi_\Gamma^1 = 0$*, for all* Γ*, any solution* $u \in V$ *of problem* $[L, B]u = [f, \varphi]$ *(L and B being homogeneous with constant coefficients) admits the following expansion:*

$$u = u_0 + \sum_{S \in \mathcal{S}, (\lambda,\mu,k) \in \Lambda_S^1(k_1,p,\beta)} c_{S,\lambda,\mu,k} \Phi_S \sigma^{S,\lambda,\mu,k}, \tag{5. 11}$$

where $u_0 \in V_\beta^{k_1+\vec{t},p}(\Omega), c_{S,\lambda,\mu,k} \in C$ *and*

$$\begin{aligned}\Lambda_S^1(k_1, p, \beta) = \{(\lambda, \mu, k) : \lambda \text{ is a eigensolution of } \mathcal{A}_S(\xi) \text{ such that} \\ Re\lambda \in]-\eta - 1; k_1 - \tfrac{2}{p} - \beta[, \mu = 1, ..., I^{S,\lambda}, k = 0, ..., N^{S,\lambda,\mu}, \\ \sigma^{S,\lambda,\mu,k} \in H_v^{\vec{m}}(\Omega)\}.\end{aligned} \tag{5. 12}$$

η *is defined as in Lemma 11.*

Moreover, there exists a constant $C > 0$ *independent of* u *such that*

$$\| u_0 \|_{V_\beta^{k_1+\vec{t},p}(\Omega)} + \sum_{S \in \mathcal{S}, (\lambda,\mu,k) \in \Lambda_S^1(k_1,p,\beta)} |c_{S,\lambda,\mu,k}| \leq C \| [f, \varphi] \|_{Y_\beta^{k_1,p}(\Omega)} \tag{5. 13}$$

PROOF Using Corollary 1.25 of [9], there exists $r \in]1, p[$, such that $f \in W^{-\overrightarrow{s},r}(\Omega)$ and $\varphi_{\Gamma,h} \in W^{-r_{\Gamma h}-\frac{1}{r},r}(\Gamma), \forall \Gamma, h$; by Lemma 11, $u \in V_\alpha^{\overrightarrow{t},r}(\Omega)$ with $\alpha = \gamma - \frac{2}{r} + \eta + 1, \forall \gamma \in]0;1[$. Since $\alpha > 0$, we obtain $[f, \varphi] \in Y_\alpha^{0,r}(\Omega) \cap Y_\beta^{k_1,p}(\Omega)$. Applying Theorem 10 in each cone C_S at $\Phi_S u$ with γ sufficiently close to 0, we obtain the result.■

REMARK 13 *Using a lifting trace Theorem we can generalize the previous result for* $\varphi_\Gamma^1 \neq 0$.

6 REGULARITY IN USUAL SOBOLEV SPACES

In this section we give without proof the decomposition of the solution in classical Sobolev spaces. Let us start by defining the injectivity modulo the polynomials initially introduced in [3].

If J is an integer, consider the polynomial spaces

$$\begin{aligned} P_J(C_S) &= \{p, p = (p_{il}) : \text{each } p_{il} \text{ is a homogeneous polynomial} \\ &\qquad \text{of degree } J + t_{il} \text{ defined on } C_i\} \end{aligned}$$

where we suppose that $p_{il} = 0$ when $J + t_{il} < 0$,

$$\begin{aligned} X_J(C_S) = \{[p, q], p = (p_{il}), q = (q_{\Gamma h}) &: \text{each } p_{il} \text{ is a homogeneous polynomial} \\ &\text{of degree } J - s_{il} \text{ defined on } C_i, \\ \text{and } q_{\Gamma h} \text{ is a homogeneous polynomial of degree } &J - r_{\Gamma h} \text{ defined on } \Gamma\}, \end{aligned}$$

with $p_{il} = 0$ if $J - s_{il} < 0$, $q_{\Gamma h} = 0$ if $J - r_{\Gamma h} < 0$.

For $S \in \mathcal{S}$, J integer and for each $[p, q] \in X_J(C_S)$, there exists a solution $w_S \in S^J(C_S)$ of the equation

$$[L_0(S), B_0(S)]w_S = [p, q], \tag{6. 1}$$

$S^J(C_S)$ is the space of functions of the form

$$r^{J+\overrightarrow{t}} \sum_{q=0}^{Q} c_q(\omega) \ln^q r, \tag{6. 2}$$

where each vectorial function c_q is in $H^{\overrightarrow{t}}(\Omega_S)$.

Now we define the injectivity modulo the polynomials:

DEFINITION 14 *Let us fix* $S \in \mathcal{S}$ *and an integer* J. *We say that the operator* $[L_0(S), B_0(S)]$ *is injective modulo the polynomials on* $S^J(C_S)$ *if and only any solution* w_S *in* $S^J(C_S)$ *of (6. 1) with a right-hand side in* $X_J(C_S)$ *is in* $P_J(C_S)$.

Now we can give the result obtained in a classical way (see [3, 14]) with the following hypotheses **(H1)** and **(H2)**.

(H1) [L,B] is injective modulo the polynomials on $S^{k_1-1}(C_S)$.

(H2) For each $\Gamma \in \mathcal{A}_S$ and $h \in \{1, ..., m_\Gamma\}$, we suppose that

$$t_{il} - m_{il} = cst = \eta_{\Gamma h},$$

for indices i and l such that $\gamma_{il} = \Gamma$ and such that $(B_{\Gamma h}^{il})_0 \neq 0$.

THEOREM 15 *Under the hypotheses (**H1**)-(**H2**), let* $p \in]1, +\infty[$ *and suppose that no eigenvalue of* $\mathcal{A}_S(\lambda)$ *lies on the line* $Re\lambda = k_1 - \frac{2}{p}$ *for all* $S \in \mathcal{S}$, *except possibly at* $k_1 - 1$ *for* $p = 2$ *and let* $[f, g] \in X^{k_1,p}(\Omega)$.

A weak solution $u \in H_v^{\overline{m}}(\Omega)$ *of the problem (3. 1)-(3. 2) admits the following decomposition*

$$u = u_0 + \sum_{S\in\mathcal{S},(\lambda,\mu,k)\in\Lambda_S^1(k_1,p,0)} c_{S,\lambda,\mu,k}\Phi_S\sigma^{S,\lambda,\mu,k} + \sum_{S\in\mathcal{S},(k,\nu)\in\Lambda_S^2(k_1,p)} d_{S,k,\nu}\Phi_S e^{S,k,\nu},$$

where $u_0 \in W^{k_1+\overrightarrow{t},p}(\Omega)$, $c_{S,\lambda,\mu,k}, d_{S,k,\nu} \in \mathbb{C}$, *and* $\Lambda_S^2(k_1,p) = \{(k,\nu) : k \in Z\cap] - \eta - 1, k_1 - \dfrac{2}{p}[$, *such that* $\mathcal{A}_S(k)$ *is not injective modulo the polynomials*$\}$.

REMARK 16 The singular functions $e^{S,k,\nu}$ coming from the polynomial resolution are in the form (5. 9) but with integer λ.

7 THE GENERAL CASE

In this section we give the decomposition for operators with variable coefficients. We suppose here that for all i and l : $m_{il} > 0$.

As usual the singular functions are defined recursively (see [3]). We first write the operators $l_i^{ll'}$ in the form

$$l_i^{ll'}(x, Dx) = \sum_{|\alpha|\le s_{il}+t_{il'}} a_\alpha(x)D_x^\alpha.$$

For $j \in \mathbb{N}, S \in \mathcal{S}$, we set

$$l_{i,j,S}^{ll'}(x, Dx) = \sum_{s_{il}+t_{il'}-j\le|\alpha|\le s_{il}+t_{il'}} \sum_{|\beta|=j+|\alpha|-(s_{il}+t_{il'})} D^\beta a_\alpha(S)\frac{x^\beta}{\beta!}D_x^\alpha.$$

Similarly, for the boundary and transmission operators $B_{\Gamma h}^{il}(x, D_x)$ in the form

$$B_{\Gamma h}^{il}(x, D_x) = \sum_{|\alpha|\le r_{\Gamma h}+t_{il}} a_\alpha(x)D_x^\alpha,$$

we set

$$B_{\Gamma h,j,S}^{il}(x, D_x) = \sum_{r_{\Gamma h}+t_{il}-j\le|\alpha|\le r_{\Gamma h}+t_{il}} \Big(\sum_{|\beta|=j+|\alpha|-(r_{\Gamma h}+t_{il})} D^\beta a_\alpha(S)\frac{x^\beta}{\beta!}D_x^\alpha\Big).$$

Finally we set

$$[L(x), B(x)]_{j,S} = [(\sum_{l'=1}^{N_i} l_{i,j,S}^{ll'}(x, D_x))_{il}, (\sum_{i\in I_\Gamma}\sum_{l=1}^{N_i} B_{\Gamma h,j,S}^{il}(x, D_x))_{\Gamma h}]. \qquad (7.\ 1)$$

Remark that the operator $[L(x), B(x)]_{0,S}$ is equal to $[L_0(S), B_0(S)]$.

DEFINITION 17 *If $\sigma^{S,\lambda,\mu,k}$ is a singular function of $[L_0(S), B_0(S)]$, we construct by iteration on j : $\sigma_0^{S,\lambda,\mu,k} = \sigma^{S,\lambda,\mu,k}$ and $\sigma_j^{S,\lambda,\mu,k}$ is a solution of*

$$[L_0(S), B_0(S)]\sigma_j^{S,\lambda,\mu,k} = -\sum_{p=0}^{j-1}[L(x), B(x)]_{j-p,S}\sigma_p^{S,\lambda,\mu,k}.$$

Analogously we define the fonctions $e_j^{S,k,\nu}$ for $(k,\nu) \in \Lambda_S^2(k_1, 2)$.

Finally we set (for the given integer k_1)

$$\tau^{S,\lambda,\mu,k} = \sum_{Re\lambda + j \leq k_1 - 1} \sigma_j^{S,\lambda,\mu,k},$$

$$E^{S,k,\nu} = \sum_{k+j \leq k_1 - 1} e_j^{S,k,\nu}.$$

THEOREM 18 *Let us suppose that $[L_0(S), B_0(S)]$ is injective modulo the polynomials on $S^{k_1-1}(C_S)$ and that no eigenvalue lies on the line $Re\lambda = k_1 - 1$ with the possible exception for $\lambda = k_1 - 1$.*

Let $u \in V$ be a solution of the problem $[L(x), B(x)] = [f, 0]$ with f in $H^{k_1 - \vec{s}}(\Omega)$. Then u admits in a neighborhood of S the following decomposition:

$$u = u_0 + \sum_{(\lambda,\mu,k)\in\Lambda_{S,\iota}^1} c_{S,\lambda,\mu,k}\Phi_S\tau^{S,\lambda,\mu,k} + \sum_{(k,\nu)\in\Lambda_S^2(\iota,2)} d_{S,k,\nu}\Phi_S E^{S,k,\nu},$$

where $u_0 \in H^{k_1 + \vec{t}}(\Omega)$, the coefficients $c_{S,\lambda,\mu,k}$ and $d_{S,k,\nu}$ being in $\mathbb{C}$.

PROOF For a fixed vertex S, we show that the principal part frozen at S is a Fredholm operator (between the usual Sobolev spaces) in C_S.Then, we use perturbation arguments with the adaptation to systems (see [3]).■

8 NUMERICAL EXAMPLES

The goal of this section is to present some computational results on the singular exponents of example 2.1, which are computed with the help of the Newton method (for other methods see [15, 16]).

EXAMPLE 19 Movement of a body constituted of 2 elastic interconnected membranes (see section 2.1). The Poisson coefficients of the two rectangular bodies Ω_1 and Ω_2 are respectively ν_1 and ν_2. We recall that α is the angle between Ω_1 and Ω_2 and we consider a vertex S common to Ω_1 and Ω_2 (see Figure 1).

In each cone $C_k, k = 1, 2$, we may suppose that S is at the origin and use polar-coordinates centered at S. We introduce the local basis

$$\mathbf{e}_r = \begin{pmatrix} \cos\omega \\ \sin\omega \end{pmatrix}, \quad \mathbf{e}_\omega = \begin{pmatrix} -\sin\omega \\ \cos\omega \end{pmatrix},$$

and write the displacement vector $\mathbf{u}^k$ in the plane of Ω_k in the form

$$\mathbf{u}^k = u_r^k(r,\omega)\mathbf{e}_r + u_\omega^k(r,\omega)\mathbf{e}_\omega.$$

The eigensolutions $((\widehat{u}_r^k, \widehat{u}_\omega^k, \widehat{u}_3^k))_{k=1,2}$ associated with the eigenvalue $\lambda - 2$ satisfy the following equations: The differential equations are

$$C_k \frac{\partial^2 \widehat{u}_r^k(\omega)}{\partial \omega^2} + (\lambda^2 - 1)\widehat{u}_r^k(\omega) + [(\lambda - 1) - C_k(\lambda + 1)]\frac{\partial \widehat{u}_\omega^k(\omega)}{\partial \omega} = 0, \quad (8.\ 1)$$

$$\frac{\partial^2 \widehat{u}_\omega^k(\omega)}{\partial \omega^2} + C_k(\lambda^2 - 1)\widehat{u}_\omega^k(\omega) + [(\lambda + 1) - C_k(\lambda - 1)]\frac{\partial \widehat{u}_r^k(\omega)}{\partial \omega} = 0, \quad (8.\ 2)$$

$$\lambda^2 \widehat{u}_3^k(\omega) + \frac{\partial^2 \widehat{u}_3^k(\omega)}{\partial \omega^2} = 0, \quad (8.\ 3)$$

for $k = 1, 2$, where we have set $C_k = \frac{1-2\nu_k}{2(1-\nu_k)}$.

The external boundary condition are:

$$\widehat{u}_r^1(\frac{\pi}{2}) = \widehat{u}_\omega^1(\frac{\pi}{2}) = \widehat{u}_3^1(\frac{\pi}{2}) = 0, \quad (8.\ 4)$$

$$\widehat{u}_r^2(\frac{\pi}{2}) = \widehat{u}_\omega^2(\frac{\pi}{2}) = \widehat{u}_3^2(\frac{\pi}{2}) = 0. \quad (8.\ 5)$$

The transmission conditions are :

$$\begin{cases} \widehat{u}_r^1(0) &= \widehat{u}_r^2(0)\,, \\ \widehat{u}_\omega^1(0) &= \cos\alpha\ \widehat{u}_\omega^2(0) + \sin\alpha\ \widehat{u}_3^2(0), \\ \widehat{u}_3^1(0) &= -\sin\alpha\ \widehat{u}_\omega^2(0) + \cos\alpha\ \widehat{u}_3^2(0), \end{cases} \quad (8.\ 6)$$

$$\begin{cases} \widehat{\sigma}_{r\omega}^1(0) &= -\widehat{\sigma}_{r\omega}^2(0), \\ \widehat{\sigma}_{\omega\omega}^1(0) &= -\cos\alpha\ \widehat{\sigma}_{\omega\omega}^2(0) - \mu_2 \sin\alpha \frac{\partial \widehat{u}_3^2(0)}{\partial \omega}, \\ \mu_1 \frac{\partial \widehat{u}_3^1(0)}{\partial \omega} &= +\sin\alpha\ \widehat{\sigma}_{\omega\omega}^2(0) - \mu_2 \cos\alpha \frac{\partial \widehat{u}_3^2(0)}{\partial \omega}, \end{cases} \quad (8.\ 7)$$

where

$$\widehat{\sigma}_{r\omega}^k(\omega) = \mu_k[\frac{\partial \widehat{u}_r^k(\omega)}{\partial \omega} + (\lambda - 1)\widehat{u}_\omega^k(\omega)],$$

$$\widehat{\sigma}_{\omega\omega}^k(\omega) = 2\mu_k[\frac{\partial \widehat{u}_\omega^k(\omega)}{\partial \omega} + \widehat{u}_r^k(\omega)] + \lambda_k[(\lambda + 1)\widehat{u}_r^k(\omega) + \frac{\partial \widehat{u}_\omega^k(\omega)}{\partial \omega}],$$

λ_k and μ_k being the corresponding Lamé coefficients.

The equations (8. 1) and (8. 2) have for solutions when $\lambda \neq 0$ and $3+\lambda-4\nu_k \neq 0$

$$\widehat{u}_r^k(\omega) = c_{1k}(\lambda - 1)\cos(\lambda + 1)\omega + c_{2k}(\lambda - 1)\sin(\lambda + 1)\omega + c_{3k}d_k(\lambda + 1)\cos(\lambda - 1)\omega - c_{4k}d_k(\lambda + 1)\sin(\lambda - 1)\omega,$$

$$\widehat{u}_\omega^k(\omega) = c_{1k}(\lambda - 1)\sin(\lambda + 1)\omega + c_{2k}(\lambda - 1)\cos(\lambda + 1)\omega - c_{3k}(\lambda + 1)\sin(\lambda - 1)\omega + c_{4k}(\lambda + 1)\cos(\lambda - 1)\omega,$$

with $d_k = \frac{3-\lambda-4\nu_k}{3+\lambda-4\nu_k}$. The solutions of equations (8. 3) have the form

$$\widehat{u}_3^k(\omega) = c_{5k}\cos\lambda\omega + c_{6k}\sin\lambda\omega.$$

Replacing these solutions in the boundary and transmission conditions we obtain a homogeneous linear system with 12 equations and 12 unknows. This system admits a non-trivial solution if and only if the corresponding determinant vanishes. The complex numbers λ which cancel this determinant (the eigenvalues) are then approximated by the Newton method. Figure 5 represents the real part of the eigenvalues when α grows from $0°$ to 360^0 and for $\nu_1 = \nu_2 = 0.17$ (thick lines corresponding to complex eigenvalues).

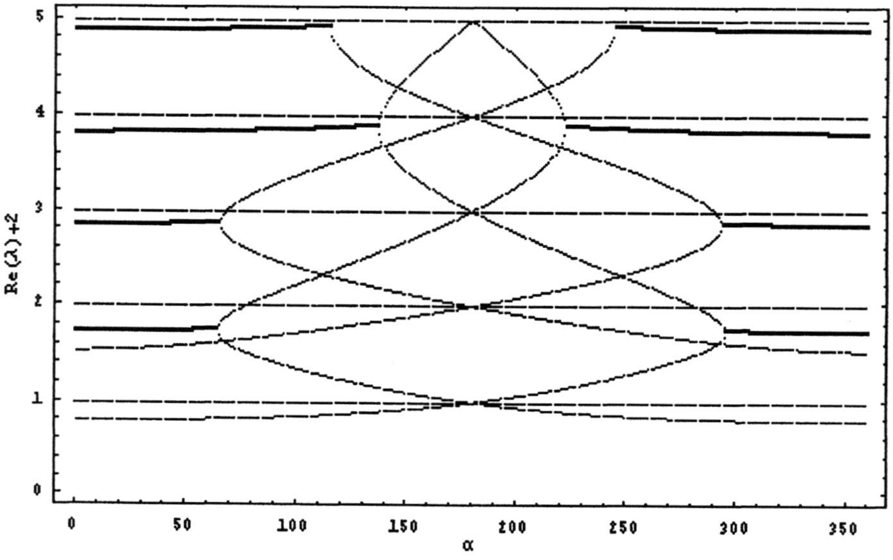

Figure 5:

9 THE DUAL SINGULAR FUNCTION EQUAL TO THE CONSTANT

The choice of V as subspace of $H_v^m(\Omega)$ is justified by Green's formula introduced in the general case. This choice is the natural one in the example presented previously. But in some cases, the natural variational space is $\widetilde{V}$. An interesting question is then to know the consequence of the use of V instead of $\widetilde{V}$. Let us clarify this point on the following example:

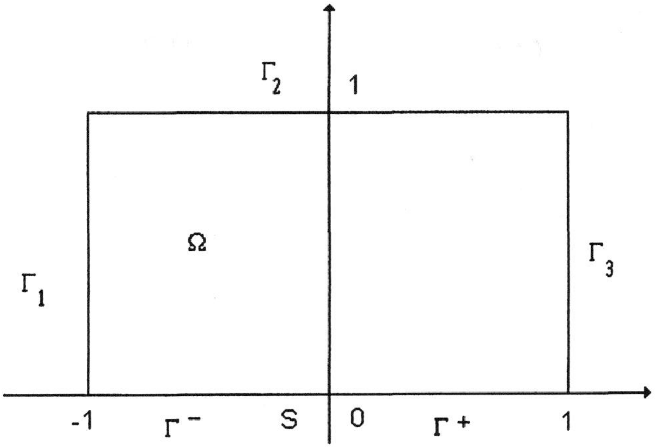

Figure 6:

EXAMPLE 20

Let $\Omega =]-1,1[\times]0,1[$ and take S as the point $(0,0)$ (see Figure 6). The considered problem is

$$\begin{aligned} \rho\Delta^2 u &= f \text{ on } \Omega, && (9.\ 1)\\ Mu &= Nu = 0 \text{ on } \Gamma^-, && (9.\ 2)\\ Mu &= Nu = 0 \text{ on } \Gamma^+, && (9.\ 3)\\ u &= \frac{\partial u}{\partial \nu} = 0 \text{ on } \Gamma_i, i = 1,2,3, && (9.\ 4) \end{aligned}$$

where the boundary operators are defined by

$$\begin{aligned} Mu &= \rho(\sigma\Delta u + (1-\sigma)\frac{\partial^2 u}{\partial \nu^2}),\\ Nu &= \rho(\frac{\partial \Delta u}{\partial \nu} + (1-\sigma)\frac{\partial^3 u}{\partial \nu \partial \tau^2}), \end{aligned}$$

with $\rho = \frac{E}{1-\sigma^2}$. ($E > 0$ and $\sigma \in]0,1[$ are respectively the Young modulus and the Poisson coefficient of the constitutive plate Ω). In this case, we have

$$\begin{aligned} V &= \{u \in H^2(\Omega) \ / \ u(S) = 0 \text{ and } u = \frac{\partial u}{\partial \nu} = 0 \text{ on } \Gamma_i, i = 1,2,3\},\\ \widetilde{V} &= \{u \in H^2(\Omega) \ / \ u = \frac{\partial u}{\partial \nu} = 0 \text{ on } \Gamma_i, i = 1,2,3\}, \end{aligned}$$

$$\begin{aligned} a(u,v) &= \rho \int_\Omega (\Delta u.\Delta \overline{v} - (1-\sigma)\{\frac{\partial^2 u}{\partial x^2}\frac{\partial^2 \overline{v}}{\partial x^2}\\ &+ \frac{\partial^2 u}{\partial y^2}\frac{\partial^2 \overline{v}}{\partial y^2} - 2\frac{\partial^2 u}{\partial x \partial y}\frac{\partial^2 \overline{v}}{\partial x \partial y}\})dxdy, \forall u,v \in H^2(\Omega). \end{aligned}$$

Let $f \in L^2(\Omega)$, the problem associated with (9. 1)-(9. 4) is to find $\widetilde{u} \in \widetilde{V}$ solution of

$$a(\widetilde{u}, \widetilde{v}) = \int_\Omega f\, \overline{\widetilde{v}}\, dxdy,\ \forall \widetilde{v} \in \widetilde{V}. \tag{9. 5}$$

By Remark 4, setting $u = \widetilde{u} - \Phi_S \widetilde{u}(0)$, $u \in V$ is solution of

$$a(u, v) = \int_\Omega g\, v\, dxdy + \int_{\Gamma^- \cup \Gamma^+} h \frac{\partial v}{\partial \nu},\ \forall v \in V, \tag{9. 6}$$

where $g = f - \Delta^2(\widetilde{u}(0)\Phi_S)$ et $h = -M(\widetilde{u}(0)\Phi_S)$. By Theorem 15, u admits the decomposition

$$u = u_{reg} + c_S \Phi_S \sigma^S, \tag{9. 7}$$

where $u_{reg} \in H^4(\Omega) \cap V$ has the optimal regularity and σ^S is the unique singular function corresponding to the vertex S given by

$$\sigma^S(r, \theta) = r^2(-c + \frac{c^2}{2} - \frac{3}{2}\cos 2\theta - \theta \sin 2\theta) + r^2 \ln r(c + \cos 2\theta), \tag{9. 8}$$

with $c = \frac{1-\sigma}{1+\sigma}$. Consequently we obtain $\widetilde{u} = \widetilde{u}_{reg} + c_S \Phi_S \sigma^S$ with $\widetilde{u}_{reg} \in H^4(\Omega)$.

It is clear that according to the general theory, one must have $c_S = 0$. This follows from the following lemma:

LEMMA 21 *The solution* $\widetilde{u} = \widetilde{u}_{reg} + c_S \Phi_S \sigma^S$ *satisfies*

$$\int_\Omega \Delta^2 \widetilde{u}\, \overline{\widetilde{v}} dxdy = a(\widetilde{u}, \widetilde{v}) + c_S k \overline{\widetilde{v}}(0), \forall \widetilde{v} \in \widetilde{V}. \tag{9. 9}$$

where $k \neq 0$.

PROOF Let $\mathcal{D}_\varepsilon$ be the half circle with center S included in Ω, with radius ε. Applying Green's formula in $\Omega \backslash \mathcal{D}_\varepsilon$, we obtain

$$\begin{aligned} \int_\Omega \Delta^2(\Phi_S \sigma^S)\overline{\widetilde{v}} = a(\Phi_S \sigma^S, \widetilde{v}) &+ \int_{\Gamma^-} (M(\Phi_S \sigma^S)\tfrac{\partial \overline{\widetilde{v}}}{\partial \nu} - N(\Phi_S \sigma^S)\overline{\widetilde{v}})d\tau \\ &+ \int_{\Gamma^+} (M(\Phi_S \sigma^S)\tfrac{\partial \overline{\widetilde{v}}}{\partial \nu} - N(\Phi_S \sigma^S)\overline{\widetilde{v}})d\tau \\ &+ \lim_{\varepsilon \to 0} \int_{\partial \mathcal{D}_\varepsilon} \mathcal{M}\sigma^S(\varepsilon, \theta)(-\tfrac{\partial \overline{\widetilde{v}}}{\partial r}(\varepsilon, \theta))\varepsilon d\theta \\ &+ \lim_{\varepsilon \to 0} \int_{\partial \mathcal{D}_\varepsilon} -\mathcal{N}\sigma^S(\varepsilon, \theta)\overline{\widetilde{v}}(\varepsilon, \theta)\varepsilon d\theta, \end{aligned}$$

$\mathcal{M}$, $\mathcal{N}$ being respectively operators of order 2 and 3. Due to the form of σ^S, it is clear that $\lim_{\varepsilon \to 0} \int_{\partial \mathcal{D}_\varepsilon} \mathcal{M}\sigma^S(\varepsilon, \theta)(-\frac{\partial \overline{v}}{\partial r}(\varepsilon, \theta))\varepsilon d\theta = 0$. Now, thanks to the Sobolev imbedding theorem, we obtain for a fixed $\alpha \in]0, 1[$ the following estimate

$$|v(\varepsilon, \theta) - v(0)| \leq cst\ \varepsilon^\alpha.$$

Thus, $\lim_{\varepsilon\to 0} \int_{\partial\mathcal{D}_\varepsilon} -\mathcal{N}\sigma^S(\varepsilon,\theta)\overline{v}(\varepsilon,\theta)\varepsilon d\theta = \lim_{\varepsilon\to 0} \int_0^\pi (h_1(\theta)+\ln\varepsilon h_2(\theta))d\theta\,\overline{\widetilde{v}}(0)$, where we have set $-\mathcal{N}\sigma^S(\varepsilon,\theta) = h_1(\theta) + \ln\varepsilon h_2(\theta)$. A computation gives $\int_0^\pi h_1(\theta)d\theta = 4(\frac{2+c}{1+c})\pi = k$ and $h_2(\theta) = 0$, so we obtain the identity (9. 9).

Notice that the function $v = 1$ is the dual function of σ^S, therefore k is different from 0 due to Theorem 3.1 of [4].■

The consequence of this lemma is the following one: Since $\widetilde{u}$ is solution of (9. 5) we readily obtain $c_S = 0$. In other words, the resolution that we obtain in using the space V as variational space (instead of $\widetilde{V}$) seems to introduce a singularity which in fact does not exist when we solve the problem in the space $\widetilde{V}$. Conversely, the solution $u \in V$ of (9. 6) with any data $g \in L^2(\Omega)$ and $h \in H^{\frac{3}{2}}(\Gamma^- \cup \Gamma^+)$ is not solution of (9. 6) when we take the test function v in $\widetilde{V}$ and this solution may have a singularity in S.

References

[1] S. Agmon, A. Douglis and L. Nirenberg, Estimates near the boundary for solutions of elliptic partial differential equations satisfying general boundary conditions I, II, Com. on Pure and Applied Math., 12, 1959, 623-727; 17, 1964, 35-92.

[2] V. A. Kondratiev, Boundary value problems for elliptic equations in domains with conical or angular points, Trans. Moscow Math. Soc., 16, 1967, 227-313.

[3] M. Dauge, Elliptic Boundary Value Problems in Corner Domains. Smoothness And Asymptotics of Solutions, Lecture Notes in Math., Vol. 1341, Springer, Berlin 1988.

[4] V. G. Maz'ya and B. A. Plameneskii, Coefficients in the asymptotics of the solutions of elliptic boundary value problems in a cone, J. of Soviet Math., 9, 1978, 750-764.

[5] J. E. Lagnese, G. Leugering and E.J.P.G. Schmidt, Modeling, analysis and control of dynamic elastic multi-link structures, Birkhäuser, Boston, 1994..

[6] H. J.-P. Morand and R. Ohayon, Interactions Fluides-Structures, RMA 13, Masson, Paris, 1992.

[7] A. Maghnouji and S. Nicaise, On a coupled problem between the plate equation and the membrane equation on polygons, Annales de la Faculté des Sciences de Toulouse, Vol. 2, n°2, 1992, 187-209.

[8] G. Lumer, Espaces ramifiés et diffusions sur les réseaux topologiques, C. R. Acad. Sc. Paris, Série A, 291, 1980, 627-630.

[9] S. Nicaise, Polygonal interface problems, Series "Methoden und Verfahren der Mathematischen Physik", 39, Peter Lang Verlag, 1993.

[10] J.-L. Lions and E. Magenes, Problèmes aux limites non homogènes et applications, T. 1, Dunod, 1968.

[11] D. Mercier and S. Nicaise, Existence results for general systems of differential equations on one-dimensional networks and prewavelets approximation, Discrete and Continuous Dynamical Systems, Vol. 4, N°2, 1998, 273-300.

[12] M. S Agranovitch and M. I Vishisk, Elliptic problems with a parameter and parabolic problems of general type, Russian Math. Surveys, 19, 1964, 53-157.

[13] V. G. Maz'ya and B.A. Plameneskii, Estimates in Lp and in Hölder classes and the Miranda-Agmon maximum principle for solutions of elliptic boundary value problems in domains with singular points on the boundary , Amer. Math. Soc.Trans., 123 , 1984, 1-56.

[14] S. Nicaise and A. M Sändig, General Interface Problems-II, Mathematical Methods in the Applied Sciences, 17, 1994, 431-450.

[15] S. Nicaise and A. M Sändig, General Interface Problems-I, Mathematical Methods in the Applied Sciences, 17, 1994, 395-429.

[16] M. Costabel, M. Dauge and Y. Lafranche, Fast semi-analytic computation of elastic edge singularities, To appear in Computer Methods in Applied Mechanics and Engineering.

About a geometrical approach to multistructures and some qualitative properties of solutions

O.M.PENKIN, Voronezh State University, Russia.
e-mail penkin@omp.vsu.ru, omp@wowmail.com

1 INTRODUCTION

In this paper we shall describe an approach to create mathematical models of certain multistructures. Roughly speaking, a multistructure is a system consisting of elements of various physical or geometrical nature. An example of a multistructure is a system consisting of finite number of strings and membranes.

The regular study of such systems began in the eighties with the work of G.Lumer, J. von Below, S. Nicaise, F. Ali Mehmeti and Ju.V. Pokornyi [11, 12, 4, 2, 18, 5].

The most part of the authors prefer to describe behaviour of multistructure in terms of an operator, defined on a linear manifold selected by transmission conditions in some Hilbert space. The detailed description of such approaches can be found in the papers mentioned above. On the basis of this approach numerous results concerning solvability of boundary value problems, the spectrum of corresponding operators etc. were obtained.

The idea to define the Laplace operator on multistructures is an old question and was introduced for instance in [11, 12, 4, 18, 1, 3, 6], where some Green's formulae were obtained. Thus it is natural to expect the validity of maximum principles for such operators. In special cases, this has be done in [13, 18, 5, 6]. Our idea is to define the Laplace – Beltrami operator on multistructures using a geometrical approach: the multistructure is embedded in $\mathbf{R}^n$, $n \geq 2$, which allows to extend in a natural way the notions of the divergence operator and the Laplace – Beltrami operator. Thus Green's formulae and maximum principles are easily deduced.

2 STRATIFIED SETS

On the next figure we see an example of a stratified set.

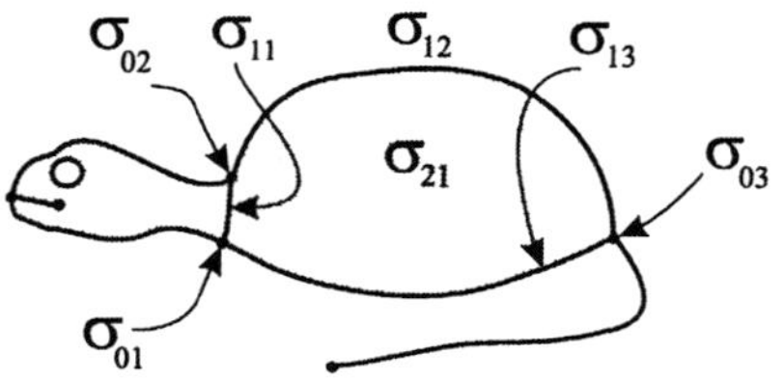

Figure 1: An example of a stratified set

In the general case a stratified set is a connected set Ω in $\mathbf{R}^n$, consisting of a finite number of smooth manifolds σ_{ki} $(k \leq n,\ i = 1, 2, \dots, \nu(k))$, which we call strata. The following conditions describing a disposition of strata in Ω are assumed to be fulfilled:

a) $\partial\sigma_{ki} = \overline{\sigma}_{ki} \setminus \sigma_{ki}$ is a union of the strata. For example $\partial\sigma_{21}$ on the above figure is equal to $\cup\sigma_{ij}$ for $i \in \{0; 1\}$ and $j \in \{1; 2; 3\}$.

b) if $\sigma_{k-1,i} \prec \sigma_{kj}$ (it means $\sigma_{k-1,i} \subset \overline{\sigma}_{kj}$) and a point $y \in \sigma_{kj}$ tends to $x \in \sigma_{k-1,i}$ then a tangent space $T_y\sigma_{kj}$ tends to some limit position $\lim\limits_{y \to x} T_y\sigma_{kj} = T_{x+}\sigma_{kj}$ and $T_{x+}\sigma_{kj} \supset T_x\sigma_{k-1,i}$.

For the flat case on the above figure these conditions are fulfilled obviously. A condition *b*) means that we exclude some "singular" situations. The next figure gives an example of such situation. In addition we assume that each set $\overline{\sigma}_{ki}$ is compact in $\mathbf{R}^n$.

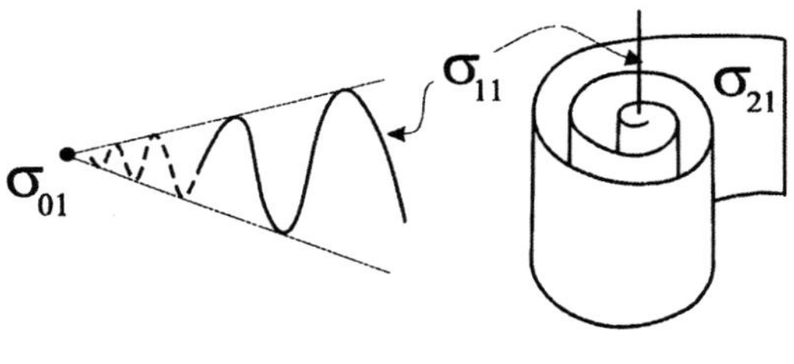

Figure 2: Impossible situations

A set Ω is not a manifold even at locally. More exactly there exists at least one point any neighborhood of which is not a manifold. Hence, it appears impossible to define local coordinates on Ω. But for each $x \in \sigma_{k-1,i}$ and for each $\sigma_{kj} \succ \sigma_{k-1,i}$ we can introduce coordinates near the point x, defining them on $\sigma_{k-1,i} \cup \sigma_{kj}$ as on the manifold with a boundary; we assume that coordinates $y^1, \dots, y^k$ are chosen in such way that $y^k \equiv 0$ on $\sigma_{k-1,i}$ and positive in the interior of σ_{kj}.

When σ_{kj} adjoins to $\sigma_{k-1,i}$ twice (see the left picture below) we define the local coordinates on a neighborhood U separately on its "left" and "right" parts.

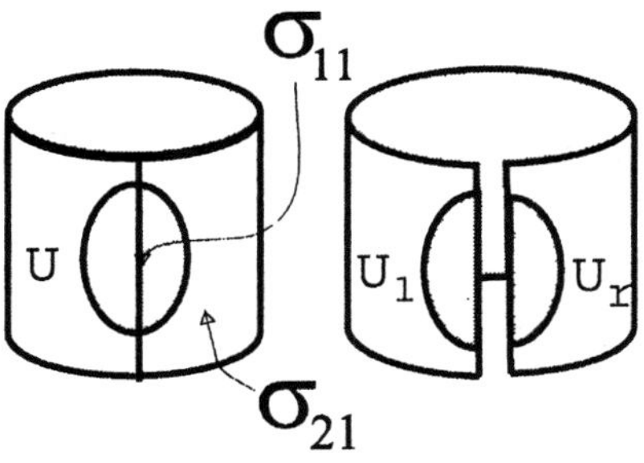

Figure 3: Separation of the neighborhood

3 DIVERGENCE OPERATOR

We shall introduce a specific definition of a divergence motivated by the following simple example.

Let us imagine a three-dimensional space divided in two open parts V_1, V_2 by means of a sufficiently smooth surface S. Besides, imagine some flow $\vec{F}$ of a hypothetical liquid with a following "strange" property - for $x \in S$ a vector $\vec{F}(x)$ lies in a tangent space T_xS, but in V_1 and V_2 a direction of $\vec{F}$ may be arbitrary (for the real liquid $\vec{F}(y)$ tends to $\vec{F}(x)$ when $y \in V_i$ tends to $x \in S$). Our purpose now is to calculate how much liquid flows through the surface of a small element $\Pi \ni x$ per unit of time (see the next picture).

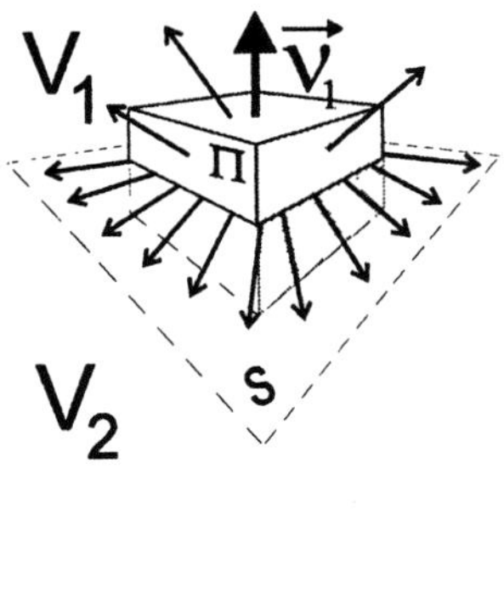

Figure 4: Tangent vector field

We assume $\vec{F}(x)$ to be proportional to m^{-2} in $V_1 \cup V_2$ and m^{-1} in S. One part of the flow, which corresponds to the boundary of $\Pi \cap S$ is approximately equal to $(\widetilde{\nabla}\vec{F})(x)\Delta s$, where Δs is a surface area of $\Pi \cap S$ and $(\widetilde{\nabla}\vec{F})(x)$ is a usual 2-dimensional divergence of the restriction of $\vec{F}(x)$ to S. The other part of the flow is approximately equal to $[(\vec{F_1} \cdot \vec{\nu}_1)(x) + (\vec{F_2} \cdot \vec{\nu}_2)(x)]\Delta s$, where $\vec{F_i}(x)$ is a limit $\lim\limits_{y \to x} \vec{F}(y)$ when y tends to x from the interior of V_i. The sum of both parts of the flow gives us

$$\Phi_\Pi(\vec{F})(x) = (\tilde{\nabla}\vec{F} + \vec{F}_1 \cdot \vec{\nu}_1 + \vec{F}_2\vec{\nu}_2)(x)\Delta s + o(\Delta s).$$

Now we define a "stratified" measure of Π as a sum $\mu(\Pi) = \Delta s + \Delta V$, where ΔV is a usual volume of Π, i.e. $\Delta V = \Delta s \cdot h$. As in the classical case let us define a 3-dimensional divergence at x by means of

$$(\nabla\vec{F})(x) = \lim_{\Pi\to x} \frac{\Phi_\Pi(\vec{F})}{\mu(\Pi)}.$$

It follows

$$(\nabla\vec{F})(x) = (\tilde{\nabla}\vec{F})(x) + \sum_{i=1}^{2}(\vec{F}_i \cdot \vec{\nu}_i)(x).$$

We can obtain something similar for the general case. Let Ω_0 be an open connected subset of Ω consisting of its strata (below we use a topology in Ω, generated by the inclusion $\Omega \subset \mathbf{R}^n$) and such that $\overline{\Omega}_0 = \Omega$. Let $\vec{F}$ be a tangent vector field defined on Ω_0. It means that for each σ_{ki} we have an implication $x \in \sigma_{ki} \Rightarrow \vec{F}(x) \in T_x\sigma_{ki}$. The divergence of a vector field $\vec{F}$ at $x \in \sigma_{k-1,i} \subset \Omega_0$ is defined by

$$(\nabla\vec{F})(x) = (\tilde{\nabla}\vec{F})(x) + \sum_{\sigma_{kj}\succ\sigma_{k-1,i}} (\vec{F}\cdot\vec{\nu})_{\overline{kj}}(x). \tag{3.1}$$

The notation $f_{\overline{kj}}$ means the following. Let f_{kj} be a usual restriction f to σ_{kj}. If f_{kj} admits an extension to $\overline{\sigma}_{kj}$ by continuity then this extension will be denoted by $f_{\overline{kj}}$. When σ_{kj} adjoins to $\sigma_{k-1,i}$ as it is shown on the Fig.3 the function $f_{\overline{kj}}$ has two values at the points of $\sigma_{k-1,i}$, but we can imagine it to be single-valued by separation of left and right parts as indicated in Fig.3.

Because we have now a notion of the divergence we can define an analog of the Laplace – Beltrami operator. After some preparations we will do it in the next section.

4 LAPLACE – BELTRAMI OPERATOR ON Ω_0

At first we present a short list of the functional spaces.

1. We write $f \in C_\sigma^m(\Omega_0)$ if for each $\sigma_{ki} \subset \Omega_0$ the restriction f_{ki} has continuous derivatives up to order m which admit an extension by continuity onto $\overline{\sigma}_{km} \setminus \partial\Omega_0$. It should be noted that a derivative $f^{(i)}$ is defined only in some chart U_α and the last condition is meaningful only when $\overline{U}_\alpha \cap \partial\sigma_{ki} \neq \emptyset$.

2. $\overline{C}_\sigma^m(\Omega_0)$ is defined like above, but now we do not exclude the points of $\partial\Omega_0$.

3. $C(\Omega)$ is a set of continuous functions on Ω.

We also use inclusions like $\vec{F} \in C_\sigma^m(\Omega_0)$ (and so on) for the vector fields on Ω_0. It means that each component of $\vec{F}$ in the local coordinates is a restriction of some function $f \in C_\sigma^m(\Omega_0)$.

Using the divergence (3.1) introduced above we can define the Laplace – Beltrami operator for a function $u \in C^2_\sigma(\Omega_0)$ by means of the formulae

$$(\Delta_p u)(x) = (\nabla(p\nabla u))(x) + \sum_{\sigma_{kj} \succ \sigma_{k-1,i}} (p\nabla u \cdot \vec{\nu})_{\overline{kj}}(x).$$

5 INTEGRAL IDENTITIES

Let $G \subset \Omega$ and μ_{ki} be a k-dimensional Lebesgue measure on σ_{ki}. The measure of G is defined by

$$\mu(G) = \sum_{\sigma_{ki}} \mu_{ki}(G \cap \sigma_{ki}). \tag{5.2}$$

We call G μ-measurable if each intersection in (5.2) is Lebesgue measurable. Using the measure defined by (5.2) we can define Lebesgue – Stieltjes integral of a measurable function (as usual f is measurable function if $f^{-1}(Q)$ is μ-measurable for any Lebesgue measurable set in $\mathbf{R}^1$). For a function $f \in C_\sigma(\Omega)$ we have

$$\int_\Omega f\, d\mu = \sum \int_{\sigma_{kj}} f\, d\mu_{kj}.$$

The following theorem is an exact analog of the classical divergence theorem. For $x \in \sigma_{k-1,i} \subset \partial\Omega_0$ we define

$$\vec{F}_\nu(x) = \sum (\vec{F} \cdot \vec{\nu})_{\overline{kj}}(x). \tag{5.3}$$

Here the sum is taken over all $\sigma_{kj} \succ \sigma_{k-1,i} \not\subset \partial\Omega_0$. The sum (5.3) is an analog of the normal projection of $\vec{F}$ at $x \in \partial\Omega_0$.

THEOREM 5.1 *Let $\vec{F} \in \overline{C}^1_\sigma(\Omega_0)$ and each $\overline{\sigma}_{ki}$ be the orientable manifold, then*

$$\int_{\partial\Omega_0} \vec{F}_\nu\, d\mu = \int_{\Omega 0} \nabla\vec{F}\, d\mu. \tag{5.4}$$

For $\partial\Omega_0 = \emptyset$ the integral in the left-hand side of (5.4) is equal to 0 by convention. Such convention is natural because one can show that the right-hand side is equal to 0 in this case.

Using theorem 5.1 we obtain, as usual, the exact analogs of the classical Green's formulae.

THEOREM 5.2 *Let $u \in \overline{C}^2_\sigma(\Omega_0)$, $v \in \overline{C}^1_\sigma(\Omega_0) \cap C(\Omega)$, then*

$$\int_{\Omega_0} v\Delta_p u\, d\mu = -\int_{\Omega_0} p\nabla v \nabla u\, d\mu - \int_{\partial\Omega_0} v(p\nabla u)_\nu\, d\mu. \tag{5.5}$$

THEOREM 5.3 *If $u, v \in \overline{C}^2_\sigma(\Omega_0) \cap C(\Omega)$, then*

$$\int_{\Omega_0} (v\Delta_p u - u\Delta_p v)\, d\mu = \int_{\partial\Omega_0} (u(p\nabla v)_\nu - v(p\nabla u)_\nu)\, d\mu. \tag{5.6}$$

If we take u such that $\Delta_p u = 0$ and assume $v \equiv p \equiv 1$ then (5.6) implies

$$\int\limits_{\partial\Omega_0} (\nabla u)_\nu \, d\mu = 0.$$

That is an analog of the classical identity

$$\int\limits_S \frac{\partial u}{\partial \nu} \, ds = 0$$

for the harmonic functions.

Let us assume $L_q u = \Delta_p u - qu$, $q \in C_\sigma(\Omega_0)$. Using the identity $vL_q u - uL_q v = v\Delta_p u - u\Delta_p v$ and assuming the conditions of the theorem 5.3 are fulfilled we obtain

$$\int\limits_{\Omega_0} (vL_q u - uL_q v) \, d\mu = \int\limits_{\partial\Omega_0} (u(p\nabla v)_\nu - v(p\nabla u)_\nu) \, d\mu. \tag{5.7}$$

6 INCOMPATIBLE INEQUALITIES

In this section we present some consequences of Green's formulae.

THEOREM 6.1 *Let Ω be orientable stratified set and $u \in C^2(\Omega_0) \cap \overline{C}^1(\Omega_0) \cap C(\Omega)$ be a nonnegative solution of the inequality $L_q u \geq 0$ in Ω_0 vanishing at $\partial\Omega_0$, satisfying $(p\nabla u)_\nu > 0$ on $\partial\Omega_0$. Besides, let $v \in C^2(\Omega_0) \cap \overline{C}^1(\Omega_0) \cap C(\Omega)$ be a positive solution of the inequality $L_q v \leq 0$ in Ω_0. Then both u and v are the solutions of the following boundary value problem:*

$$L_q w = 0, \tag{6.8}$$

$$w\Big|_{\partial\Omega_0} = 0. \tag{6.9}$$

As a consequence, the inequalities $L_q u > 0$, $L_q v \leq 0$ or $L_q u \geq 0$, $L_q v < 0$ are incompatible.

It is well known that the inequality $\Delta u \geq 0$, where Δ is the Laplace – Beltrami operator, defined on a compact Riemannian manifold, admits only the constant solutions (see [10]). The following theorem shows that the stratified set Ω_0 without boundary (i.e. $\Omega = \Omega_0$) is an analog of the compact Riemannian manifold.

THEOREM 6.2 *Let Ω be orientable, $\partial\Omega_0 = \emptyset$ (i.e. $\Omega_0 = \Omega$). Then the inequality $\Delta_p u \geq 0$ in $C^2(\Omega_0) \cap C(\Omega)$ admits only constant solutions.*

Moreover, one can prove a more general result:

THEOREM 6.3 *Let $q \geq 0$ and $\partial\Omega_0 = \emptyset$. If an inequality $L_q u \geq 0$ has a positive solution $u \in C^2(\Omega_0) \cap C(\Omega)$, then $u \equiv const$ on Ω and $q \equiv 0$.*

COMMENT 6.1 *One can prove that some conditions in theorem 6.1 may be omitted (orientability of Ω and positivity of $(p\nabla u)_\nu$), but the corresponding proof cannot be based only on Green's formulae.*

7 WEAK MAXIMUM PRINCIPLE

Now we assume q to be nonnegative.

In the classical case it is impossible for a nonconstant solution of the inequality

$$(\Delta_p u - qu)(x) \geq 0 \quad (x \in \Omega_0) \tag{7.10}$$

to have a local nonnegative maximum in the interior of Ω_0. But in our case it is not true; for example, a solution of the equation $\Delta u = 0$ may be constant in some n-dimensional stratum with $n \geq 1$, but $u \not\equiv const$ entirely in Ω_0. As a consequence we need some new formulation of a strong maximum principle (see the next section). Nevertheless, a weak maximum principle holds in the classical form.

THEOREM 7.1 *Let u be a solution of (7.10). Then*

$$\sup_{\Omega_0} u \leq \sup_{\partial\Omega_0} u^+ \quad (u^+ = \max\{u; 0\}).$$

For 2-dimensional case (i.e. when the maximal dimension of the strata in Ω is equal two) this assertion was proved in [15], but that proof could not be extended to the general case. The proof based on the scheme in [8] is general but rather complicated.

As a consequence of the last theorem we obtain the uniqueness theorem for the boundary value problem

$$\Delta_p u - qu = f,$$

$$u\Big|_{\partial\Omega_0} = \phi,$$

and the continuous dependence of its solutions on ϕ and f.

8 STRONG MAXIMUM PRINCIPLE

If u achieves a local maximum at $\xi \in \Omega_0$ and $u \not\equiv const$ in any neighborhood of ξ then ξ is called a point of nontrivial maximum of u. We say also that u has a nontrivial maximum at ξ. Next assertion is an analog of the classical strong maximum principle.

THEOREM 8.1 *Any solution $u \in C^2(\Omega_0)$ of the inequality (7.10) hasn't a nontrivial and nonnegative local maximum in Ω_0.*

For a 2-dimensional case this assertion was proved in [7]. The considerations of that paper are applicable to the general case. In spite of the fact that we use some classical ideas, the result is very difficult. We based our proof on the analog of Hopf lemma (see [9]) about the normal derivative of the solution (7.10) at the point of its maximum, which lies at the boundary. In our case an analog of the normal derivative is an expression $(p\nabla u)_\nu$.

THEOREM 8.2 *Let u be a solution of (7.10) which has at $\xi \in \partial\Omega_0$ nonnegative and nontrivial maximum. Then $(p\nabla u)_\nu < 0$.*

It is rather trivial to prove a strong maximum principle for the strict inequality $\Delta_p u - qu > 0$. Moreover, we can present a more exotic assertion.

THEOREM 8.3 *Let $\Delta_p u - qu > 0$. Than u has no point of nontrivial descent in Ω_0.*

A point $\xi \in \sigma_{k-1,i}$ is called a point of descent for a function u if $u(x) \leq u(\xi)$ when ξ lies in $\sigma_{k-1,i} \cup (\bigcup_{\sigma_{kj} \succ \sigma_{k-1,i}} \sigma_{kj})$ near a point ξ. The word "nontrivial" in the formulation means that $u \not\equiv const$ in any neighborhood of ξ.

It should be noted that the descent points are admissible for the solutions of (7.10). In fact let us consider the following example (see the next figure)

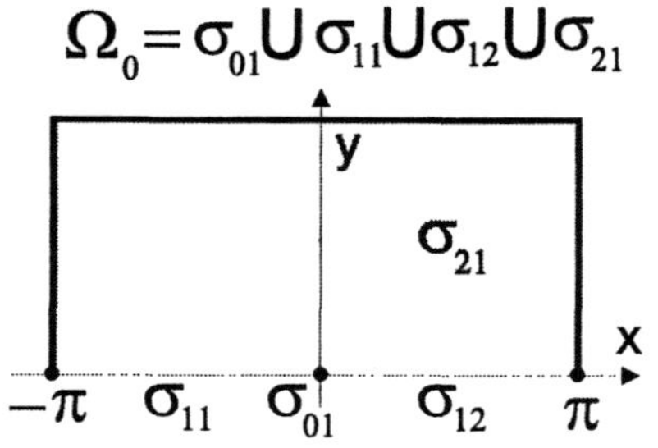

Figure 5: To an example of a descent point

The function $u(x,y) = \exp y \sin x$ is a solution of $\Delta u = 0$. It is not a classical Laplace equation; for example on σ_{11} and σ_{12} the relation $\Delta u = 0$ is equivalent to

$$\frac{\partial^2 u}{\partial x^2} + \frac{\partial u}{\partial y} = 0.$$

At $\sigma_{01} = (0;0)$ we have

$$\frac{\partial u}{\partial x}(0+;0) - \frac{\partial u}{\partial x}(0-;0) + \frac{\partial u}{\partial y}(0;0) = 0.$$

The point $(0;0)$ is a point of nontrivial descent of u and lying in the interior of Ω_0 (a boundary $\partial\Omega_0$ is drawn by the bold line).

References

[1] F. Ali Mehmeti, Regular solutions of transmission and interaction problems for wave equations, Math. Meth. Appl. Sc., 11, 1989, 665-685.

[2] F. Ali Mehmeti, Nonlinear waves in networks, Mathematical Research, 80, Akademie Verlag, 1994. F. Ali Mehmeti and S. Nicaise, Some realizations of interaction problems, Lecture Notes in Pure and Appl. Math., vol. 135, 1991, 15-27.

[3] F. Ali Mehmeti and S. Nicaise, Nonlinear interaction problems, Nonlinear Analysis: Theory, methods and applications, 20, 1992, 27-61.

[4] J. von Below, Classical solvability of linear parabolic equations on networks, J. Differential Equation, 72, 1988, 316-337

[5] J. von Below, Parabolic Network Equations, Habilitation Thesis, Eberhard-Karls-Universität Tübingen, 1993

[6] J. von Below and S. Nicaise, Dynamical interface transition with diffusion in ramified media, Commun. in Partial Differential Equations, 21, 1996, 255-279.

[7] A.A. Gavrilov, O.M. Penkin, About strong maximum principle for elliptic inequality on 2-dimensional stratified sets, Differential Equations, 2000 (to appear)

[8] D. Gilbarg, N.S. Trudinger, Elliptic Partial Differential Equations of Second Order, Springer Verlag, 1983.

[9] E. Hopf, Elementare Bemerkungen uber die Lösungen partieller Differentialgleihungen zweiter Ordnung vom elliptischen Typus, Sitz. Ber. Preuss. Akad. Wissensch. Berlin, Math.-Phys. Kl. 19, 1927, 147-152.

[10] K. Jano and S. Bochner, Curvature and Betti numbers, Princeton University Press, 1953.

[11] G. Lumer, Espaces ramifiés et diffusion sur les réseaux topologiques, C. R. Acad. Sc. Paris, Série A, 291, 1980, 627–630.

[12] S. Nicaise, Le laplacien sur les réseaux deux dimensionnels polygonaux topologiques, J. Math. Pures et Appl., 67, 1988, 93-113

[13] S. Nicaise, Elliptic operators on elementary ramified spaces, Integral Eq. and Op. Theory, 11, 1988, 230-257.

[14] S. Nicaise, Polygonal interface problems, Methoden und Verfahren der mathematischen Physic, 39, Peter Lang Verlag, Frankfurt a. M., 1993

[15] O.M. Penkin, On the weak maximum principle for an elliptic equation on a two-dimensional cell complex, Differential equations, vol. 33, No 10, 1997, 1410-1414.

[16] O.M. Penkin, On the Maximum Principle for an Elliptic Equation on a Two-Dimensional Cell Complex, Doklady Mathematics, vol. 55, No 1, 1997, 91-94.

[17] O.M. Penkin, Yu.V. Pokornyi, Differential Inequalities with Elliptic Operator on Composite Manifolds, Doklady Mathematics, vol. 57, No 3, 1998, 421-423.

[18] Yu.V. Pokornyi, O.M. Penkin, Comparison theorems for equations on graph, Differential equations, vol. 25, No 7, 1989, 802-809.

Study of a Vibration Problem for a Perforated Plate with Fourier Boundary Conditions

J.-M. SAC-ÉPÉE * & J. SAINT JEAN PAULIN **
Université de Metz, Département de Mathématiques, Ile du Saulcy, 57045 Metz, France.
* jmse@poncelet.sciences.univ-metz.fr ** sjpaulin@poncelet.sciences.univ-metz.fr

ABSTRACT

We consider a thin plate, periodically perforated by cylindrical holes, the axes of which are perpendicular to the plane of the domain. On a horizontal cross section, we denote by ε the size of the period, and by e the thickness of the plate. On this domain, we consider an elasticity problem with Fourier boundary conditions. First, we pass to the limit in ε, then in e. Next, we pass to the limit in the reverse order. In each of these asymptotic processes, we consider two cases depending on the value of the data on the lateral boundary of the holes.

1. PRESENTATION AND NOTATIONS.

We consider a problem defined on a thin plate periodically perforated. More precisely, the geometry of the domain is defined as follows.

Set $\omega = (0, L_1) \times (0, L_2)$, with $L_1/L_2 \in \mathbb{Q}$. We assume that ω is periodically covered by cells homothetic to a cell of reference $\mathcal{Y} = (-1/2, 1/2) \times (-1/2, 1/2)$, the ratio being ε:1. Let s (resp. $\mathcal{Y}^* = \mathcal{Y} \setminus s$) be the part of $\mathcal{Y}$ corresponding to the hole (resp. to the material). If we choose ε such that $N_\varepsilon^1 = L_1/\varepsilon$ and $N_\varepsilon^2 = L_2/\varepsilon$ are integers, the domain ω is perforated periodically with period ε by holes which do not intersect $\partial\omega$. Let ω_ε be the domain corresponding to the cross-section of the material, and set $s_\varepsilon = \omega \setminus \omega_\varepsilon$. Thus, s_ε corresponds to the cross-section of the holes. Denote ∂s_ε its boundary. We define the perforated cylindrical plate by $\Omega_{e\varepsilon} = \omega_\varepsilon \times (-e/2, e/2)$, and the set of the holes contained in $\Omega_e = \omega \times (-e/2, e/2)$ by $S_{e\varepsilon} = s_\varepsilon \times (-e/2, e/2)$. Hence, $\partial S_{e\varepsilon}$ is the lateral boundary of the holes. We set also $S_e = s \times \big(-e/2, e/2\big), \Gamma_{e\varepsilon}^{\pm} = \omega_\varepsilon \times \{\pm e/2\}$ and $\Gamma_e^0 = \partial\omega \times (-e/2, e/2)$. Finally, we set $Y = \mathcal{Y} \times \big(-e/2, e/2\big)$ and $Y^* = \mathcal{Y}^* \times \big(-e/2, e/2\big)$. When $e = 1$, we drop subindex e, thus $\Omega = \Omega_1$, and so on.

The problem we intend to study is inspired by Eljendy [10]. We consider the vibration problem of an elastic material with linearized elasticity constants a_{ijkh}. The displacement $u^{e\varepsilon}$ is the solution of the following system:

$$
\begin{aligned}
& e^2\frac{\partial^2 u_i^{e\varepsilon}}{\partial t^2} - \frac{\partial}{\partial x_j}\left(a_{ijkh}\frac{\partial u_k^{e\varepsilon}}{\partial x_h}\right) = F_i^{e\varepsilon} && \text{in } \Omega_{e\varepsilon}\times(0,T)\\
& u_i^{e\varepsilon} = 0 && \text{on } \Gamma_e^0\times(0,T)\\
& a_{i3kh}\frac{\partial u_k^{e\varepsilon}}{\partial x_h} n_3 = 0 && \text{on } \Gamma_{e\varepsilon}^{\pm}\times(0,T)\\
& a_{i\alpha kh}\frac{\partial u_k^{e\varepsilon}}{\partial x_h} n_\alpha + \lambda e^2 \varepsilon u_i^{e\varepsilon} + \gamma e^2\varepsilon \frac{\partial u_i^{e\varepsilon}}{\partial t} = G_i^{e\varepsilon} && \text{on } \partial S_{e\varepsilon}\times(0,T)\\
& u_i^{e\varepsilon}(0) = u^{\varepsilon 0} \text{ and } \frac{\partial u_i^{e\varepsilon}}{\partial t}(0) = u^{\varepsilon 1} && \text{in } \Omega_{e\varepsilon}.
\end{aligned}
\tag{1.1}
$$

Here, $F^{e\varepsilon}$ is a body force, and $G^{e\varepsilon}$ is a surface force independent of x_3. The functions $u^{\varepsilon 0}$ and $u^{\varepsilon 1}$ are the initial displacement and velocity, and $n_\alpha = cos(n, x_\alpha)$, where n is the unit exterior normal. The constants λ and γ are positive and related to a damping effect. The a_{ijkh} satisfy the usual symmetry and coercivity assumptions:

$$
\begin{aligned}
& \exists \alpha > 0 \mid a_{ijkh}\xi_{ij}\xi_{kh} \geq \alpha \xi_{ij}\xi_{ij} \quad \forall (\xi_{ij})_{i,j} \in \mathbb{R}^9 \text{ such that } \xi_{kh} = \xi_{hk}\ \forall k, h,\\
& a_{ijkh} = a_{khij} = a_{jikh} \qquad \forall i, j, k, h.
\end{aligned}
$$

For repeated indices, summation convention is understood. We set $G_i^e(\frac{x_1}{\varepsilon}, \frac{x_1}{\varepsilon}, t) = G_i^{e\varepsilon}(x_1, x_2, t)$, and we assume that

$$
\begin{aligned}
& G_i^e \text{ is } \mathcal{Y}\text{-periodic}, G_i^e \in L^\infty\big((0,T); L^2(\partial S_e)\big), \frac{\partial G_i^e}{\partial t} \in L^\infty\big((0,T); L^2(\partial S_e)\big),\\
& F^{e\varepsilon} \in L^2\big((0,T); (L^2(\Omega_{e\varepsilon}))^3\big), F^{e\varepsilon} \to F^e \text{ strongly in } L^2\big((0,T); L^2(\Omega_e)\big),\\
& u^{\varepsilon 0} \in W_{e\varepsilon}, \quad u^{\varepsilon 1} \in [L^2(\Omega_{e\varepsilon})]^3,
\end{aligned}
\tag{1.2}
$$

where $V_{e\varepsilon} = \{v \in H^1(\Omega_{e\varepsilon}), v = 0 \text{ on } \Gamma_e^0\}$, $W_{e\varepsilon} = (V_{e\varepsilon})^3$ and where ∂S_e refers to the vertical boundaries of S_e.

Then, using Galerkin's method and Gronwall's lemma, we check that Problem (1.1) has a unique solution $u^{e\varepsilon}$ such that $u^{e\varepsilon} \in L^\infty\big((0,T); W_{e\varepsilon}\big)$ and $\partial u^{e\varepsilon}/\partial t \in L^\infty\big((0,T); \big(L^2(\Omega_{e\varepsilon})\big)^3\big)$.

Our aim is to study the asymptotic behaviour of this plate when the two parameters e and ε are small. In Section 2, we establish some basic lemmas. In Section 3, we pass to the limit first in ε and then in e. In Section 4, we pass to the limit in the reverse order.

2. BASIC LEMMAS.

In the course of the computations, some difficulties arise from the presence of surface integrals, originating from the Fourier conditions. To solve this problem, we use some lemmas derived from Cioranescu-Donato [6], the main point being to transform surface integrals on the boundary of the holes into volume integrals of auxiliary functions on the perforated domain. Therefore, we need extension operators for perforated domains, with some kind of uniform estimate with respect to ε. These extension operators are obtained from two-dimensional ones by using the fact that the perforated domain is cylindrical. The proofs are similar to those of Brizzi-Chalot [2], so we do not detail them.

Denote $|E|$ the measure of a set E. For any function $h \in L^\infty(0,T;L^2(\partial S_e))$ and t fixed, set

$$C_h(t) = \frac{1}{|Y^*|}\int_{\partial S_e} h(\sigma,t)d\sigma, \quad \text{and} \quad \Delta_y = \frac{\partial^2}{\partial y_1^2} + \frac{\partial^2}{\partial y_2^2} + \frac{\partial^2}{\partial y_3^2}\,.$$

We define the auxiliary function ψ_h as the solution of the following problem (in which t is a parameter):

$$\begin{aligned}
&-\Delta_y \psi_h = C_h && \text{in } Y^*,\\
&\frac{\partial \psi_h}{\partial n} = h(y_1,y_2,t) && \text{on } \partial S_e,\\
&\frac{\partial \psi_h}{\partial n} = 0 && \text{in } \mathcal{Y}^* \times \{\pm\frac{e}{2}\},\\
&\psi_h \ \mathcal{Y}\text{-periodic with a zero average.}
\end{aligned}$$

This function will be used below with different values of h to derive the homogenized equations. First, we have the following result (see [2] and integrate in t).

LEMMA 2.1. *For t fixed, consider $h(y_1,y_2,t) \in L^2(\partial S_e)$. Then, for all $v \in H^1(\Omega_{e\varepsilon})$, such that $v_{|\Gamma_0} = 0$, one has*

$$\begin{aligned}
&\varepsilon\int_{\partial S_{e\varepsilon}} h(\frac{x_1}{\varepsilon},\frac{x_2}{\varepsilon},t)v(x_1,x_2,x_3)d\sigma = \\
&= \varepsilon\int_{\Omega_{e\varepsilon}} (\frac{\partial \psi_h}{\partial y_i})(\frac{x_1}{\varepsilon},\frac{x_2}{\varepsilon},t)\frac{\partial v}{\partial x_i}dx + C_h(t)\int_{\Omega_{e\varepsilon}} v(x_1,x_2,x_3)dx.
\end{aligned}$$

DEFINITION 2.2. As in Cioranescu-Donato [6], we define on $V_{e\varepsilon}$ the continuous linear form $\mu_h^\varepsilon(t)$ by

$$\forall\varphi \in V_{e\varepsilon}, \ \langle \mu_h^\varepsilon(t),\varphi\rangle = \varepsilon\int_{\partial S_{e\varepsilon}} h(x_1/\varepsilon, x_2/\varepsilon, t)\,\varphi(x_1,x_2,x_3)\,d\sigma.$$

We can establish the following lemmas:

LEMMA 2.3. *Under the hypothesis of Lemma 2.1, we have*

$$\mu_h^\varepsilon(t) \rightarrow \mu_h(t) \qquad \textit{strongly in } H^{-1}(\Omega_e),$$

where $\langle \mu_h(t),\varphi\rangle = \mu_h(t)\int_{\Omega_e}\varphi dx$, *with* $\mu_h(t) = \frac{1}{|Y|}\int_{\partial s\times(-e/2,e/2)} h(y,t)\,d\sigma.$

LEMMA 2.4. *We assume that* $h(y_1, y_2, t) \in L^2(s \times (-\frac{e}{2}, \frac{e}{2}))$ *and that*

$$\frac{1}{|\partial s \times (-\frac{e}{2}, \frac{e}{2})|} \int_{\partial s \times (-e/2, e/2)} h(\sigma, t) d\sigma = 0.$$

Then, the function $\varepsilon^{-1} \mu_h^{\varepsilon}(t)$ *tends to zero weakly in* $H^{-1}(\Omega_e)$.

The proof of Lemma 2.3 follows the ideas of Brizzi-Chalot [2] and that of Lemma 2.4 the ideas of Cioranescu-Donato [6], so we omit them. We will pass to the limit in problem (1.1) when the two parameters tend to zero independently.

3. THE CASE WHERE THE PERIOD IS MUCH SMALLER THAN THE THICKNESS.

Mathematically, this corresponds to the case where we make first $\varepsilon \to 0$ with e fixed, and then $e \to 0$ in the homogenized problem.

3.1. The Parameter $\varepsilon \to 0$ in Problem (1.1), while the Parameter e is Fixed.

This is a homogenization process. Set $V_e = \{v \in H^1(\Omega_e), v = 0 \text{ on } \Gamma_e^0\}$. We prove

THEOREM 3.1. *Assume that* $F_i^{e\varepsilon} \to F_i^e$ *strongly in* $L^2((0,T); L^2(\Omega_e))$. *Then, there exists* $P^\varepsilon \in \mathcal{L}(L^\infty((0,T); V_{e\varepsilon}), L^\infty((0,T); V_e))$ *such that*

$$\varepsilon\,(P^\varepsilon u_i^{e\varepsilon}) \rightharpoonup U_i^e \text{ weakly in } L^p((0,T); V_e) \quad 1 \leq p \leq \infty,$$

where U^e *satisfies the following homogenized problem:*

$$\begin{aligned} e^2 \frac{|Y^*|}{|Y|} \frac{\partial^2 U_i^e}{\partial t^2} - \frac{\partial}{\partial x_j}\left(q_{ijkh} \frac{\partial U_k^e}{\partial x_h}\right) &= \\ = \mu_{G_i^e} - \lambda e^2 \frac{|\partial S|}{|Y|} U_i^e - \gamma e^2 \frac{|\partial S|}{|Y|} \frac{\partial U_i^e}{\partial t} &\text{ in } \Omega_e \times (0,T), \end{aligned} \tag{3.1}$$

and the boundary conditions

$$\begin{aligned} &q_{i3kh} \frac{\partial U_k^e}{\partial x_h} n_3 = 0 && \text{on } \Gamma_e^{\pm} \times (0,T), \\ &U_i^e = 0 && \text{on } \Gamma_e^0 \times (0,T), \\ &U_i^e(0) = u_i^0 \quad \text{and} \quad \frac{\partial u_i^e}{\partial t}(0) = u_i^1, \end{aligned} \tag{3.2}$$

where $\Gamma_e^{\pm} = \omega \times \{\pm e/2\}$.

The homogenized coefficients q_{ijkh} *are defined by*

$$q_{ijkh} = \frac{1}{|\mathcal{Y}|} \int_{\mathcal{Y}^*} \left(a_{ijkh} - a_{\ell\beta kh} \frac{\partial \chi_\ell^{ij}}{\partial y_\beta}\right) dy, \tag{3.3}$$

and the functions χ^{ij} *satisfy*

$$
\begin{aligned}
&-\frac{\partial}{\partial y_\beta}\big(a_{i\alpha k\beta}\frac{\partial \chi_i^{\ell m}}{\partial y_\alpha} - a_{\ell m k\beta}\big) = 0 && \text{in } \mathcal{Y}^*,\\
&\big(a_{i\alpha k\beta}\frac{\partial \chi_i^{\ell m}}{\partial y_\alpha} - a_{\ell m k\beta}\big)n_\beta = 0 && \text{on } \partial s,\\
&\chi^{\ell m}\ \ \mathcal{Y}^*\text{-periodic with a zero average.}
\end{aligned}
\tag{3.4}
$$

In the particular case where the constants satisfy

$$
\|C_{G_i^e}\|^2_{L^\infty(0,T)} + \|C_{\frac{\partial G_i^e}{\partial t}}\|^2_{L^\infty(0,T)} = 0, \tag{3.5}
$$

we have $u_i^e = 0$, and we establish the more precise following result:

THEOREM 3.2. *Assume that* $F_i^{e\varepsilon} \to F_i^e$ *strongly in* $L^p\big((0,T); L^2(\Omega_\varepsilon)\big)$, *and that (3.5) holds. Then,*

$$
P^\varepsilon u_i^{e\varepsilon} \rightharpoonup u_i^e \ \text{ weakly in } L^p\big((0,T); V_e\big) \quad 1 \le p \le \infty,
$$

where the limit u^e *is solution of the homogenized equation*

$$
\begin{aligned}
e^2\frac{|Y^*|}{|Y|}\frac{\partial^2 u_i^e}{\partial t^2} - \frac{\partial}{\partial x_j}\big(q_{ijkh}\frac{\partial u_k^e}{\partial x_h}\big) &=\\
= \frac{|Y^*|}{|Y|}F_i^e - e^2\lambda\frac{|\partial S|}{|Y|}u_i^e - e^2\gamma\frac{|\partial S|}{|Y|}\frac{\partial u_i^e}{\partial t} &\quad \text{in } \Omega_e\times(0,T).
\end{aligned}
\tag{3.6}
$$

The limit u^e *satisfies the same boundary conditions as in (3.2) and the homogenized coefficients* q_{ijkh} *are still defined by (3.3), (3.4).*

PROOF OF THEOREM 3.2. Multiplying Problem (1.1) by the test function $\partial u_i^{e\varepsilon}/\partial t$ (see Sac-Épée [11] for details), we have

$$
\|\,|\frac{\partial u_i^{e\varepsilon}}{\partial t}| + |\gamma_{ij}(u^{e\varepsilon})|\,\|^2_{L^\infty((0,T);L^2(\Omega_{e\varepsilon}))} \le \frac{C_1}{\varepsilon^2}\big[\varepsilon^2 + \|C_{G_i^e}\|^2_{L^\infty(0,T)} + \|C_{\frac{\partial G_i^e}{\partial t}}\|^2_{L^\infty(0,T)}\big],
$$

where $\gamma_{ij}(u^{e\varepsilon}) = \frac{1}{2}\big(\frac{\partial u_i^{e\varepsilon}}{\partial x_j} + \frac{\partial u_j^{e\varepsilon}}{\partial x_i}\big)$ is the strain tensor. The term $C_{\frac{\partial G_i^e}{\partial t}}$ comes from the fourth relation in (1.1) after multiplication by $\frac{\partial u^{e\varepsilon}}{\partial t}$ and integration by parts in time. Then, using assumption (3.5), we have

$$
\|\frac{\partial u_i^{e\varepsilon}}{\partial t}\|_{L^\infty((0,T);L^2(\Omega_{e\varepsilon}))} + \|\gamma_{ij}(u^{e\varepsilon})\|_{L^\infty((0,T);L^2(\Omega_{e\varepsilon}))} \le K. \tag{3.7}
$$

The a priori estimates (3.7) are obtained in a domain $\Omega_{e\varepsilon}$ depending on the parameter ε. It is important to obtain estimates in a domain independent of ε in order to derive weak convergence results. This can be done by extending the solution $u^{e\varepsilon}$ to the whole of Ω_e. It is desirable to control the ε dependence of such an extension operator. This is why it is built first in the cell of reference Y, and then extended to Ω_e by using the periodic structure of $\Omega_{e\varepsilon}$. For details about such extension operators, we refer the reader to Cioranescu-Saint Jean Paulin [7], to Duvaut [9], and to Brizzi-Chalot [2]. These methods enable us to define a suitable extension $P^\varepsilon u^{e\varepsilon}$ and we have, up to a subsequence

$$P^\varepsilon u_i^{e\varepsilon} \rightharpoonup u_i^e \text{ weakly in } L^p\big((0,T);V_e\big), \qquad 1 \le p \le \infty.$$

Let us introduce the notation

$$\xi_{ij}^{e\varepsilon} = a_{ijkh}\gamma_{kh}(u^{e\varepsilon}).$$

Then, we have, up to a subsequence

$$\widetilde{\xi_{ij}^{e\varepsilon}} \rightharpoonup \xi_{ij}^e \text{ weakly} \star \text{ in } L^\infty\big((0,T);L^2(\Omega)\big),$$

where $\sim$ denotes the extension by 0 to the whole of Ω. We multiply Problem (1.1) by a test function $\varphi \in \mathcal{D}\big([0,T[;\mathcal{D}(\Omega_e)\big)$, and we make $\varepsilon \to 0$. In passing to the limit, we use Lemma 2.3, with the two choices $h = e^2\lambda$ and $h = e^2\gamma$ and Lemma 2.4 with $h = G_i^e$. We obtain (see Sac-Épée [11] for details)

$$\begin{aligned} &e^2\frac{|Y^*|}{|Y|}\frac{\partial^2 u_i^e}{\partial t^2} - \frac{\partial \xi_{ij}^e}{\partial x_j} = \frac{|Y^*|}{|Y|}F_i^e - e^2\frac{|\partial S|}{|Y|}\lambda u_i^e - e^2\frac{|\partial S|}{|Y|}\gamma\frac{\partial u_i^e}{\partial t} \quad \text{in } \Omega_e \times (0,T), \\ &u_i^e(0) = u_i^0, \quad \frac{\partial u_i^e}{\partial t}(0) = u_i^1. \end{aligned} \tag{3.8}$$

Now, the difficulty is to write ξ_{ij}^e as a function of u_i^e. Let us consider the solution $W^{\ell m}$ of the adjoint problem (where δ_{ij} is the Kronecker symbol)

$$\begin{aligned} &-\frac{\partial}{\partial y_\beta}\big(a_{i\alpha k\beta}\frac{\partial W_i^{\ell m}}{\partial y_\alpha} + \delta_{m3}\, a_{\ell 3k\beta}\big) = 0 \qquad \text{in } \mathcal{Y}^*, \\ &\big(a_{i\alpha k\beta}\frac{\partial W_i^{\ell m}}{\partial y_\alpha} + \delta_{m3}\, a_{\ell 3k\beta}\big)\, n_\beta = 0 \qquad \text{on } \partial s, \\ &W_i^{\ell m} - \delta_{m\lambda}\delta_{i\ell}y_\lambda \quad \mathcal{Y}^* \text{ - periodic with a zero average.} \end{aligned} \tag{3.9}$$

Set

$$W_i^{\ell m\varepsilon} = P^\varepsilon\Big(\varepsilon W_i^{\ell m}\big(\frac{x_1}{\varepsilon},\frac{x_2}{\varepsilon}\big)\Big) \text{ and } w_{k\beta}^{\ell m\varepsilon} = a_{i\alpha k\beta}^\varepsilon\frac{\partial W_i^{\ell m\varepsilon}}{\partial x_\alpha}.$$

Then, it is easy to check that $w^{\ell m\varepsilon}$ is solution of

$$\begin{aligned} &-\frac{\partial}{\partial x_\beta}\big(w_{k\beta}^{\ell m\varepsilon} + \delta_{m3}a_{\ell 3k\beta}\big) = 0 \quad \text{in } \Omega_{e\varepsilon}, \\ &\big(w_{k\beta}^{\ell m\varepsilon} + \delta_{m3}a_{\ell 3k\beta}\big)n_\beta = 0 \quad \text{on } \partial S_{e\varepsilon}. \end{aligned} \tag{3.10}$$

A difficulty arises from the non homogeneous boundary condition satisfied by $u^{e\varepsilon}$ on $\partial S_{e\varepsilon}$. Adapting an idea of Cioranescu-Donato [6], let us introduce the solution ζ of the elasticity problem

$$\begin{aligned} &-\frac{\partial}{\partial y_\alpha}\Big(a_{i\alpha k\beta}\frac{\partial \zeta_k}{\partial y_\beta}\Big) = 0 && \text{in } \mathcal{Y}^*, \\ &a_{i\alpha k\beta}\frac{\partial \zeta_k}{\partial y_\beta} n_\alpha = G_i^e && \text{on } \partial s, \\ &\zeta \ \ \mathcal{Y}^* - \text{ periodic with a zero average.} \end{aligned} \tag{3.11}$$

We also define

$$\gamma_{i\alpha} = \begin{cases} a_{i\alpha k\beta}\dfrac{\partial \zeta_k}{\partial y_\beta} & \text{in } \mathcal{Y}^* \\ 0 & \text{in } s, \end{cases} \qquad \begin{aligned} &\gamma_{i\alpha}^\varepsilon(x_1,x_2,x_3,t) = \gamma_{i\alpha}\big(\tfrac{x_1}{\varepsilon},\tfrac{x_2}{\varepsilon},t\big) \\ &\zeta^\varepsilon(x) = \varepsilon\big(P^\varepsilon\zeta\big)\big(\tfrac{x_1}{\varepsilon},\tfrac{x_2}{\varepsilon}\big). \end{aligned}$$

Then, we have

$$\begin{aligned} &e^2\frac{\partial^2 u_i^{e\varepsilon}}{\partial t^2} - \frac{\partial}{\partial x_\alpha}\big(\xi_{i\alpha}^{e\varepsilon} - \gamma_{i\alpha}^{e\varepsilon}\big) - \frac{\partial \xi_{i3}^{e\varepsilon}}{\partial x_3} = F_i^{e\varepsilon} && \text{in } \Omega_{e\varepsilon}\times(0,T), \\ &\big(\xi_{i\alpha}^{e\varepsilon} - \gamma_{i\alpha}^{e\varepsilon}\big)n_\alpha = -\,\varepsilon\, e^2\lambda u_i^{e\varepsilon} - \varepsilon\, e^2\gamma\frac{\partial u_i^{e\varepsilon}}{\partial t} && \text{on } \partial S_{e\varepsilon}\times(0,T). \end{aligned} \tag{3.12}$$

According to the energy method (see Bensoussan, Lions and Papanicolaou [1], Tartar [12]), let us multiply Problem (3.8) by the test function $\varphi(-P^\varepsilon u_k^{e\varepsilon} + \zeta_k^\varepsilon)$, Problem (3.9) by the test function $\varphi W_i^{\ell m\varepsilon}$, with $\varphi \in \mathcal{D}(]0,T[, \mathcal{D}(\Omega_e))$, and add the results. Making $\varepsilon \to 0$ yields (see Sac-Épée [11])

$$\xi_{ij}^e = \Big(\frac{1}{|\mathcal{Y}|}\int_{\mathcal{Y}^*}\gamma_{ij}dy\Big) + \Big[\frac{1}{|\mathcal{Y}|}\int_{\mathcal{Y}^*}\Big(a_{ijkh} - a_{\ell\beta kh}\frac{\partial \chi_\ell^{ij}}{\partial y_\beta}\Big)dy\Big]\frac{\partial u_k^e}{\partial x_h}, \tag{3.13}$$

with

$$\begin{aligned} \chi_i^{\ell\lambda} &= -W_i^{\ell\lambda} + y_\lambda\,\delta_{i\ell}, \\ \chi_i^{\ell 3} &= -W_i^{\ell 3}. \end{aligned}$$

Define q_{ijkh} by (3.3). In Problem (3.8), replace ξ_{ij}^e by its expression given by (3.13). This yields the homogenized problem (3.1). The tensor q_{ijkh} is positive definite, hence the existence and uniqueness of the solution of (3.1). □

PROOF OF THEOREM 3.1. Due to the assumptions on the constants $C_{G_i^e}$ and $C_{\frac{\partial G_i^e}{\partial t}}$, relation (3.4) has now to be replaced by

$$\varepsilon\Big\|\frac{\partial u_i^{e\varepsilon}}{\partial t}\Big\|_{L^\infty\left((0,T);L^2(\Omega_{e\varepsilon})\right)} + \varepsilon\big\|\gamma_{ij}(u^{e\varepsilon})\big\|_{L^\infty\left((0,T);L^2(\Omega_{e\varepsilon})\right)} \le K.$$

We do not detail the rest of the proof since it is easier than that of Theorem 3.2. Indeed, the main difficulty in the proof of Theorem 3.2 derived from the weak convergence of $\varepsilon^{-1}\mu_h^\varepsilon$.

In the case of Theorem 3.1, this problem disappears, because if we multiply Problem (1.1) by ε, we can use the strong convergence of μ_h^ε to its limit, and it remains to apply the energy method. □

3.2. The Parameter e Tends to Zero in the Homogenized Problem (3.1).

Since the domain Ω_e depends on e, we transform it into a fixed domain by means of dilatation in the x_3 direction. In the aim of using usual plate techniques, we introduce the change of variables

$$z_\alpha = x_\alpha,\ \ \alpha = 1,\ 2 \qquad \text{and} \qquad z_3 = x_3/e,$$

and the change of functions

$$\phi_\alpha^e(z_1, z_2, ez_3) = e\phi_\alpha(z_1, z_2, z_3) \qquad \text{and} \qquad \phi_3^e(z_1, z_2, ez_3) = \phi_3(z_1, z_2, z_3). \quad (3.14)$$

As usual, latin indices take values in $1, 2, 3$ and greek indices in $1, 2$.

THEOREM 3.3. *We make the assumptions of Theorem 3.2. Moreover, we assume that* $F_3 \rightharpoonup F_3^*$ *weakly* $\star$ *in* $L^\infty\big((0,T); L^2(\Omega)\big)$. *Then, we have*

$$e^2 u_i \rightharpoonup u_i^* \ \ \textit{weakly} \star \textit{ in } \ L^\infty\big((0,T); V(\Omega)\big),$$

where $V(\Omega) = \{v \in H^1(\Omega), v_{|\Gamma^0} = 0\}$. *Moreover*

$$u_3^* = U_3^*(z_1, z_2, t) \qquad \textit{and} \qquad u_\alpha^* = -z_3 \frac{\partial U_3^*}{\partial z_\alpha} + U_\alpha^*(z_1, z_2, t), \quad (3.15)$$

which is the usual form in thin plates problems. The function U_3^* *is solution of*

$$\begin{aligned}
&\frac{|Y^*|}{|Y|}\frac{\partial^2 U_3^*}{\partial t^2} + \frac{1}{12} P_{\tau\eta\theta\rho} \frac{\partial^4 U_3^*}{\partial z_\tau \partial z_\eta \partial z_\theta \partial z_\rho} = \frac{|Y^*|}{|Y|} \int_{-\frac12}^{\frac12} f_3^* dz_3 - \\
&\qquad - \frac{|\partial S|}{|Y|}\lambda U_3^* - \frac{|\partial S|}{|Y|}\gamma \frac{\partial U_3^*}{\partial t} \qquad \textit{in } \omega \times (0,T), \\
&U_3^* = 0 \ \ \textit{and} \ \ \frac{\partial U_3^*}{\partial n} = 0 \qquad \textit{on } \partial\omega \times (0,T), \\
&U_3^*(z_1, z_2, 0) = 0 \ \ \textit{and} \ \ \frac{\partial U_3^*}{\partial t}(z_1, z_2, 0) = \int_{-\frac12}^{\frac12} U_3^{*1} dz_3 \ \ \textit{in } \omega \times \{0\},
\end{aligned} \quad (3.16)$$

with

$$P_{\alpha\beta\theta\rho} = q_{\alpha\beta\theta\rho} - q_{\alpha\beta j3}\, d_{ij}\, q_{i3\theta\rho} \ \ \textit{and} \ \ (d_{ij})_{ij} = (q_{i3j3})_{ij}^{-1}.$$

The function U_α^* *is zero.*

PROOF OF THEOREM 3.3. Using the techniques of plate theory (see Caillerie [3], Ciarlet [4], Ciarlet-Destuynder [5]), we write down (3.6) with the new function u defined in (3.14), we multiply its first two equations of by $e\frac{\partial u_\tau}{\partial t}$ and its third equation by $\frac{\partial u_3}{\partial t}$. Integrating from 0 to t and using the coercivity of a_{ijkh}, we obtain the two relations

$$e^3 \| \frac{\partial u_\alpha}{\partial t} \|_{L^\infty\left((0,T);L^2(\Omega)\right)} + e^2 \| \frac{\partial u_3}{\partial t} \|_{L^\infty\left((0,T);L^2(\Omega)\right)} \leq C, \tag{3.17}$$

$$\begin{aligned} e^2 \|\gamma_{\alpha\beta}(u)\|_{L^\infty((0,T);L^2(\Omega))} + e\|\gamma_{\alpha 3}(u)\|_{L^\infty((0,T);L^2(\Omega))} + \\ + \|\gamma_{33}(u)\|_{L^\infty((0,T);L^2(\Omega))} \leq C. \end{aligned} \tag{3.18}$$

Set

$$\sigma_{ij} = q_{ij\alpha\beta}(u) + 2e^{-1}q_{ij\alpha 3}(u) + e^{-2}q_{ij33}\gamma_{33}(u).$$

Consequently, one has, up to a subsequence,

$$e^2 u^e \rightharpoonup u^* \text{ weakly in } L^\infty\left((0,T); V(\Omega)\right),$$

where u^* has the form (3.15), and also

$$e^2 \sigma_{ij} \rightharpoonup \sigma^*_{ij} \text{ weakly in } L^\infty\left((0,T); L^2(\Omega)\right).$$

Let us now characterize U_3^* and U_α^*. We use the following moments, introduced in Caillerie [3]:

$$M_{\alpha\beta} = e^2 \int_{-\frac{1}{2}}^{\frac{1}{2}} z_3 \sigma_{\alpha\beta} dz_3, \qquad Q_\alpha = e \int_{-\frac{1}{2}}^{\frac{1}{2}} \sigma_{\alpha 3} dz_3.$$

Let $\theta_\tau \in \mathcal{D}\big([0,T[, H_0^1(\omega)\big)$. We multiply the first two equations of Problem (3.1) by $e\, z_3\, \theta_\tau(z_1, z_2, t)$, and we get, when e tends to zero

$$-\frac{\partial M^*_{\tau\eta}}{\partial z_\eta} + Q^*_\tau = 0 \text{ in } \omega \times (0,T), \tag{3.19}$$

where $M^*_{\tau\eta}$ and Q^*_τ are the limits of $M^e_{\tau\eta}$ and Q^e_τ respectively. Multiplying the third equation of Problem (3.1) by $\theta_3 \in \mathcal{D}\big([0,T[, H_0^1(\omega)\big)$, we get, when e tends to zero

$$\frac{\partial^2 U_3^*}{\partial t^2} - \frac{\partial Q_\eta^{*\varepsilon\delta}}{\partial z_\eta} = \frac{|Y^*|}{|Y|}\Big(\int_{-\frac{1}{2}}^{\frac{1}{2}} F_3^* dz_3\Big) - \frac{|\partial S|}{|Y|}\lambda\, U_3^* - \frac{|\partial S|}{|Y|}\gamma\, \frac{\partial U_3^*}{\partial t}. \tag{3.20}$$

The replacement of Q^*_η in (3.20) by its value obtained from (3.19) yields the equation characterizing U_3^*. Equations on boundaries are obtained easily from the weak formulation. The arguments concerning U_α^* are similar. □

REMARK 3.4. This method can also be applied without any additional difficulty to the solution of Problem (3.1). □

4. THE CASE WHERE THE THICKNESS IS MUCH SMALLER THAN THE PERIOD.

Mathematically, this corresponds to the case where we make first $e \longrightarrow 0$, with ε fixed, and then $\varepsilon \longrightarrow 0$ in the limit two dimensional problem.

4.1. The Parameter e Tends to Zero in Problem (1.1) and the Parameter ε is Fixed.

As we did before, we transform the domain $\Omega_{e\varepsilon}$ into a fixed domain, and we use the same changes of functions and variables as those that were required to study the dependence in e in Theorem 3.3. Then, Problem (1.1) becomes

$$
\begin{aligned}
& e^3\frac{\partial^2 u_\tau^\varepsilon}{\partial t^2} - e\frac{\partial}{\partial z_\eta}\sigma_{\tau\eta}^\varepsilon - \frac{\partial}{\partial z_3}\sigma_{\tau 3}^\varepsilon = f_\tau^\varepsilon && \text{on } \Omega_\varepsilon \times (0,T),\\
& e^2\frac{\partial^2 u_3^\varepsilon}{\partial t^2} - e\frac{\partial}{\partial z_\eta}\sigma_{3\eta}^\varepsilon - \frac{\partial}{\partial z_3}\sigma_{33}^\varepsilon = f_3^\varepsilon && \text{on } \Omega_\varepsilon \times (0,T),\\
& \sigma_{i3}^\varepsilon n_3 = 0 && \text{on } \Gamma_\varepsilon^\pm \times (0,T),\\
& \sigma_{\tau\eta}^\varepsilon n_\eta = g_\tau^\varepsilon - \lambda\varepsilon e^2 u_\tau^\varepsilon - \gamma\varepsilon e^2\frac{\partial u_\tau^\varepsilon}{\partial t} && \text{on } \partial S_\varepsilon \times (0,T),\\
& \sigma_{3\eta}^\varepsilon n_\eta = g_3^\varepsilon - \lambda\varepsilon e u_3^\varepsilon - \gamma\varepsilon e\frac{\partial u_3^\varepsilon}{\partial t} && \text{on } \partial S_\varepsilon \times (0,T),\\
& u^\varepsilon = 0 && \text{on } \Gamma^0 \times (0,T),\\
& u_\alpha^\varepsilon(0) = e^{-1}u_\alpha^{\varepsilon 0},\ u_3^\varepsilon(0) = u_3^{\varepsilon 0},\ \frac{\partial u_\alpha^\varepsilon}{\partial t}(0) = e^{-1}u_\alpha^{\varepsilon 1},\ \frac{\partial u_3^\varepsilon}{\partial t}(0) = u_\alpha^{\varepsilon 1},\\
& \sigma_{ij}^\varepsilon = a_{ij\alpha\beta}\gamma_{\alpha\beta}(u^\varepsilon) + 2e^{-1}a_{ij\alpha 3}(u^\varepsilon) + e^{-2}a_{ij33}\gamma_{33}(u^\varepsilon),
\end{aligned}
\tag{4.1}
$$

with

$$f_\tau^\varepsilon = eF_\tau^\varepsilon,\ f_3^\varepsilon = F_3^\varepsilon,\ g_\tau^\varepsilon = G_\tau^\varepsilon,\ g_3^\varepsilon = e^{-1}G_3^\varepsilon,\ \Omega_\varepsilon = \omega_\varepsilon \times (-\frac{1}{2},\frac{1}{2}),\ S_\varepsilon = s_\varepsilon \times (-\frac{1}{2},\frac{1}{2}).$$

Let e tend to zero in Problem (4.1). We get

THEOREM 4.1. *Assume that*

$$f_3^\varepsilon \rightharpoonup f_3^{*\varepsilon} \textit{ weakly in } L^2((0,T);L^2(\Omega_\varepsilon)) \textit{ and } eg_3^\varepsilon \rightharpoonup g_3^{*\varepsilon} \textit{ weakly in } L^2((0,T;L^2(\partial S_\varepsilon)).$$

Then, we have

$$e^2 u_i^\varepsilon \rightharpoonup u_i^{*\varepsilon} \textit{ weakly} \star \textit{ in } L^\infty((0,T);V(\Omega_\varepsilon),$$

where

$$u_3^{*\varepsilon} = U_3^{*\varepsilon}(z_1,z_2,t) \textit{ and } u_\alpha^{*\varepsilon} = -z_3\frac{\partial U_3^{*\varepsilon}}{\partial z_\alpha}(z_1,z_2,t) + U_\alpha^{*\varepsilon}(z_1,z_2,t).$$

Moreover, $U_3^{*\varepsilon}$ satisfies the problem

$$
\begin{aligned}
&\frac{\partial^2 U_3^{*\varepsilon}}{\partial t^2} + \frac{1}{12} b_{\tau\eta\theta\rho} \frac{\partial^4 U_3^{*\varepsilon}}{\partial z_\tau \partial z_\eta \partial z_\theta \partial z_\rho} = \int_{-\frac{1}{2}}^{\frac{1}{2}} f_3^{*\varepsilon} dz_3 && \text{in } \omega_\varepsilon \times (0,T);\\
&-\frac{\partial}{\partial z_\tau}\Big(b_{\eta\tau\theta\rho} \frac{\partial^2 U_3^{*\varepsilon}}{\partial z_\theta \partial z_\rho}\Big) n_\eta + \lambda\varepsilon U_3^{*\varepsilon} + \gamma\varepsilon \frac{\partial U_3^{*\varepsilon}}{\partial t} = g_3^{*\varepsilon} && \text{on } \partial s_\varepsilon \times (0,T);\\
&b_{\tau\eta\theta\rho} \frac{\partial^2 U_3^{*\varepsilon}}{\partial z_\theta \partial z_\rho} n_\eta = 0 && \text{on } \partial s_\varepsilon \times (0,T); \qquad (4.2)\\
&U_3^{*\varepsilon} = 0 \;\; \text{and} \;\; \frac{\partial U_3^{*\varepsilon}}{\partial n} = 0 && \text{on } \partial\omega \times (0,T);\\
&U_3^{*\varepsilon}(z_1,z_2,0) = 0 \;\text{and}\; \frac{\partial U_3^{*\varepsilon}}{\partial t}(z_1,z_2,0) = \int_{-\frac{1}{2}}^{\frac{1}{2}} U_3^{*\varepsilon 1} dz_3 = V_3^{*\varepsilon 1} && \text{in } \omega \times \{0\}.
\end{aligned}
$$

The function $U_3^{*\varepsilon 1}$ is defined by

$$e^2 u_3^{\varepsilon 1} \rightharpoonup U_3^{*\varepsilon 1} \quad \text{weakly in } L^2(\omega),$$

and

$$b_{\alpha\beta\theta\rho} = a_{\alpha\beta\theta\rho} - a_{\alpha\beta j3}\, d_{ij}\, a_{i3\theta\rho} \quad \text{with} \quad (d_{ij})_{ij} = (a_{i3j3})_{ij}^{-1}.$$

The function $U_\alpha^{*\varepsilon}$ is the function zero.

The proof of Theorem 4.1 is based on the same plate theory method as that of Theorem 3.3, so we omit it (see Sac-Épée [11] for details). □

4.2. The Parameter ε Tends to Zero in Problem (4.2).

THEOREM 4.2. *Assume that $f_3^{*\varepsilon} \to f_3^*$ strongly in $L^\infty\big((0,T); L^2(\Omega)\big)$. Then, there exists an operator $P^\varepsilon \in \mathcal{L}\Big(L^\infty\big((0,T); H^2(\Omega_\varepsilon)\big), L^\infty\big((0,T); H^2(\Omega)\big)\Big)$ such that*

$$\varepsilon P^\varepsilon(U_3^{*\varepsilon}) \rightharpoonup U_3^{**} \quad \text{weakly} \star \text{ in } L^\infty\big((0,T); H_0^2(\Omega)\big),$$

where the function U_3^{**} satisfies

$$\frac{|\mathcal{Y}^*|}{|\mathcal{Y}|} \frac{\partial^2 U_3^{**}}{\partial t^2} + \frac{\partial^2}{\partial z_\tau \partial z_\eta}\Big(q_{\tau\eta\theta\rho} \frac{\partial^2 U_3^{**}}{\partial z_\theta \partial z_\rho}\Big) = \mu_{g_3^*} \qquad \text{in } \omega \times (0,T), \tag{4.3}$$

$$
\begin{aligned}
&U_3^{**} = 0 \quad \text{and} \quad \frac{\partial U_3^{**}}{\partial n} = 0 && \text{on } \partial\omega \times (0,T),\\
&U_3^{**}(0) = 0 \quad \text{and} \quad \frac{\partial U_3^{**}}{\partial t}(0) = V_3^{*1} && \text{in } \omega \times \{0\}.
\end{aligned} \tag{4.4}
$$

The homogenized coefficients are given by

$$q_{\tau\eta\theta\rho} = \frac{1}{|\mathcal{Y}^*|} \int_{\mathcal{Y}^*} \Big(b_{\tau\eta\theta\rho} - b_{\alpha\beta\theta\rho} \frac{\partial^2 \chi^{\tau\eta}}{\partial z_\alpha \partial z_\beta}\Big)\, dy, \tag{4.5}$$

the function $\chi^{\tau\eta}$ *being solution of*

$$
\begin{aligned}
&- b_{\theta\rho\alpha\beta}\frac{\partial^4(-\chi^{\tau\eta}+y^{\tau\eta})}{\partial y_\alpha \partial y_\beta \partial y_\theta \partial y_\rho} = 0 && \text{in } \mathcal{Y}^*,\\
&b_{\theta\rho\alpha\beta}\frac{\partial^3(-\chi^{\tau\eta}+y^{\tau\eta})}{\partial y_\alpha \partial y_\theta \partial y_\rho} n_\eta = 0 && \text{on } \partial s,\\
&b_{\theta\rho\alpha\beta}\frac{\partial^2(-\chi^{\tau\eta}+y^{\tau\eta})}{\partial y_\theta \partial y_\rho} n_\tau = 0 && \text{on } \partial s,\\
&\chi^{\tau\eta} \quad \mathcal{Y}\text{ - periodic with a zero average,}
\end{aligned}
\tag{4.6}
$$

with

$$
y^{11} = \frac{1}{2}y_1^2, \quad y^{12} = y^{21} = \frac{1}{2}y_1 y_2, \quad y^{22} = \frac{1}{2}y_2^2.
$$

PROOF OF THEOREM 4.2. Set $\mathcal{V}(\omega_\varepsilon) = \{v \in H^2(\omega_\varepsilon), v = \dfrac{\partial v}{\partial n} = 0 \text{ on } \partial\omega\}$. Multiplying system (4.2) by the test function $\frac{\partial U_3^{*\varepsilon}}{\partial t}$ and integrating from 0 to t, we derive

$$
\begin{aligned}
\|\frac{\partial U_3^{*\varepsilon}}{\partial t}\|^2_{L^\infty((0,T);L^2(\omega_\varepsilon))} + \|U_3^{*\varepsilon}\|^2_{L^\infty((0,T);\mathcal{V}(\omega_\varepsilon))} \leq \\
\leq \frac{C}{\varepsilon^2}(\varepsilon^2 + \|C_{g_3^*}\|^2_{L^\infty(0,T)} + \|C_{\frac{\partial g_3^*}{\partial t}}\|^2_{L^\infty(0,T)}).
\end{aligned}
\tag{4.7}
$$

Hence,

$$
\varepsilon\,\|\frac{\partial U_3^{*\varepsilon}}{\partial t}\|^2_{L^\infty((0,T);\mathcal{V}(\omega_\varepsilon))} + \varepsilon\,\|U_3^{*\varepsilon}\|^2_{L^\infty((0,T);\mathcal{V}(\omega_\varepsilon))} \leq K.
$$

Set $\xi^\varepsilon_{\tau\eta} = \dfrac{1}{12} b_{\tau\eta\theta\rho}\dfrac{\partial^2(\varepsilon U_3^{*\varepsilon})}{\partial z_\theta \partial z_\rho}$, and denote $\widetilde{\xi^\varepsilon_{\tau\eta}}$ its extension by 0 in s_ε. Combining the methods of Duvaut [9] and Brizzi-Chalot [2], we define a suitable extension operator P^ε and, up to a subsequence,

$$
\begin{aligned}
&P^\varepsilon(\varepsilon\, U_3^{*\varepsilon}) \rightharpoonup U_3^{**} \quad \text{weakly} \star \text{ in } L^\infty((0,T);H_0^2(\omega)),\\
&\widetilde{\xi^\varepsilon_{\tau\eta}} \rightharpoonup \xi^*_{\tau\eta} \quad \text{weakly} \star \text{ in } L^\infty((0,T);L^2(\omega)).
\end{aligned}
$$

Multiplying (4.2) by $\varphi \in \mathcal{D}([0,T[,\mathcal{D}(\omega))$, integrating by parts and making ε tend to zero, we obtain

$$
\frac{|\mathcal{Y}^*|}{|\mathcal{Y}|}\frac{\partial^2 U_3^{**}}{\partial t^2} + \frac{\partial^2 \xi^*_{\tau\eta}}{\partial z_\tau \partial z_\eta} = \mu_{g_3^*} - \mu_\chi U_3^{**} - \mu_\gamma \frac{\partial U_3^{**}}{\partial t} \quad \text{in } \omega \times (0,T). \tag{4.8}
$$

We now use the energy method (see Bensoussan-Lions-Papanicolaou [1], Tartar [12]) in order to express $\xi^*_{\tau\eta}$ in terms of U_3^{**}. Let us consider the adjoint problem

$$
\begin{aligned}
&b_{\theta\rho\tau\eta}\frac{\partial^4 W^{\alpha\beta}}{\partial y_\tau \partial y_\eta \partial y_\theta \partial y_\rho} = 0 && \text{in } \mathcal{Y}^*,\\
&b_{\theta\rho\tau\eta}\frac{\partial^3 W^{\alpha\beta}}{\partial y_\tau \partial y_\theta \partial y_\rho} n_\eta = 0 \quad \text{and} \quad b_{\theta\rho\tau\eta}\frac{\partial^2 W^{\alpha\beta}}{\partial y_\theta \partial y_\rho} n_\tau = 0 && \text{on } \partial s,\\
&W^{\alpha\beta} - y^{\alpha\beta} \quad \mathcal{Y} \text{ periodic with a zero average.}
\end{aligned}
$$

We introduce the functions

$$W^{\varepsilon,\alpha\beta}(z) = \varepsilon^2 P^\varepsilon(W^{\alpha\beta}(\frac{z}{\varepsilon})) \text{ and } \chi^{\alpha\beta} = -W^{\alpha\beta} + y^{\alpha\beta}.$$

Let us use $\xi^\varepsilon_{\tau\eta}$ in Problem (4.2), and multiply this system by $\varphi W^{\varepsilon,\alpha\beta}$ with $\varphi \in \mathcal{D}([0,T[,\mathcal{D}(\omega))$. Let us also multiply by $\varepsilon\varphi U_3^{*\varepsilon}$ the system satisfied by $W^{\varepsilon,\alpha\beta}$. Passing to the limit in ε, we derive the claimed homogenized equation. The other equations of the homogenized problem are obtained by using other appropriate test functions. □

In the particular case where the constants satisfy

$$\|C_{g_3^*}\|_{L^\infty(0,T)} + \|C_{\frac{\partial g_3^*}{\partial t}}\|_{L^\infty(0,T)} = 0, \tag{4.9}$$

we have $U_3^{**} = 0$ and we establish the more precise following result.

THEOREM 4.3. *Assume that $f_3^{*\varepsilon} \to f_3^*$ strongly in $L^\infty((0,T);L^2(\Omega))$ and that (4.9) holds. Then,*

$$P^\varepsilon(U_3^{*\varepsilon}) \rightharpoonup U_3^* \quad \textit{weakly} \star \textit{ in } L^\infty((0,T);H_0^2(\Omega)),$$

where U_3^* satisfies the equation

$$\begin{aligned}\frac{|\mathcal{Y}^*|}{|\mathcal{Y}|}\frac{\partial^2 U_3^*}{\partial t^2} + \frac{\partial^2}{\partial z_\tau \partial z_\eta}(q_{\tau\eta\theta\rho}\frac{\partial^2 U_3^*}{\partial z_\theta \partial z_\rho}) &= \\ = \frac{|Y^*|}{|Y|}\int_{-\frac{1}{2}}^{\frac{1}{2}} f_3^* dz_3 - \frac{|\partial S|}{|Y|}\lambda U_3^* - \frac{|\partial S|}{|Y|}\gamma\frac{\partial U_3^*}{\partial t} &\quad \textit{in } \omega \times (0,T),\end{aligned}$$

and the same boundary conditions as in (4.4).

PROOF OF THEOREM 4.3. Using (4.9) in the a priori estimate (4.7), we have

$$\|\frac{\partial U_3^{*\varepsilon}}{\partial t}\|_{L^\infty\left((0,T);L^2(\omega_\varepsilon)\right)} + \|U_3^{*\varepsilon}\|_{L^\infty\left((0,T);\mathcal{V}(\omega_\varepsilon)\right)} \leq K.$$

For the rest of the proof, the methods are quite similar to those used in the proof of Theorem 3.1. For more details, we refer the reader to Sac-Épée [11]. □

COMMENTS. It would be interesting to compare numerically the limit coefficients $\frac{1}{12}P_{\tau\eta\theta\rho}$ and $q_{\tau\eta\theta\rho}$ which appear respectively in (3.16) and (4.3). We conjecture that they are different because on simpler problems, they are known to be different. This is the case for instance for gridworks satisfying a stationary thermal problem with homogeneous Neumann conditions on the boundary of the holes (see Cioranescu-Saint Jean Paulin [8]). If our conjecture holds, this means that when $\varepsilon << e$, one has to take the limit problem (3.16) and when $e << \varepsilon$, one has to take the limit problem (4.3).

REFERENCES

[1] A. BENSOUSSAN, J.-L. LIONS and G. PAPANICOLAOU, *Asymptotic analysis for periodic structures*, North-Holland, Amsterdam, 1978.

[2] R. BRIZZI and J.-P. CHALOT, Homogénéisation dans des ouverts à frontière fortement oscillante, Thèse de spécialité, Nice, 1978.

[3] D. CAILLERIE, Etude de quelques problèmes de perturbations en théorie de l'élasticité et de la conduction thermique, Thèse d'état, Paris, 1982.

[4] Ph. CIARLET, A justification of the Von Karman equations, Arch. Rat. Mech. Anal. 73 (1980), 349-389.

[5] Ph. CIARLET and Ph. DESTUYNDER, A justification of the two-dimentional linear plate model, J. Mécanique, 18 (1979), 315-344.

[6] D. CIORANESCU and P. DONATO, Homogénéisation du problème de Neumann non homogène dans des ouverts perforés, Asymptotic Analysis 1 (1988) 115-138.

[7] D. CIORANESCU and J. SAINT JEAN PAULIN, Homogenization in open sets with holes, J. Math. Pures et Appl. 65 (1986), 403-422.

[8] D. CIORANESCU and J. SAINT JEAN PAULIN, *Homogenization of Reticulated Structures*, Applied Mathematical Sciences 136, Springer New-York1999.

[9] G. DUVAUT, Comportement macroscopique d'une plaque perforée périodiquement, *Singular Perturb. Bound. Layer Theory*, Proc. Conf. Lyon 1976, Lecture Notes Math. 594, Springer-Verlag 1977, 131-145.

[10] A. ELJENDY, Boundary stabilization of the wave equation, Preprint 143, Jyvaskyla, May 1992.

[11] J.-M. SAC-ÉPÉE, Étude de problèmes d'homogénéisation définis sur des plaques minces perforées périodiquement, Thèse de l'Université de Metz, Metz, 1994. For details, see http://poncelet.sciences.univ-metz.fr/~jmse/these.html

[12] L. TARTAR, Quelques remarques sur l'homogénéisation. in *Functional Anal. and Num. Anal.*, Proc. Japan -France Seminar, ed. Fujita, Soc. for the Promotion of Science, Japan 1978, 469-482.

Singular perturbations with non-smooth limit and finite element approximation of layers for model problems of shells

J. SANCHEZ-HUBERT Laboratoire de Mécanique, Université de Caen, Caen, France

E. SANCHEZ PALENCIA Laboratoire de Modélisation en Mécanique, Université Paris VI, Paris, France

1 Introduction

We consider perturbation problems depending on a small parameter $\varepsilon \in (0,1]$ for $\varepsilon > 0$ which are variational problems in the energy space V. The energy space of the limit problem as $\varepsilon \searrow 0$ is $V_a \supset V$. The corresponding duals are such that $V_a' \subset V'$. When the right hand side $f \in V_a'$ classical singular perturbation theory shows that the solution u^ε converges to the solution u^0 of the limit problem. We are here concerned with the case $f \in V'$ but $f \notin V_a'$. This situation often appears in shell theory where the space V_a' is, in some cases, so small that it does not contain usual loadings. In this case obviously the limit problem does not make sense as a variational problem. Nevertheless it may be shown [3] that the solutions u^ε converge in some abstract topology which goes out of the variational framework. In usual partial differential equations, the convergence may not take place in the distribution sense. Two examples of such a situation in one space dimension are handled in Sect.3. In the first one (section 3.1) the limit exists in the distribution sense but the limit solution is not variational. In the second one (section 3.2) there is no limit in the usual sense (even as distributions). In any case when $f \notin V_a'$ the energy of the solution u^ε tends to infinity as $\varepsilon \searrow 0$ (see theorem 2). It will be seen from examples that in such cases the energy is concentrated in boundary or internal layers. In fact a part of the present work is concerned with the asymptotic description of the boundary and internal layers using the classical formal method of matched asymptotic expansions (see for instance [17]). We shall not give here a complete description of the asymptotic expansions which may be found in [10] and [9].

Concerning thin shell theory, it is well known [15] that they may be classified in inhibited and non inhibited ones; this is equivalent to saying that the middle surface with kinematic boundary conditions is geometrically rigid or not. In the inhibited case, denoting by ε the relative thickness of the shell, we are in the above situation (see [15], Sect.VI.1.4). For $\varepsilon > 0$ the energy space V is, roughly speaking, $H^1 \times H^1 \times H^2$ where H^1 deals with the two tangential components and H^2 with the normal one of the displacement vector. The limit problem is associated with the membrane approximation where the flexion terms are neglected. The corresponding energy space V_a is the completion of V with the membrane energy norm. It is known [15], Chap.VII that the structure of V_a depends highly on the shape and on the kinematic boundary conditions. It turns out that the space V_a' is very small so that $f \in V_a'$ implies smoothness conditions which in general are not satisfied in usual physical problems. Descriptions of this situation for hyperbolic and parabolic middle surfaces may be found in [15], Sect.VII.2.4 and VII.4.2 respectively. Moreover, in certain cases the space V_a' is so small that it does not contain the space $\mathcal{D}$ of the test functions of distributions (see [12], [13], [6]). Accordingly, it is known that the numerical approximation of shell problems exhibits particular features for very small ε (see [4], [16]). The exact solution u^ε becomes less and less smooth as $\varepsilon \searrow 0$ developing singularities which propagate along the characteristic curves [8]. As a consequence the finite element convergence $h \searrow 0$ is not uniform with respect to ε so that the smaller ε, the smaller h must be taken to get a good approximation (see [7], [6]).

In Sect.4 of this paper, instead of general shells, we shall only deal with a model problem in two variables x_1, x_2 and $u^\varepsilon = (u_1^\varepsilon, u_2^\varepsilon)$ where u_1^ε play the roleof the tangential components and u_2^ε the normal one. The problem in Sect.4 is parabolic so that it is a model of geometrically rigid developable surfaces. In Sect.4.1 we give a criterion for $f \in V_a'$ which is not satisfied for piecewise constant "forces" f bearing a discontinuity along a characteristic curve $x_2 = const.$ The corresponding solution u^ε develops an internal layer which is described in Sect.4.3. It appears that the energy concentrates in the internal layer with thickness of order $\mathcal{O}(\eta)$, $\eta = \varepsilon^{\frac{1}{3}}$ so that the layer is described in terms of the inner variables (x_1, y_2) with

$$y_2 = \frac{x_2 - Const.}{\eta} \tag{1.1}$$

The asymptotic behavior of u^ε is such that out of the layer it is of order

$\mathcal{O}(1)$ but inside the layer the components u_1^ε and u_2^ε tend to infinity as η^{-1} and η^{-2} respectively:

$$\begin{cases} u_1^\varepsilon = \frac{1}{\eta} U_1^\eta (x_1, y_2) = \frac{1}{\eta} U_1^0 (x_1, y_2) + \cdots \\ u_2^\varepsilon = \frac{1}{\eta^2} U_2^\eta (x_1, y_2) = \frac{1}{\eta^2} U_2^0 (x_1, y_2) + \cdots \end{cases} \tag{1.2}$$

In addition the study of the structure of the layer shows that the leading terms satisfy a constraint so that the corresponding equations involve a Lagrange multiplier.

Moreover as a consequence of the structure of the layer we consider the numerical approximation using anisotropic meshes (i.e. with triangles elongated in the tangential direction) for the finite element approximation. We recall that error estimates for anisotropic meshes were obtained in [2] for problems with a singularity along an edge and in [1] for convection - diffusion problems depending on a parameter ε with a limit solution exhibiting a discontinuity along a characteristic curve. We may compare [2] with [14] which used an isotropic adapted mesh to handle the same problem. Error estimates for anisotropic meshes in the model problem of Sect.4 are considered in sections 5 and 6. More precisely, we only consider the layer region, not the whole domain, and we take advantage of the structure of u^ε given by the formal asymptotic expansion procedure (2) and we replace u^ε by the leading term of the expansions (2) in order to handle a simple but essentially correct description of u^ε. Knowing the description in the (x_1, y_2) variables we use classical error estimates for isotropic meshes (i.e. satisfying the classical non-flattening condition) [5] in the (x_1, y_2) variables with polynomials of degrees k_1 and k_2 for u_1^ε and u_2^ε respectively. Coming back to the (x_1, x_2) variables we get the corresponding estimates for an anisotropic mesh the dimensions of the triangles being H and ηH in the tangential and normal directions respectively. It should be noticed that these estimates may also be obtained directly (and we checked them in certain cases) from error estimates for anisotropic interpolation theory [1], but our method shows more explicitly the influence of the asymptotic structure.

In Sect.5 we only consider the interpolation error. It appears that the relative error estimates for $u_1^\varepsilon (x_1, x_2)$ and $u_2^\varepsilon (x_1, x_2)$ are exactly the same as the ones for $U_1^0 (x_1, y_2)$ and $U_2^0 (x_1, y_2)$ with the corresponding isotropic meshes.

The problem of the error estimates for the Galerkin approximation is addressed in Sect.6 . The advantage of an anisotropic mesh follows from the comparison with classical estimates for an isotropic (in (x_1, x_2)) mesh: for a given error the number of triangles of the anisotropic mesh is $\eta = \varepsilon^{\frac{1}{3}}$ times the corresponding number for an isotropic mesh. Moreover, the local structure of the layer which involves a constraint evokes the possibility of improving the estimates by using special locking - free finite elements.

Numerical experiments for the model problem of Sect.4 may be bound in [9]. These experiments include several cases of loading and exhibit the corresponding patterns of layers (boundary and internal layers). Both isotropic and anisotropic meshes are used. The advantage of a refined anisotropic mesh appears clearly in certain cases.

2 Singular perturbations

Let V be a real Hilbert space, $b(u,v)$ a continuous, coercive and symmetric bilinear form on V and $a(u,v)$ a continuous and symmetric bilinear form on V. Let the form a satisfy the conditions

$$a(v,v) \geq 0, \quad a(v,v) = 0 \Rightarrow v = 0 \tag{2.3}$$

so that b is a scalar product on V with norm equivalent to that of V and $a(v,v)^{\frac{1}{2}}$ defines a norm on V. Let us denote by V_a the completion of V with this norm, then

$$V \subset V_a \tag{2.4}$$

with dense and continuous inclusion. Correspondingly, the dual spaces satisfy

$$V_a' \subset V' \tag{2.5}$$

We consider the family of problems with parameter $\varepsilon \in (0,1]$:

Problem 1 Let $f \in V'$, find $u^\varepsilon \in V$ satisfying

$$a(u^\varepsilon, v) + \varepsilon^2 b(u^\varepsilon, v) = \langle f, v \rangle \quad \forall v \in V \tag{2.6}$$

Obviously the solution of (6) exists and is unique. The energy of the solution is defined by

$$E(u^\varepsilon) = \frac{1}{2}\left[a(u^\varepsilon, u^\varepsilon) + \varepsilon^2 b(u^\varepsilon, u^\varepsilon)\right] \tag{2.7}$$

We note that in usual cases the expression of the energy is an integral on some domain so that we may define the energy on a part of that domain.

If $f \in V_a'$ then the limit problem: find $u \in V_a$ satisfying

$$a(u,v) = \langle f, v \rangle \quad \forall v \in V_a \tag{2.8}$$

has a unique solution.

We then have the classical result (see for instance [14], p. 196)

Theorem 1 *Let $f \in V_a'$ be fixed. Let u^ε and u be the solutions of (6) and (8) respectively. Then*

$$u^\varepsilon \to u \quad \text{in } V_a \text{ strongly} \tag{2.9}$$

and $E(u^\varepsilon)$ remains bounded.

Moreover we have (see [6] or [9])

Theorem 2 *Let $f \in V'$ be fixed. Let u^ε be the solution of (6), then*

1. The necessary and sufficient condition for $E(u^\varepsilon)$ to remain bounded is that $f \in V_a'$
2. If $f \notin V_a'$, then $E(u^\varepsilon)$ tends to infinity as $\varepsilon \searrow 0$.

Remark 1 *Often in shell theory (non well inhibited shells [15]) V_a' is a very small space so that usual loadings lead in the situation of theorem 2, part 2.*■

Remark 2 *In order give a sense to equations and boundary conditions, integrations by parts are done using a pivot space H analogous to L^2 which is identified to its dual. Usually $V \subset H \subset V'$ but it may happen that $V_a \not\subset H$ and $V_a' \not\supset H$. In shell theory this is the definition of non well inhibited.*■

Remark 3 *We shall see in certain examples with $f \notin V_a'$ that the energy (which tends to infinity) accumulates in boundary or internal layers. In other words, the energy out of the layers is asymptotically small with respect to the energy inside the layers.*■

Remark 4 *When $f \notin V_a'$, the limit problem (8) does not make sense as a variational problem in V_a but it may happen that it does in the Lions and Magenes theory [11].*■

3 Examples in one space dimension

In this section we consider some simple examples which are developed in [10]. We take the space $V = H_0^2(0,1)$ and the form

$$b(u,v) = \int_0^1 u''(x)\, v''(x)\, dx \tag{3.10}$$

with the completion space

$$V_a = H_0^1(0,1) \tag{3.11}$$

for the form

$$a(u,v) = \int_0^1 u'(x)\, v'(x)\, dx \tag{3.12}$$

Then, taking $H \equiv H' = L^2(0,1)$, we have

$$V' = H^{-2}(0,1), \quad V'_a = H_0^{-1}(0,1) \tag{3.13}$$

we shall consider two cases of loadings $f \notin V'_a$.

3.1 First example

In the previous situation, we consider $f = \delta'_{\frac{1}{2}}$ where $\delta_{\frac{1}{2}}$ denotes the Dirac distribution at $x = \frac{1}{2}$. We note that $\delta'_{\frac{1}{2}} \in H^{-2}$ and even $\delta'_{\frac{1}{2}} \in \Xi^{-2}$ of [11]. So, the problem is

$$\begin{cases} \left(-\dfrac{d^2}{dx^2} + \varepsilon^2 \dfrac{d^4}{dx^4}\right) u^\varepsilon = \delta'_{\frac{1}{2}} \\ u^\varepsilon(0) = u^\varepsilon(1) = \dfrac{du^\varepsilon}{dx}(0) = \dfrac{du^\varepsilon}{dx}(1) = 0 \end{cases} \tag{3.14}$$

Moreover, the limit problem

$$\begin{cases} -\dfrac{d^2u}{dx^2} = \delta'_{\frac{1}{2}} \\ u(0) = u(1) = 0 \end{cases} \tag{3.15}$$

has a unique solution in the sense of Lions - Magenes:

$$u(x) = \begin{cases} x & \text{for} \quad 0 < x < \frac{1}{2} \\ x - 1 & \text{for} \quad \frac{1}{2} < x < 1 \end{cases} \tag{3.16}$$

There is an internal layer in the vicinity of $x = \frac{1}{2}$ which is described by an inner asymptotic expansion the leading term of which is

$$v^0(y) = \begin{cases} -\frac{1}{2}e^y + \frac{1}{2} & \text{for } y < 0 \\ \frac{1}{2}e^{-y} - \frac{1}{2} & \text{for } y > 0 \end{cases} \tag{3.17}$$

in terms of the inner variable $y = \varepsilon^{-1}\left(x - \frac{1}{2}\right)$.

It is easily seen that the energy out of the layer is $\mathcal{O}(1)$ and in the layer (17) of order $\mathcal{O}(\varepsilon^{-1})$. So, the total energy tends to infinity according to theorem 2, part 2 and it accumulates in the layer.

We note that in the finite element approximation for small ε, a refinement of the mesh in the vicinity of $x = \frac{1}{2}$ allows to have a good description of the layer. Without this refinement we only have a good approximation of u^ε out of the layer.

3.2 Second example

Let us now consider $f = x^{-p-2} + (1-x)^{-p-2}$ for some $p \in \left(0, \frac{1}{2}\right)$ so that $f \in H^{-2}$, $f \notin H^{-1}$, $f \notin \Xi^{-2}$. It is easily seen that the limit problem

$$\begin{cases} -\dfrac{d^2u}{dx^2} = x^{-p-2} + (1-x)^{-p-2} \\ u(0) = u(1) = 0 \end{cases} \tag{3.18}$$

has no solution. Indeed, from the equation, it follows that $u \cong x^{-p} + (1-x)^{-p} + Ax + B$ and the boundary conditions cannot be satisfied. In fact, the formal asymptotic expansions are:

$$\begin{array}{c} \text{Outer expansion: } u^{\varepsilon}(x) = \varepsilon^{-p}C + \mathcal{O}(1) \\ \text{Inner expansion (near } x=0\text{): } u^{\varepsilon} = \varepsilon^{-p}v^0(y) + \cdots \quad y = \dfrac{x}{\varepsilon} \end{array} \tag{3.19}$$

where v^0 is the unique solution of

$$\begin{cases} \left(-\dfrac{d^2}{dy^2} + \dfrac{d^4}{dy^4}\right) v^0 = y^{-p-2} \\ v^0(0) = \dfrac{dv^0}{dy}(0) = 0 \end{cases} \tag{3.20}$$

which is bounded on $(0, +\infty)$. Of course there is a symmetric layer near $x = 1$.

It appears that the energy is $\mathcal{O}(\varepsilon^{-2p})$ out of the layers and of order $\mathcal{O}(\varepsilon^{-2p-1})$ in the layers. In order to obtain a good finite element approximation for small ε the mesh step must be very fine inside the layers. Otherwise the boundary layers are badly computed as well as the region out of the layers which depends of them. It will be noticed that the refinement out of the layer is not necessary.

4 A model problem of shells

The model problem $P(\varepsilon)$ is defined as follows. Let us consider the domain $\Omega = (0, \pi) \times (0, \pi)$ of the plane $x = (x_1, x_2)$. The whole boundary $\partial\Omega$ is "clamped".

The configuration space V is

$$V = H_0^1(\Omega) \times H_0^2(\Omega) \tag{4.21}$$

We consider the two bilinear forms

$$a(u, v) = \int_{\Omega} \left[(\partial_1 u_1)(\partial_1 v_1) + (\partial_2 u_1 - u_2)(\partial_2 v_1 - v_2)\right] dx \tag{4.22}$$

$$b(u,v) = \int_\Omega \sum_{|\alpha|\leq 2} \partial_\alpha u_2 \partial_\alpha v_2 dx \equiv (u_2, v_2)_2 \tag{4.23}$$

Let f be an element of V' (the dual of V), the problem $P(\varepsilon)$ writes:

$$\left\{ \begin{array}{l} \text{Find } u^\varepsilon \in V \text{ such that } \forall v \in V \\ a\left(u^\varepsilon, v\right) + \varepsilon^2 b\left(u^\varepsilon, v\right) = \langle f, v \rangle_{V'V} \end{array} \right. \tag{4.24}$$

This problem is somewhat classical. It is equivalent to the system of equations

$$\left\{ \begin{array}{l} -\Delta u_1^\varepsilon + \partial_2 u_2^\varepsilon = f_1 \\ -\partial_2 u_1^\varepsilon + u_2^\varepsilon + \varepsilon^2 \left(\Delta^2 u_2^\varepsilon - \Delta u_2^\varepsilon + u_2^\varepsilon\right) = f_2 \end{array} \right. \tag{4.25}$$

with the "clamping" boundary conditions corresponding to $u^\varepsilon \in V$ and where

$$\langle f, v \rangle_{V'V} = \int_\Omega f_i v_i dx \tag{4.26}$$

We denote by V_a the completion of V with the norm $a\left(u,u\right)^{\frac{1}{2}}$. We note that this space is somewhat abstract so that its description is not easy as well as that of its dual V_a'; consequently it is useful to get a criterion to characterize the loadings $f \in V_a'$.

4.1 Criterion for $f \in V_a'$

In this subsection, we shall consider functionals f defined in (26) by functions f_i (on Ω).

The form a defined in (22) is associated with the two quantities

$$\left\{ \begin{array}{l} \gamma_1\left(u\right) \equiv \partial_1 u_1 \\ \gamma_2\left(u\right) \equiv \partial_2 u_1 - u_2 \end{array} \right. \tag{4.27}$$

Let us define the injective application γ by (27)

$$V \xrightarrow{\gamma} \left(L^2\right)^2$$

this application may be continued by continuity to V_a to an isomorphism from V_a onto its range X which is clearly a closed subspace of $\left(L^2\right)^2$.

Let us now consider the functional

$$l_f(v) = \int_\Omega \left(f_1 v_1 + f_2 v_2\right) dx \tag{4.28}$$

defined on V. Then we have [9]:

Theorem 3 *The functional defined by (26) may be extended by continuity to V_a (that is equivalent to $f \in V_a'$) if, and only if, there exists $T = (T^1, T^2) \in$*

$(L^2)^2$ *such that*

$$\int_{\Omega} (f_1 v_1 + f_2 v_2)\, dx = \left(T^1, \gamma_1(v)\right)_{L^2} + \left(T^2, \gamma_2(v)\right)_{L^2} \ \forall v \in V. \quad (4.29)$$

It is easily seen that from the boundary conditions on V (29) is equivalent to the existence of $(T^1, T^2) \in (L^2)^2$ such that

$$\begin{cases} f_1 = -\partial_1 T_1 - \partial_2 T_2 \\ f_2 = -T_2 \end{cases} \quad (4.30)$$

Obviously, if f is a smooth function there exist (T_1, T_2) satisfying (30) so that $f \in V_a'$. Oppositely, if f_2 is piecewise smooth with a discontinuity along $x_2 = C$ then $T_1 \in L^2$ cannot exist because the first equation (30) contains a term in $\delta(x_2 - C)$.

We shall take in the sequel

$$\begin{cases} f_1 \equiv 0 \\ f_2 = \begin{cases} 0 \text{ if } x_2 < \frac{\pi}{2} \\ 1 \text{ if } x_2 > \frac{\pi}{2} \end{cases} \end{cases} \quad (4.31)$$

so that $f \notin V_a'$

4.2 Remarks on the limit problem

Clearly the limit problem is

$$\begin{cases} -\Delta u_1 + \partial_2 u_2 = f_1 \\ -\partial_2 u_1 + u_2 = f_2 \end{cases} \quad (4.32)$$

with $u_1 = 0$ on $x_1 = 0$ and $x_1 = \pi$.

This problem is equivalent to

$$-\partial_1^2 u_1 = f_1 - \partial_2 f_2 \quad (4.33)$$

$$u_2 = f_2 + \partial_2 u_1 \quad (4.34)$$

Under this form it is obvious that the limit problem is essentially equivalent to (33) for u_1. This equation is an elliptic one with respect to x_1 with parameter x_2. In Ω it is parabolic with double characteristics $x_2 = const.$ As x_2 appears as a parameter, f_1 and f_2 may be chosen to be distributions of x_2 with values in an appropriate space for the variable x_1. Consequently, when f_2 is not sufficiently smooth with respect to x_2, the equations (32) and the boundary conditions keep a sense in a more general framework which is not that of the variational problem (8).

Denoting by $-A$ the Laplace operator in x_1 with Dirichlet condition on $x_1 = 0$ and $x_1 = \pi$ and taking f given by (31) it is easily seen that the solution is

$$\begin{cases} u_1(\bullet, x_2) = -\left(A^{-1}1\right)\delta\left(x_2 - \frac{\pi}{2}\right) \\ u_2(\bullet, x_2) = f_2(\bullet, x_2) - \left(A^{-1}1\right)\delta'\left(x_2 - \frac{\pi}{2}\right) \end{cases} \tag{4.35}$$

where 1 denote the function of x_1 equal to 1. It is clear that the solution (35) exhibits a very strong singularity along $x_2 = \frac{\pi}{2}$. For small ε, u^ε develops an internal layer in the vicinity of $x_2 = \frac{\pi}{2}$ which converges to the singular solution (35).

4.3 Internal layer in the vicinity of $x_2 = \frac{\pi}{2}$

We consider the case when the loading f is given by (31). In order to find the appropriate scaling to describe the layer, we use a method based on the structure of the sinusoidal solutions of the system (see [9]). As the characteristics of the limit problem are normal to the vector $(0,1)$ there exist solutions of the form

$$u(x_1, x_2) = v e^{i(\xi_1 x_1 + \xi_2 x_2)} \tag{4.36}$$

with $\xi_1 = 0, \xi_2 \neq 0$. Let us search, for $\varepsilon > 0$, solutions of the form

$$u^\varepsilon(x_1, x_2) = v^\varepsilon e^{i\xi_1 + \mu x_2} \tag{4.37}$$

with $\xi_1 \neq 0$ and $|\mu| \to +\infty$ (i.e. such that $(\xi_1 \text{ (real)}, \xi_2 \equiv -i\mu)$ tends to be proportional to $(0,1)$). The solutions (37) are sinusoidal in x_1 with wave length of order $\mathcal{O}(1)$ and very fast variations in x_2. By substitution of (37) in the homogeneous system associated with (32) it is easily seen that

$$\mu \cong e^{\frac{ik\pi}{3}} \xi_1^{\frac{1}{3}} \varepsilon^{-\frac{1}{3}} \tag{4.38}$$

It then appears that the solutions with $\xi_1 = \mathcal{O}(1)$, $\mu = \mathcal{O}\left(\varepsilon^{-\frac{1}{3}}\right)$ have a characteristic length of variation in the x_2 direction of order $\mathcal{O}\left(\varepsilon^{\frac{1}{3}}\right)$; this corresponds to a layer of thickness $\mathcal{O}(\eta)$,

$$\eta = \varepsilon^{\frac{1}{3}} \tag{4.39}$$

It also follows from the above considerations that the corresponding scaling of u^ε is

$$\frac{u_2^\varepsilon}{u_1^\varepsilon} = \mathcal{O}\left(\eta^{-1}\right). \tag{4.40}$$

consequently we search in the layer expansions of the form

$$\begin{cases} u_1^\varepsilon(x_1,x_2) = \theta(\eta)\,U_1^\eta(x_1,y_2) = \theta(\eta)\,U_1^0(x_1,y_2) + \cdots \\ u_2^\varepsilon = \eta^{-1}\theta(\eta)\,U_2^\eta(x_1,y_2) = \eta^{-1}\theta(\eta)\,U_2^0(x_1,y_2) + \cdots \end{cases} \tag{4.41}$$

where

$$y_2 = \frac{x_2 - \frac{\pi}{2}}{\eta} \tag{4.42}$$

and $\theta(\eta)$ is undetermined for the time being.

By substituting (41) and (42) in the variational formulation (24) we obtain the corresponding variational formulation in the domain transformed by (42). At the limit $\eta \searrow 0$ the domain Ω becomes the strip $D = (0,\pi) \times (-\infty, +\infty)$. The variational formulation in D is then

$$\begin{aligned} &\theta(\eta) \int_D \Big\{ \partial_1 U_1^\eta \partial_1 V_1^\eta + \frac{1}{\eta^2}\left(D_2 U_1^\eta - U_2^\eta\right)\left(D_2 V_1^\eta - V_2^\eta\right) \\ &+ \left[D_2^2 U_2^\eta D_2^2 V_2^\eta + \cdots\right]\Big\}\, dx_1 dy_2 = \int_D \Big\{\Big[f_1\left(x_1, \frac{\pi}{2}\right) + \cdots\Big] V_1^\eta \\ &+ \frac{1}{\eta}\Big[f_2\left(x_1,\frac{\pi}{2}\right) + \eta y_2 \partial_2 f_2\left(x_1, \frac{\pi}{2}\right) + \cdots\Big] V_2^\eta \Big\}\, dx_1 dy_2 \end{aligned} \tag{4.43}$$

where $\partial_1 = \frac{\partial}{\partial x_1}$, $D_2 = \frac{\partial}{\partial y_2}$. Moreover, the asymptotic expansion (41) must match with the outer expansion as $|y_2| \to +\infty$. But from the structure of the singularity in (35) it clearly appears that $\theta(\eta)$ must tend to infinity as $\eta \searrow 0$ so that the matching of the leading terms U_1^0, U_2^0 implies that they tend to zero as $|y_2| \to +\infty$. Then the function θ is chosen such that the layer problem in D for the leading terms has a finite not vanishing right hand side. This gives (see [9] for details) $\theta = \eta^{-1}$ and the constraint

$$D_2 U_1^0 - U_2^0 = 0 \tag{4.44}$$

as well as the variational limit problem for U_1^0

$$\int_D \left\{\partial_1^2 U_1^0 V_1^0 + D_2^3 U_1^0 D_2^3 V_1^0\right\} dx_1 dy_2 = \int_{D+} D_2 V_1^0 dx_1 dy_2 \tag{4.45}$$

where $D^+ = (0,\pi) \times (0, +\infty)$ (analogously $D^- = (0,\pi) \times (-\infty, 0)$).

In each region D^+ and D^- we obtain the equation

$$-\partial_1^2 U_1^0 - D_2^6 U_1^0 = 0 \tag{4.46}$$

and the boundary conditions (taking on $x_2 = \frac{\pi}{2}$ the quantities V_1^0, $D_2V_1^0$ and $D_2^2V_1^0$ continuous) become the interface conditions

$$\begin{cases} [|D_2^3U_1^0(x_1,0)|] = 0 \\ [|D_2^4U_1^0(x_1,0)|] = 0 \\ [|D_2^5U_1^0(x_1,0)|] = 1 \end{cases} \tag{4.47}$$

where $[|\bullet|]$ denotes the jump across the interface. Obviously, we shall also prescribe

$$\left[\left|U_1^0\right|\right] = \left[\left|D_2U_1^0\right|\right] = \left[\left|D_2^2U_1^0\right|\right] = 0. \tag{4.48}$$

Moreover, the function U_1^0 satisfies the boundary conditions

$$U_1^0(0,y_2) = U_1^0(\pi,y_2) = 0 \tag{4.49}$$

and the matching conditions

$$U_1^0(x_1,y_2) \underset{y_2\to\pm\infty}{\to} 0 \tag{4.50}$$

The solution is then searched in the form

$$U_1^0(x_1,y_2) = \sum_{n=1}^{+\infty} a_n(y_2)\sin nx_1 \tag{4.51}$$

We search in each region D^+ and D^- a solution with coefficients a_n^+ and a_n^- respectively, which satisfy

$$-a_n^{\pm(6)}(y_2) + n^2a_n^{\pm}(y_2) = 0 \tag{4.52}$$

and are bounded in their corresponding domain of definition, so that

$$\begin{cases} a_n^+(y_2) = \sum\limits_{k=2}^{4} c_k^+ \exp\left[n^{\frac{1}{3}}\exp\left(\frac{ik\pi}{3}\right)y_2\right] \\ a_n^-(y_2) = \sum\limits_{k=0}^{3} c_k^- \exp\left[n^{\frac{1}{3}}\exp\left(\frac{i(2k+1)\pi}{3}\right)y_2\right] \end{cases} \tag{4.53}$$

Moreover, they must satisfy the interface conditions

$$\begin{aligned} &[|a_n|] = [|a_n^1|] = \cdots = \left[\left|a_n^{(4)}\right|\right] = 0 \\ &\left[\left|a_n^{(5)}\right|\right] = \Phi_n \end{aligned} \tag{4.54}$$

where Φ_n is the coefficient of $\sin nx_1$ in the Fourier expansion of the function unity. The coefficients and the structure of the layer are determined.

5 Interpolation error for anisotropic finite elements in the layers

In order to disclose the essential trends of finite element approximation in the layers, we shall consider the model problem of section 4 in a domain D_L which is the domain D truncated at $|y_2| = L$ for sufficiently large L (note that this modification is not essential because of the exponential decreasing of the solution at infinity). We shall also neglect unessential terms, in particular we shall retain only the leading terms in the layer. In this context, we shall also assume that U_2^0 and $D_2 U_2^0$ vanish for $|y_2| = L$. More precisely, we shall consider the domain

$$D_L = (0, \pi) \times (-L, L) \tag{5.55}$$

in the (x_1, y_2) variables and the corresponding domain

$$D_{\eta L} = (0, \pi) \times (-\eta L, \eta L) \tag{5.56}$$

in the (x_1, x_2) variables. We recall that $y_2 = \frac{x_2 - \frac{\pi}{2}}{\eta}$, $\eta = \varepsilon^{\frac{1}{3}}$.

The variational formulation of the problem for u^ε in the domain $D_{\eta L}$ writes

$$\int_{D_{\eta L}^+} \left[\left(\partial_2 u_1^\varepsilon - u_2^\varepsilon \right) \left(\partial_2 v_1 - v_2 \right) + \partial_1 u_1^\varepsilon \partial_1 v_1 + \varepsilon^2 \partial_2^2 u_2^\varepsilon \partial_2^2 v_2 \right] dx_1 dx_2$$

$$= \int_{D_{\eta L}} f_2 \left(x_1, \frac{\pi}{2} + 0 \right) v_2 dx_1 dx_2 \tag{5.57}$$

Now from

$$\begin{cases} u_1^\varepsilon (x_1, x_2) = \frac{1}{\eta} U_1^\eta (x_1, y_2) \approx \frac{1}{\eta} U_1^0 (x_1, y_2) \\ u_2^\varepsilon (x_1, x_2) = \frac{1}{\eta^2} U_2^\eta (x_1, y_2) \approx \frac{1}{\eta^2} U_2^0 (x_1, y_2) \end{cases} \tag{5.58}$$

follows the variational problem for U^η in the domain D_L^+ :

$$\int_{D_L^+} \left\{ \frac{1}{\eta^2} \left(D_2 U_1^\eta - U_2^\eta \right) \left(D_2 V_1^\eta - V_2^\eta \right) + \partial_1 U_1^\eta \partial_1 V_1^\eta + D_2^2 U_2^\eta D_2^2 V_2^\eta \right\} dx_1 dy_2$$

$$= \int_{D_L^+} f_2 \left(x_1, \frac{\pi}{2} + 0 \right) V_2^\eta dx_1 dy_2 \tag{5.59}$$

We note that the problem (57) is a singular perturbation problem whereas the problem (59) is a penalty one. This is the reason why we had the constraint (44) at the leading term in the layer.

Let us consider the two functions U_1^η, U_2^η belonging to $H^1(D_L)$ and $H^2(D_L)$ respectively and their interpolation approximations U_{1H}^η, U_{2H}^η obtained in the classical framework [5] with polynomials of degrees k_1 and k_2

and an isotropic (i.e. satisfying the non-flattening condition) mesh of step $\mathcal{O}(H)$. Let us denote by δ_H the interpolation error. We classically have

$$\|\delta_H U_1^\eta\|_1 \leq C H^{k_1} \tag{5.60}$$

$$\|\delta_H U_2^\eta\|_2 \leq C H^{k_2 - 1} \tag{5.61}$$

where C denotes generic constants independent of the parameters. Due to the approximation (58) U^η is considered as independent of η so that the norm was taken into account in the constants C of (60), (61). It then obviously follows

$$\|\delta_H (\partial_1 U_1^\eta)\|_0 \leq C H^{k_1} \tag{5.62}$$

$$\left\|\delta_H \left(D_2^2 U_2^\eta\right)\right\|_0 \leq C H^{k_2 - 1} \tag{5.63}$$

$$\|\delta_H (D_2 U_1^\eta - U_2^\eta)\|_0 \leq C H^{k_1} + C H^{k_2 - 1} \leq C H^{\inf(k_1, k_2 - 1)} \tag{5.64}$$

It should be noticed that estimates (62) and (63) are probably optimal, but this is not the case for (64) when $k_2 - 1 < k_1$, as its term in U_2^η was estimated in H^2 instead of L^2.

From the above estimates we easily obtain, using (42) and (58) for u^ε and u_H^ε, the corresponding ones:

$$\|\delta_H (\partial_1 u_1^\varepsilon)\|_0 \leq C H^{k_1} \eta^{-\frac{1}{2}} \tag{5.65}$$

$$\left\|\delta_H \left(\partial_2^2 u_2^\varepsilon\right)\right\|_0 \leq C H^{k_2 - 1} \eta^{-\frac{7}{2}} \tag{5.66}$$

$$\|\delta_H (\partial_2 u_1^\varepsilon - u_2^\varepsilon)\|_0 \leq C H^{\inf(k_1, k_2 - 1)} \eta^{-\frac{3}{2}} \tag{5.67}$$

which is concerned with an anisotropic mesh with mesh steps $\mathcal{O}(H)$ in x_1 and $\mathcal{O}(\eta H)$ in x_2.

We note that the errors tend to infinity as $\eta \searrow 0$ but the functions themselves obviously satisfy

$$\|\partial_1 u_1^\varepsilon\|_0 \cong \eta^{-\frac{1}{2}}, \ \left\|\partial_2^2 u_2^\varepsilon\right\|_0 \cong \eta^{-\frac{7}{2}}, \ \|\partial_2 u_1^\varepsilon - u_2^\varepsilon\|_0 \cong \eta^{-\frac{3}{2}} \tag{5.68}$$

so that we have obtained the

Proposition 4 *The estimates for the relative errors for u^ε (see (65)-(67) and (68))with the anisotropic mesh with steps H and ηH in x_1 and x_2 respectively are the same as the relative error estimates (62) - (64) for $U^0(x_1, y_2)$ with the isotropic mesh of step H in x_1 and y_2.*

6 Error estimates for the Galerkin approximation with anisotropic finite elements

Let u^ε_{HG} be the Galerkin approximation with the finite elements described in sect.5, i.e. u^ε_{HG} is the solution of (59) in the corresponding space V_H. Let $\delta_{HG}u^\varepsilon = u^\varepsilon - u^\varepsilon_{HG}$ be the corresponding error. According to the classical minimizing property of the Galerkin approximation, the energy norm of $\delta_{HG}u^\varepsilon$ is majorized by that of $\delta_H u^\varepsilon$ so that, using the interpolation estimates (65)-(67), and on account of $\varepsilon = \eta^3$ we have

$$\begin{aligned}\|\delta_{HG}\left(\partial_1 u_1^\varepsilon\right)\|_0^2 + \|\delta_{HG}\left(\partial_2 u_1^\varepsilon - u_2^\varepsilon\right)\|_0^2 + \varepsilon^2 \left\|\delta_{HG}\left(\partial_2^2 u_2^\varepsilon\right)\right\|_0^2 \\ \le CH^{2k_1}\eta^{-1} + CH^{2\inf(k_1,k_2-1)}\eta^{-3} + \varepsilon^2 CH^{2(k_2-1)}\eta^{-7} \\ \le CH^{2\inf(k_1,k_2-1)}\eta^{-3}\end{aligned} \tag{6.69}$$

from which the errors of the significative terms are

$$\|\delta_{HG}\left(\partial_1 u_1^\varepsilon\right)\|_0 \le CH^{\inf(k_1,k_2-1)}\eta^{-\frac{3}{2}} \tag{6.70}$$

$$\left\|\delta_{HG}\left(\partial_2^2 u_2^\varepsilon\right)\right\|_0 \le CH^{\inf(k_1,k_2-1)}\eta^{-\frac{9}{2}} \tag{6.71}$$

Remark 5 *On account of (68), we note that (70) and (71) give for the relative error an estimate which is η^{-1} times the standard interpolation estimate for a couple of functions in H^1 and H^2 in a fixed domain with polynomials of degrees k_1 and k_2 respectively.*

We note that (70) and (71) may also to be obtained in the following way: we first perform the Galerkin approximation for $U^\eta(x_1, y_2)$ and we write it in terms of u^ε. The advantage of such a method is to exhibit the structure of the problem for U^η as a penalty one allowing us to improve the convergence by using finite elements avoiding the locking.

Indeed, the Galerkin approximation for the penalty problem (59) in the variables x_1, y_2 with (62) - (64) gives

$$\begin{aligned}\|\delta_{HG}\left(\partial_1 U_1^\eta\right)\|_0^2 + \frac{1}{\eta^2}\|\delta_{HG}\left(D_2 U_1^\eta - U_2^\eta\right)\|_0^2 + \left\|\delta_{HG}\left(D_2^2 U_2^\eta\right)\right\|_0^2 \\ \le C\frac{1}{\eta^2}H^{2\inf(k_1,k_2-1)}\end{aligned} \tag{6.72}$$

from which we immediately get

$$\|\delta_{HG}\left(\partial_1 U_1^\eta\right)\|_0 \le C\eta^{-1}H^{\inf(k_1,k_2-1)} \tag{6.73}$$

and

$$\left\|\delta_{HG}\left(D_2^2 U_2^\eta\right)\right\|_0 \leq C\eta^{-1} H^{\inf(k_1,k_2-1)} \tag{6.74}$$

which in account of (58) and (42) give again (70) and (71).

Remark 6 *As we saw, the problem (59) for* $U^\eta(x_1, y_2)$ *is a penalty one so that the locking phenomenon may appear. The above obtained estimates (73) and (74) hold true in any case. Nevertheless, in principle, these estimates may be improved using finite elements avoiding the locking, i.e. consistent with the limit subspace (44). In this case, according to our simplification (58),* $D_2 U_1^\eta - U_2^\eta \approxeq D_2 U_1^0 - U_2^0 = 0$. *The "consistency with the limit subspace (44) means that the approximation also vanishes as well as (64). Then the estimates are*

$$\begin{cases} \left\|\delta_{HG}\left(\partial_1 U_1^\eta\right)\right\|_0 \leq C H^{\inf(k_1,k_2-1)} \\ \left\|\delta_{HG}\left(D_2^2 U_2^\eta\right)\right\|_0 \leq C H^{\inf(k_1,k_2-1)} \end{cases} \tag{6.75}$$

instead of (73) and (74) which give for u^ε

$$\begin{cases} \left\|\delta_{HG}\left(\partial_1 u_1^\varepsilon\right)\right\|_0 \leq C\eta^{-\frac{1}{2}} H^{\inf(k_1,k_2-1)} \\ \left\|\delta_{HG}\left(\partial_2^2 u_2^\varepsilon\right)\right\|_0 \leq C\eta^{-\frac{7}{2}} H^{\inf(k_1,k_2-1)} \end{cases} \tag{6.76}$$

On account of (68) the relative error is the same as the standard interpolation estimate (compare with Remark5).■

We then have

Proposition 5 *The Galerkin finite element approximation for* $u^\varepsilon(x_1, x_2)$ *with an anisotropic mesh of steps* $\mathcal{O}(H)$ *and* $\mathcal{O}(\eta H)$ *in the directions of* x_1 *and* x_2 *respectively satisfy in general the estimates (70) and (71). Moreover, using finite elements which avoid the locking these estimates may be improved to become (76).*

Our aim is to compare the previous error estimates with the corresponding ones for an isotropic mesh in (x_1, x_2). We then consider an isotropic mesh with step h in (x_1, x_2) in the domain $D_{\eta L}$. We classically [5] have the interpolation estimates

$$\|\delta_h u_1^\varepsilon\|_1 \leq C h^{k_1} \left|u_1^\varepsilon\right|_{k_1+1} \tag{6.77}$$

$$\|\delta_h u_2^\varepsilon\|_2 \leq C h^{k_2-1} \left|u_2^\varepsilon\right|_{k_2+1} \tag{6.78}$$

where obviously

$$\begin{cases} \left|u_1^\varepsilon\right|_{k_1+1} \approxeq \left[\left(\frac{1}{\eta\times\eta^{k_1+1}}\right)^2 \eta\right]^{\frac{1}{2}} = \eta^{-k_1-\frac{3}{2}} \\ \left|u_2^\varepsilon\right|_{k_2+1} \approxeq \left[\left(\frac{1}{\eta^2\times\eta^{k_2+1}}\right)^2 \eta\right]^{\frac{1}{2}} = \eta^{-k_2-\frac{5}{2}} \end{cases} \tag{6.79}$$

so that (77) becomes

$$\|\delta_h u_1^\varepsilon\|_1 \le C h^{k_1} \eta^{-k_1-\frac{3}{2}} = C\left(\frac{h}{\eta}\right)^{k_1} \eta^{-\frac{3}{2}} \tag{6.80}$$

Analogously, from (78) we have

$$\|\delta_h u_2^\varepsilon\|_2 \le C h^{k_2-1} \eta^{-k_2-\frac{3}{2}} = C\left(\frac{h}{\eta}\right)^{k_2-1} \eta^{-\frac{7}{2}} \tag{6.81}$$

It immediately follows

$$\|\delta_h (\partial_1 u_1^\varepsilon)\|_0 \le C\left(\frac{h}{\eta}\right)^{k_1} \eta^{-\frac{3}{2}} \tag{6.82}$$

$$\left\|\delta_h \left(\partial_2^2 u_2^\varepsilon\right)\right\|_0 \le C\left(\frac{h}{\eta}\right)^{k_2-1} \eta^{-\frac{7}{2}} \tag{6.83}$$

Moreover, by the triangle inequality we have :

$$\|\delta_h (\partial_2 u_1^\varepsilon - u_2^\varepsilon)\|_0 \le \|\delta_h (\partial_2 u_1^\varepsilon)\|_0 + \|\delta_h u_2^\varepsilon\|_0 \tag{6.84}$$

Now, in order to estimate the L^2 norm of $\delta_h u_2^\varepsilon$, using the Poincaré inequality in the domain $D_{\eta L}$ of thickness $\mathcal{O}(\eta)$, on account of the fact that $\delta_h u_2^\varepsilon$ and its derivative $\partial_2 \delta_h u_2^\varepsilon$ vanish at $|x_2| = \eta L$ (see our assumption at the beginning of Sect.5) we obtain

$$\|\delta_h u_2^\varepsilon\|_0 \le C\eta \|\partial_2 \delta_h u_2^\varepsilon\|_0 \le C\eta^2 \|\delta_h u_2^\varepsilon\|_2 \tag{6.85}$$

Then (84) gives

$$\|\delta_h (\partial_2 u_1^\varepsilon - u_2^\varepsilon)\|_0 \le C \|\delta_h (\partial_2 u_1^\varepsilon)\|_0 + C\eta^2 \|\delta_h u_2^\varepsilon\|_2 \tag{6.86}$$

and, from (80) and (81),

$$\|\delta_h (\partial_2 u_1^\varepsilon - u_2^\varepsilon)\|_0 \le C\left(\frac{h}{\eta}\right)^{k_1} \eta^{-\frac{3}{2}} + C\left(\frac{h}{\eta}\right)^{k_2-1} \eta^{-\frac{3}{2}} \le C\left(\frac{h}{\eta}\right)^{\inf(k_1,k_2-1)} \eta^{-\frac{3}{2}} \tag{6.87}$$

Let now u_{hG}^ε be the Galerkin approximation for the h isotropic mesh and $\delta_{hG} u^\varepsilon = u^\varepsilon - u_{hG}^\varepsilon$ the corresponding error. In the same way as we obtained (69), but using (82), (83) and (87), we get

$$\|\delta_{hG} (\partial_1 u_1^\varepsilon)\|_0^2 + \|\delta_{hG} (\partial_2 u_1^\varepsilon - u_2^\varepsilon)\|_0^2 + \varepsilon^2 \left\|\delta_{hG} \left(\partial_2^2 u_2^\varepsilon\right)\right\|_0^2$$
$$\le C\eta^{-3}\left(\frac{h}{\eta}\right)^{2k_1} + C\eta^{-3}\left(\frac{h}{\eta}\right)^{2\inf(k_1,k_2-1)} + C\eta^{-1}\left(\frac{h}{\eta}\right)^{2(k_2-1)}$$

$$\leq C\eta^{-3}\left(\frac{h}{\eta}\right)^{2\inf(k_1,k_2-1)} \tag{6.88}$$

so that the errors of the significative terms are

$$\|\delta_{hG}\left(\partial_1 u_1^\varepsilon\right)\|_0 \leq C\eta^{-\frac{3}{2}}\left(\frac{h}{\eta}\right)^{\inf(k_1,k_2-1)} \tag{6.89}$$

$$\|\delta_{hG}\left(\partial_2^2 u_2^\varepsilon\right)\|_0 \leq C\eta^{-\frac{9}{2}}\left(\frac{h}{\eta}\right)^{\inf(k_1,k_2-1)} \tag{6.90}$$

Comparing with (70) and (71), we observe that, taking

$$h = \eta H \tag{6.91}$$

the estimates (70) and (71) are the same as (89) and (90) respectively. Moreover, in this case the area of the triangles are in the ratio

$$\frac{\text{Area of isotropic triangle}}{\text{Area of anisotropic triangle}} = \frac{h^2}{\eta H^2} = \eta. \tag{6.92}$$

Proposition 6 *The error estimate of the Galerkin approximation for an anisotropic mesh $(H, \eta H)$ is given by (70) and (71). The analogous error estimate for an isotropic mesh (h, h) is given by (89) and (90). Taking H and h linked by (91) the number of triangles of the anisotropic mesh is $\eta = \varepsilon^{\frac{1}{3}}$ times the number of triangles of the isotropic one and the estimate for the anisotropic mesh is the corresponding one for the isotropic one.*

Acknowledgment The authors are indebted to S. Nicaise for bringing their attention to estimates for anisotropic meshes.

[1] Apel T. & Lube G. *Anisotropic mesh refinement in stabilized Galerkin methods.* Numer. Math. vol.74, pp. 261-282, 1996.

[2] Apel T. & Nicaise S. *Elliptic problems in domains with edges: anisotropic regularity and anisotropic finite element meshes.* In Partial Differential Equations and Functional Analysis, in memory of P. Grisvard, Ed. J. Céa, D. Chesnais, G. Geymonat, J.L. Lions, Birkhauser, Boston, pp. 207-220, 1996.

[3] Caillerie D. *Étude générale d'un type de problèmes raides et de perturbation singulière.* Compt. Rend. Acad. Sc. Paris, série I, tome 323, pp. 835-840, 1996.

[4] Chapelle D. & Bathe K.J., *Fundamental considerations for the finite element analysis of shell structures,* Computers & Structures, vol. 66, N^0 1, pp. 19-36, 1998.

[5] Ciarlet P.G. & Raviart P.A. *General Lagrange and Hermite interpolation in* $\mathbb{R}^n$ *with applications to finite element method.* Arch. Rational Mech. Anal., vol. 46, pp.177-199, 1972.

[6] Gérard P. & Sanchez Palencia É. *Sensitivity phenomena for certain thin elastic shells with edges*, Math. Meth. in Appl. Sciences, vol. 23, p. 379-399, 2000 .

[7] Karamian P. *Nouveaux résultats numériques concernant les coques minces hyperboliques inhibées: cas du paraboloïde hyperbolique*, Compt. Red. Acad. Sci., Paris, série IIb, vol 326, pp. 755-760, 1998.

[8] Karamian P., *Réflexion des singularités dans les coques hyperboliques inhibées*, Compt.Rend. Acad. Sci., Paris, série IIb, vol. 326, pp. 609-614, 1998.

[9] Karamian P., Sanchez-Hubert J & Sanchez Palencia É. *A model problem for boundary layers of thin elastic shells.* Model Math. Anal. Num, vol. 34, p. 1-30, 2000.

[10] Leguillon D., Sanchez-Hubert J. & Sanchez Palencia É., *Model problem of singular perturbation without limit in the space of finite energy and its computation*, Compt. Rend. Acad. sci. Paris, série IIb, pp.485-492, 1999

[11] Lions J.L. & Magenes E. *Problèmes aux limites non homogènes et applications. vol.I, Dunod, Paris, 1968*

[12] Lions J.L. & Sanchez Palencia É, *Problèmes sensitifs et coques élastiques minces*, in Partial Differential Equations and Functional Analysis, in memory of P. Grisvard, Ed. J. Céa, D. Chesnais, G. Geymonat, J.L. Lions, Birkhauser, Boston, pp. 207-220, 1996

[13] Lions J.L. & Sanchez Palencia É, *Sur quelques espaces de la théorie des coques et la sensitivité.* In Homogenization and applications to material sciences, ed. D. Cioranescu, A. Damlamian, P. Donato, Gakkotosho, pp. 271-278, 1995.

[14] Medina, Picasso & Rappaz. *Error estimates and adaptive finite elements for nonlinear diffusion- convection problems,* Math. Meth. in Appl. Sciences, vol. 6, N°5, pp. 689-712, 1996.

[15] Sanchez-Hubert J. & Sanchez Palencia É., *Coques élastiques minces. Propriétés asymptotiques,* Masson, Paris, 1997

[16] Sanchez-Hubert J. & Sanchez Palencia É., *Pathological phenomena in computation of thin elastic shells*, Transactions Can. Soc. Mech. Engin., vol. 2, n^0 4B, pp. 435-446, 1999.

[17] Van Dyke M. *Perturbation methods in fluid mechanics*, Academic Press, New-York, 1964.

Modelling of a thin piezoelectric shell coupled with a distributed electronic circuit by distributed piezoelectric transducers.

G. SENOUCI and M. LENCZNER Equipe de Mathématiques de Besançon, Université de Franche-Comté, Besançon, France

1 INTRODUCTION

This work is devoted to mathematical and numerical modelling of a coupled system including an elastic thin shell, distributed piezoelectric patches and a distributed electronic circuit. The formulation of the circuit is general, it can take into account passive devices and voltage to voltage amplifiers. Two numerical simulations based on a finite elements method are presented. They are related to two different distributed electronic circuits which are both stabilizing shell vibrations. In our mind, the shell vibration stabilization problem is an intermediary step before to study the question of sound transmission control through a such an thin elastic shell. This last question is a key for noise reduction in all terrestrial and non terrestrial vehicles.
Before to go into details, let us summarize the contributions of this work.

A variational formulation is given for the global system in the static case. It takes into account mechanical as well as electric phenomena. The use of variational formulation is not usual for the electronic circuit. It is based on a previous work, by M. Lenczner and G. Senouci-Bereksi [L2], devoted to homogenization of active electric circuits. Let us remark that variational framework has been intensively developped for partial differential equations posed on networks, see for example [A1], [B1], [N1] and [L1].

A mathematical framework based on variational methods and graph theory is used in order to state conditions of existence and uniqueness of the solution of the global system in the static case. They are formulated in terms of circuit topology. The main difficulty comes from the voltage to voltage amplifiers. Our statements are particular to this case, but the general approach developed here can be extended to other classes of active devices.

The Reisner-Mindlin two-dimensional shell model of the coupled system is derived. Conditions for existence and uniqueness of the solution are also stated in the static case.

In view of active control applications, a new model for collocated actuators and sensors is proposed. It is based on the assumption that the metallization of each piezoelectric patch is divided in two intricately parts, the first one being a sensor when the second one is an actuator.

Numerical simulations of the evolution problem related to the model with collocated sensors and actuators are presented. They are based on a finite element method. Two simulations for two examples of distributed electric circuits acting as vibration dampers are discussed. The first one is local when the second one is non local.

The paper is organized in three parts. First, the three-dimensional piezoelectric thin shell model is stated. Sufficient conditions for existence and uniqueness of solution are announced. In the second part, the two-dimensional models are presented. The third part is devoted to the numerical simulation.

Section 2. Equations of mechanics of continuum media are often written in the framework of variational formulations. By another way, electric circuits are modelized using algebraic equations, see A. Recski [R1] for example. Since we are interested on the coupling between electric circuits and continuum media, we develop a framework adapted to both phenomena.

A variational formulation of electric circuit including passive and voltage to voltage amplifiers has already been introduced in [L3]. Here, we extend this formulation to the coupling between an electric network and a piezoelectric shell covered by a metallization. This leads to a system :

$$a(u,v) + b_1(v,p) =< f, v >$$

$$b_2(u,q) =< g, q >,$$

where the bilinear symmetric form $a(.,.)$ is defined on both the electric circuit and the mechanical structure when $b_1(.,.)$ and $b_2(.,.)$ are defined on the electric network only. This formulation is not symmetric because the bilinear forms $b_1(.,.)$ and $b_2(.,.)$ are different. The lack of symmetry is due to the presence of active amplifiers.

The electric potential and the current that are the "electric unknowns" of this variational formulation are respectively affine and constant on each branch of the circuit. These conditions result directly from the classical equations of electronic circuit.

Conditions for existence and uniqueness of the solution of such an abstract problem have been derived in C. Bernardi, C. Canuto and Y. Maday [B2] under the form of some inf-sup conditions. Here, we focus our attention on the derivation of graph interpretations of these conditions. They are mainly related to the location of the various devices in the network : voltage sources, current sources, resistors, amplifiers inputs and outputs, and earth.

The main point of interest, in our approach, is that it provides a tool which is well suited for a global analysis of the coupled problem. It allows to use generic methods based on the framework of variational formulations. For example, using this approach, we already have derived an homogenized model for such a coupling. This work is based on the homogenization of electric circuits presented in [L3], and will be reported in a forthcoming paper.

Our work may be extended to electric networks including other active devices such as current to current amplifiers, voltage to current amplifiers, current to voltage amplifiers, diodes and operational amplifiers, as well as passive devices such as capacitors and inductors.

Finally, it may also be extended to electric potential belonging in H^1 and current in L^2. In this case, the graph conditions insuring existence and uniqueness of the solution should be reconsidered, when our homogenization method applies directly without any change (assuming sufficient a priori estimates).

Section 3. This section is devoted to a simplified shell model (ie a two-dimensional shell model) that is used in our numerical simulation. Using the above formulation of the coupled system, we state a two-dimensional thin piezoelectric shell model. This piezoelectric shell has a single layer. Its derivation is based on the Reisner-Mindlin kinematics, for mechanical displacements, and on a similar assumption for the electric potential. This last one was suggested by a previous work of E. Canon and M. Lenczner [C3] where a piezoelectric thin plate model was derived using an asymptotic method. The conditions of existence and uniqueness of the solution of the two-dimensional model are the same as those already used for the three-dimensional model.

By another way, several piezoelectric thin shell models have already been derived, see for example S. Tzou [T1], N. Rogacheva [R2, R3], A. Saidi [S1], M. Bernadou and D. Haenel [B5] and D. Haenel [H1]. They are based on various kinematics. Concerning the piezoelectric part, our model is close to those in [S1] and [H1] which have been derived during the same period. In comparison with these papers, our contribution lies in the coupling of the piezoelectric thin shell model with a general electric circuit.

Finally, the choice of a particular configuration of the electrode covering the piezoelectric patch leads to a second model. We assume that it is divided in two parts, and that these two parts are intricately shaped. This conditions is interpreted as an assumption on the strain tensor. For control applications, one part is devoted to sensing when the second part is devoted to actuation. The resulting model is called the thin shell model with collocated actuators and sensors.

Section 4. The above models have been extended to some multilayered thin piezoelectric shell models including many distributed transducers being connected to a distributed electric circuit and also to the evolution problem. Here, we consider the particular case where each transducer is a collocated sensor and actuator modelled in section 3. In order to avoid too many notations, we do not report in this paper the detailed statement of the model. The reader interested by more details can refer to [C3] where a thin elastic plate including many piezoelectric transducers was considered. Let us quote that the statement of the variational formulation does not present new difficulties (even for a general electric circuit); however, the mathematical properties of existence and uniqueness of solution of such a model are still not proved. This is an additional reason why we do not have detailed it in section 3.

The numerical simulation is based on the degenerate finite element introduced by S. Ahmad, B.M. Irons and O.C. Zienkiewicz [A2] for elastic thin shells, and improved by C. Gelin [G1] for the prevention of the locking effect. Here, we extend this method to the case of multilayer piezoelectric thin shells. In addition, our simulation takes into account the presence of a distributed electric network. This last may be local, in the sense that it does not connect piezoelectric patches together, or non local, in the sense that each piezoelectric patch is connected with its closer neighbors.

The numerical simulation is related to the case of a quarter of cylinder, subjected to an uniform and oscillating pressure on one of its side. Two particular choices of electric network are proposed. The first is local when the second is non local. Both of them are designed in order to stabilize the shell vibrations. The operation principle of these systems is the following. Voltages are measured on each piezoelectric sensor, so they constitute the input of the distributed circuit. Its outputs are also voltages that are differentiated with respect to the time variable, and applied to actuators. The circuits operate as finite difference operators from the input to the output. In the local case, it acts as the identity operator, when in the non local case, it acts as a discrete second order derivative (with respect to space). Since input voltages are a linear combination of strains, both the local and the non local circuits produce a term proportional to the time derivative of displacements in the shell equation. So, their effect is to stabilize shell vibrations.

Our idea with the non local circuit is to build approximations of distributed stabilization or control laws having properties that cannot be reached using local circuits.

2 MODELING OF THE THREE-DIMENSONAL ELECTRO-PIEZO-MECHANICAL SYSTEM

In the following, Greek indices $\alpha, \beta, ...$ will belong to the set $\{1, 2\}$, and Latin indices $i, j, ...$ will belong to the set $\{1, 2, 3\}$. The usual Einstein summation convention will be adopted. Following [B5] and [G1], curvilinear coordinates in $\mathbb{R}^3$ are now defined. Let $(\mathbf{O}, \mathbf{e}_1, \mathbf{e}_2, \mathbf{e}_3)$ be a reference system in $\mathbb{R}^3$. Any point $\mathbf{M}$ (see figure 2.1) in this space is referred as $\mathbf{OM} = x^i\mathbf{e}_i$ and it may also be referred to a curvilinear coordinate system $\langle \xi^1, \xi^2, \xi^3 \rangle$ such that $x^i = x^i(\xi^1, \xi^2, \xi^3)$ with $det(\frac{\partial x^i}{\partial \xi^j}) \neq 0$. Two different local basis are associated at any point $\mathbf{M}$. The covariant basis is given by $(\mathbf{g}_i = \frac{\partial x^j}{\partial \xi^i}\mathbf{e}_j)_i$ and the contravariant one $(\mathbf{g}^j)$ which is the adjoint basis of $(\mathbf{g}_i)$ is such that $\mathbf{g}^i\mathbf{g}_j = \delta^i_j$. The contravariant and the covariant components of tensors are associated to these basis. These two sets of components of a first order tensor $\mathbf{v}$ are defined by $\mathbf{v} = v_i\mathbf{g}^i = v^i\mathbf{g}_i$. The relation between the contravariant and the covariant components of a tensor is $v^i = g^{ij}v_j$ and $v_i = g_{ij}v^j$. The tensors (g^{ij}) and (g_{ij}) being called the metric tensors. The covariant and contravariant basis are not orthonormal, so Christoffel's symbols Γ^k_{ij} are introduced for the computation of differential of tensors. Let $\mathbf{v} = v^i\mathbf{g}_i$, so that

$$\frac{\partial\, \mathbf{v}}{\partial \xi^j} = \mathbf{v}_{,j} = v^i_{,j}\mathbf{g}_i + v^i\mathbf{g}_{i,j},$$

then defining

$$\Gamma^k_{ij} = \mathbf{g}^k\mathbf{g}_{i,j} = g^{kl}\mathbf{g}_l\mathbf{g}_{i,j},$$

leads to

$$\mathbf{g}_{i,j} = \Gamma^k_{ij}\mathbf{g}_k,$$

so that

$$\mathbf{v}_{,j} = v_{i|j}\mathbf{g}^i = v^i_{|j}\mathbf{g}_i,$$

where

$$v_{i|j} = v_{i,j} - \Gamma^k_{ij}v_k \text{ and } v^i_{|j} = v^i_{,j} + \Gamma^i_{kj}v^k.$$

Finally, the volume element formulated in the curvilinear coordinates is $d\Omega = g^{\frac{1}{2}}d\xi^1 d\xi^2 d\xi^3$, with $g^{\frac{1}{2}} = (\mathbf{g}_1 \times \mathbf{g}_2).\mathbf{g}_3$.

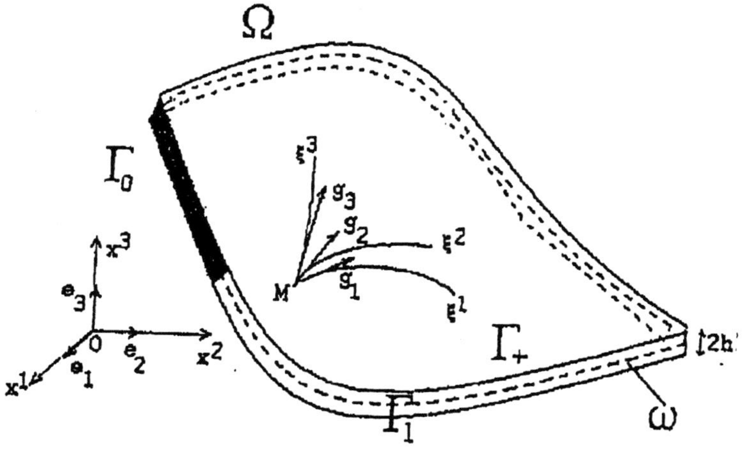

Figure 1: Local basis

1 The three-dimensional piezoelectric shell

Let Ω be the three-dimensional domain occupied by a piezoelectric shell. We denote by Γ_+ and Γ_- its upper and lower faces, $\Gamma_\pm = \Gamma_+ \cup \Gamma_-$ and Γ its lateral boundary. The lateral boundary Γ is divided in two parts Γ_0 and Γ_1. We assume that mes(Γ_0) does not vanish. The body is submitted to volume loading of density (f^i) and surface loading of density (r^i) applied on $\Gamma_\pm \cup \Gamma_1$. The boundary Γ_0 is assumed to be clamped. We denote by σ^{ij} the stress tensor components ($\boldsymbol{\sigma} = \sigma^{ij}\mathbf{g}_i \otimes \mathbf{g}_j$), D^i the electric displacement components ($\mathbf{D} = D^i\mathbf{g}_i$), n_i the components of the external normal vector to the boundary ($\mathbf{n} = n_i\mathbf{g}^i$), E_i the components of the electric field ($\mathbf{E} = E_i\mathbf{g}^i$), ϕ the electric potential, u_i the components of the displacement vector ($\mathbf{u} = u_i\mathbf{g}^i$), ϵ_{ij} the components of the strain tensor. The electric field is defined by $E_i(\phi) = -\phi_{|i} = -\phi_{,i}$, and $\epsilon_{ij}(\mathbf{u}) = \frac{1}{2}(u_{i|j} + u_{j|i})$. The static behavior of a piezoelectric body is governed by the following equations. The elastostatic and electrostatic equations written in Ω are $-\sigma^{ij}|_j = f^i$ and $D^i|_i = 0$. The natural boundary conditions satisfied by mechanical stresses on $\Gamma_\pm \cup \Gamma_1$ and by electric displacements on Γ are $\sigma^{ij}n_j = r^i$, $D^i n_i = 0$. The clamping condition on Γ_0 is $\mathbf{u} = 0$. Concerning the electric conditions imposed on $\Gamma_\pm$, one assumes that the lower side Γ_- is covered by a metallization and connected to the earth, ie $\phi = 0$ on Γ_-. The upper part Γ_+ is also covered by a metallization. It is connected to one branch of the electric circuit (see figure 3). The electric potential and the current flowing from Γ_+ satisfy ϕ =constant on Γ_+ and

$$i = -\frac{d}{dt}\int_{\Gamma_+} D^i n_i d\Gamma.$$

Since this section is devoted to the statement of the static model, the time derivative in the above expression will be dropped out. However, it will be taken into account in the section related to the numerical simulation when the dynamic model will be considered. The stress tensor and the electric displacements vector are expressed in terms of the strain tensor and the electric field :

$$\begin{aligned} \sigma^{ij}(\mathbf{u},\phi) &= R^{ijkl}\epsilon_{kl}(\mathbf{u}) - d^{mij}E_m(\phi), \\ D^i(\mathbf{u},\phi) &= d^{ikl}\epsilon_{kl}(\mathbf{u}) + c^{ij}E_j(\phi), \end{aligned}$$

where the coefficients R^{ijkl}, d^{mij} and c^{ij} are the components of the tensors of stiffness, piezoelectricity and permittivity respectively. They satisfy some symmetry and ellipticity conditions :

$$R^{ijkl} = R^{jikl} = R^{jilk} = R^{klij}, \; d^{mij} = d^{mji}, \; c^{ij} = c^{ji},$$

$$\begin{aligned} R^{ijkl}\epsilon_{ij}\epsilon_{kl} &\geq \alpha \sum_{i,j=1}^{3} (\epsilon_{ij})^2, \quad \forall(\epsilon_{ij}) \in \mathbb{R}^9 \text{ such that } (\epsilon_{ij}) = (\epsilon_{ji}) \\ \text{and } c^{ij}w_i w_j &\geq \beta \sum_{i=1}^{3} (w_i)^2, \quad \forall(w_i) \in \mathbb{R}^3. \end{aligned}$$

Here α and β are some strictly positive constants.

2 The electric circuit

For basic definitions and properties relatives to electric circuits, we refer to [V1]. An electric network is composed of vertices (or nodes) and edges (or branches). Vertices are linked by edges. The set of edges is denoted by Θ. Mathematically, Θ is a network in $\mathbb{R}^3$. We denote by σ_0 the subset of vertices linked to the earth (i.e. where the electric potential is equal to zero). The network Θ is divided in five disjoints parts Θ_0, Θ_1, Θ_2, Θ_3 and Θ_4 occupied respectively by the voltage sources, the current sources, the resistors, the input of the amplifiers and the output of the amplifiers. The manifold Θ is supposed to be parameterized. This parameterization defines a positive sense for each edge. We name s_e^+ and s_e^- the vertices belonging to an edge $e \in \Theta$ such that $s_e^+ \to s_e^-$ is the positive sense. The set of edges arriving in a positive (respectively negative) sense at a vertices s is denoted by Θ_s^+ (respectively Θ_s^-). The function L defined on Θ is constant on each edge e and is defined by $L(\mathbf{x}) = |e|$ (length of the edge e) for every $\mathbf{x} \in e$. The potential ψ which is usually defined on the vertices, is extended as an affine function, still denoted by ψ, on each edge $e \in \Theta$. Let us define the sets $\mathbb{P}^0(\Theta)$ or $(\mathbb{P}^0(\Theta_k))_{k=0..4}$ (respectively $\mathbb{P}^1(\Theta)$) of functions constant on each edge $e \in \Theta$ or $(e \in \Theta_k)_{k=0..4}$ (respectively affine on each edge $e \in \Theta$ and continuous on Θ). As it is usual in electronics, the current i and the voltage v are some distributed fields belonging to $\mathbb{P}^0(\Theta)$. Its turn out that the electric potential is also a distributed field, it belongs to $\mathbb{P}^1(\Theta)$. The tangential derivative of a function ψ defined on Θ is denoted by $\nabla_\tau \psi$.

The voltage Kirchhoff law is stated on each edge $e \in \Theta$:

$$v_{|e} = \phi(s_e^+) - \phi(s_e^-),$$

or equivalently

$$-L\nabla_\tau \phi = v \text{ on } \Theta. \tag{2. 1}$$

The Kirchhoff law for currents is stated for each vertex s as

$$\sum_{e \subset \Theta_s^+} i_{|e} - \sum_{e \subset \Theta_s^-} i_{|e} = 0.$$

It can be equivalently written under a weak formulation

$$\int_\Theta i(\mathbf{x}) \nabla_\tau \psi(\mathbf{x}) dl(\mathbf{x}) = 0 \text{ for all } \psi \in \mathbb{P}^1(\Theta) \text{ such that } \psi = 0 \text{ on } \sigma_0. \tag{2. 2}$$

The voltages, currents and electric potentials are imposed respectively on Θ_0, Θ_1 and σ_0 to be equal to the voltage sources $v_d \in \mathbb{P}^0(\Theta_0)$, the current sources $i_d \in \mathbb{P}^0(\Theta_1)$ and 0 on σ_0 :

$$v = v_d \text{ on } \Theta_0, i = i_d \text{ on } \Theta_1 \text{ and } \phi = 0 \text{ on } \sigma_0. \tag{2. 3}$$

The sign of v_d and of i_d on an edge e depends on the orientation of e. An impedance $1/g \in \mathbb{P}^0(\Theta_2)$ is associated to Θ_2, which means that v and i are linked by the constitutive linear equation on Θ_2

$$i = gv \text{ on } \Theta_2. \tag{2. 4}$$

We assume that $g \geq g_{\min} > 0$. A voltage to voltage amplifier is a device which imposes two equations between currents and voltages of two edges. The set Θ_3 and Θ_4 are respectively the sets of amplifier's inputs and outputs. Each input edge $e_3^l \in \Theta_3$ is associated to an unique output edge $e_4^l \in \Theta_4$ where l varies from one to the number of amplifiers. The constitutive relations of the voltage to voltage amplifier are for each l,

$$v_{|e_4^l} - k_l v_{|e_3^l} = 0 \text{ and } i_{|e_3^l} = 0,$$

where $k_l \in \mathbb{R}^*$ is the amplification coefficient. The edges e_3^l and e_4^l are respectively called the input and the output of the amplifier. Since this equation applies for each amplifier, we consider that $k \in \mathbb{P}^0(\Theta_3)$ and we write the amplifier constitutive equations as follows

$$v_{|\Theta_4} - k_l v_{|\Theta_3} = 0 \text{ and } i_{|\Theta_3} = 0. \tag{2. 5}$$

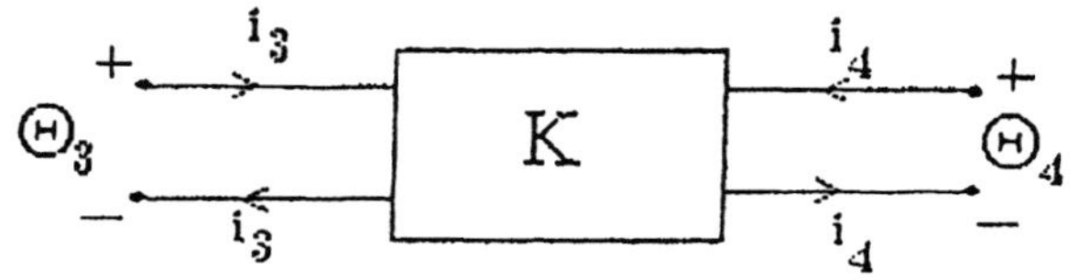

Figure 2: Voltage to voltage amplifier

3 Connection between the electric circuit and the piezoelectric shell

The electric network is connected to the upper side Γ_+ of the piezoelectric shell. The upper face Γ_+ corresponds to a node of the network Θ. The piezoelectric shell it self is not considered as a part of Θ, because, it is already a part of Ω. We consider that the lower side Γ_- is connected to the earth, and does not correspond to any node of Θ.

4 Variational formulation

The space of admissible functions for the model is

$$W_{ad}(v_d) = H^1_{\Gamma_0}(\Omega)^3 \times \Psi_{ad}(v_d)$$

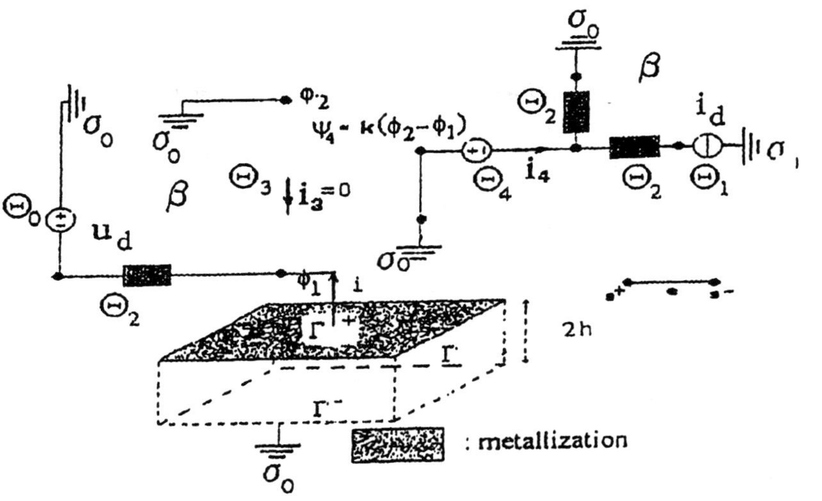

Figure 3: A circuit connected to a piezoelectric transducer

where $\Psi_{ad}(v_d)$ is the set of admissible electric potential. It is defined by

$$\begin{aligned}\Psi_{ad}(v_d) &= \{(\psi, j) \in H^1(O) \times \mathbb{P}^0(\Theta_4), \psi_{|\Theta} \in \mathbb{P}^1(\Theta), \psi_{|\Gamma_+} \in \mathbb{P}^0(\Gamma_+),\\ \psi &= 0 \text{ on } \sigma_0 \cup \Gamma_- \text{ and } -L\nabla_\tau \psi = v_d \text{ on } \Theta_0\},\end{aligned}$$

where $v_d \in \mathbb{P}^0(\Theta_0)$ and $O = \overset{o}{(\overline{\Omega \cup \Theta})}$. The set $\mathbb{P}^0(\Gamma_+)$ is the set of functions constants on Γ_+. Usual regularity of the data $(f^i)_i \in L^2(\Omega)^3$, $r^i \in L^2(\Gamma_\pm \cup \Gamma_1)^3$ is required. Thus $(\mathbf{u},\phi, i) \in W_{ad}(v_d)$ is solution of the variational formulation : for every $(\mathbf{v},\psi, j) \in W_{ad}(0)$,

$$\begin{aligned}a((\mathbf{u},\phi),(\mathbf{v},\psi)) + b_1(\psi, i) &= l(\mathbf{v},\psi)\\ b_2((\mathbf{u},\phi), j) &= 0.\end{aligned} \tag{4. 6}$$

Here

$$\begin{aligned}a((\mathbf{u},\phi),(\mathbf{v},\psi)) &= \int_\Omega (R^{ijkl}\epsilon_{kl}(\mathbf{u}) - d^{mij}E_m(\phi))\epsilon_{ij}(\mathbf{v})\\ &\quad +(d^{ikl}\epsilon_{kl}(\mathbf{u}) + c^{im}E_m(\phi))E_i(\psi)d\Omega + \int_{\Theta_2} Lg\nabla_\tau\phi\nabla_\tau\psi dl(\mathbf{x}),\end{aligned}$$

$$\begin{aligned}b_1(\psi, i) &= \int_{\Theta_4} i\nabla_\tau\psi dl(\mathbf{x}),\\ b_2((\mathbf{u},\phi), j) &= \int_{\Theta_4} \nabla_\tau\phi\ j\ dl(x) - \int_{\Theta_3} k\nabla_\tau\phi\ j\ dl(\mathbf{x})\end{aligned}$$

and

$$l(\mathbf{v},\psi) = -\int_{\Theta_1} i_d \nabla_\tau \psi dl(\mathbf{x}) + \int_\Omega f^i v_i d\Omega + \int_{\Gamma_\pm \cup \Gamma_1} r^i v_i dS.$$

We adopt the rule that j takes the same value on the input e_3^l and on the output e_4^l of an amplifier. In the following assumptions, Ω is treated as an edge e of the generalized circuit O. For example, $\mathbb{P}^0(O)$ is the set of functions being constant on each edge of Θ and on Ω. In addition, for a given function ψ such that $(\psi,.) \in \Psi_{ad}(0)$, $\nabla_\tau \psi$ is also defined in Ω by

$$L\nabla_\tau \psi = \psi_{|\Gamma+} - \psi_{|\Gamma-}$$

where L represents the shell thickness.

Definition 1 *(i) A path is a sequence of edges where the end of an edge is connected to the beginning of the following one.*

(ii) A circuit is a path where the beginning of the first edge is connected to the end of the last one. For this definition, all vertices belonging to $\sigma_0 \cup \Gamma_-$ (the earth) are considered as one. The circuits are denoted by the letter β.

The notion of circuit plays a central role in the following conditions because natural estimates on $\mathbb{P}^1$ functions can be associated to each circuit.
In order to check the conditions in [B2], we will introduce the following linear system. For $v \in \mathbb{P}^0(\Theta_3)$ such that

$$\int_{\beta \cap \Theta_3} v \, dl(\mathbf{x}) = 0 \text{ for each circuit } \beta \text{ of } \Theta_0 \cup \Theta_3 \cup \Theta_4, \tag{4. 7}$$

we need to construct a solution $\eta \in \mathbb{P}^0(O - \Theta_1)$, relative to v, of the linear system :

$$\begin{aligned} \eta_{|\Xi} &= 0 \\ \eta_{|e_4^l} - k\eta_{|e_3^l} &= v_{|e_3^l} \text{ for every } l \\ \int_\beta \eta \, dl(\mathbf{x}) &= 0 \text{ for each circuit } \beta \text{ of } O - \Theta_1, \end{aligned} \tag{4. 8}$$

where Ξ is a subset of $O - \Theta_1$.

Consider the class of subsets $X \subset O - \Theta_1$ such that equations in (4. 8) are independent when $\Xi = (O - \Theta_1) - X$.

Definition 2 *We say that X is minimal for the independency of equations in (4. 8) if for any $X^* \subset X$ (with $X^* \neq X$), equations in (4. 8) are not independent when $\Xi = (O - \Theta_1) - X^*$.*

Remarks : (i) For a given set of equations (4. 8), the minimal set X is not unique.
(ii) Every minimal set have the same cardinal (see A. Recksi [R1]).
(iii) There exist algorithms for building up such minimal set X. See [R1].

Assumptions

(H1) There exist $\overline{O}_2 \subset O_2$ and $\tilde{O}_2 = O_2 - \overline{O}_2$ such that the two following conditions are fulfilled :
(i) There exists a minimal set $X = \tilde{O}_2 \cup \Theta_3 \cup \Theta_4$ for the independency of equations in (4. 8) such that $\Xi = (O - \Theta_1) - X = \Theta_0 \cup \overline{O}_2$.
(ii) For every $v \in \mathbb{P}^0(\Theta_3)$ verifying the compatibility condition (4. 7), the linear system (4. 8) has at most one solution $u \in \mathbb{P}^0(O - \Theta_1)$.

Remark : Here, $O_2 = \Theta_2 \cup \Omega$. It can be proved that (H1)(i) is equivalent to the existence of the solution of (4. 8). Therefore, (i) and (ii) imply that (4. 8) has one and only one solution. That is, the system (4. 8) has as many equations as many unknowns.

Let us consider such a minimal set X. For $e \in X$, $X^* = (X - e)$ is not a minimal set, that is, equations (4. 8) are not independent when $\Xi = (O - \Theta_1) - X^*$. After deleting some equations in (4. 8) (excepted the equation $u_{|e} = 0$), the remaining equations can be independent.

Definition 3 *(i) One says that a subset E of dependent equations of (4. 8) with $\Xi = (O - \Theta_1) - X^*$, is minimal with respect to e, when, after deleting any equation, the remaining equations are independent and when the number of equations in E is equal to the number of edges involved in E plus one.*
(ii) The set of edges involved in a minimal set of dependent equations is called the minimal set of edges linked with e and is denoted by $Z(e)$.

Remarks : (i) In the above definition (i), the subset necessary contains the equation $u_{|e} = 0$, otherwise equations would be independent.
(ii) The definition of minimal subset of dependent equations leads to existence of solution of system E. When the number of equations in E is equal to the number of edges involved in E plus one, the solution is unique.
(iii) The definition of $Z(e)$ implies that $\eta_{|e}$ is a unique linear combination of $(\eta_{|e'})_{e' \in Z(e) - \{e\}}$. Therefore,

$$|\eta|_e \leq C |\eta|_{Z(e) - \{e\}}.$$

(H2) Let us consider a real $\alpha_0 \in \mathbb{R}$. One can choose a function $\alpha \in \mathbb{P}^0(O)$, constant on each circuit β, such that for each $e \in \tilde{O}_2$, there exists a minimal set $Z(e)$ of edges, linked with e, such that $\alpha_{|Z(e) \cap \Theta_3} = \alpha_0$ and $\alpha_{|Z(e) \cap \tilde{O}_2} = 1$.

The aim of the third assumption is to interpret the following condition : There exists a positive constant C such that for any $(\psi, 0) \in \Psi_{ad}(0)$ satisfying $B_1(\psi) = 0$ we have

$$|\nabla_\tau \psi|_{\Theta_3} \leq C |\nabla_\tau \psi|_{\overline{O}_2}.$$

(H3) Every edge $e \in \Theta_3$ belongs to a circuit $\beta \subset \{e\} \cup \Theta_0 \cup \overline{O}_2 \cup \Theta_4$.

The following assumption (H4) means that there exists a positive constant C such that for every $(\psi, 0) \in \Psi_{ad}(0)$ we have $|\nabla_\tau \psi|_{\Theta_1} \leq C|\nabla_\tau \psi|_{O-\Theta_1}$. It implies the continuity of the linear form

$$l(\psi) = \int_{\Theta_1} i_d \nabla_\tau \psi dl(\mathbf{x})$$

with respect to the semi norm $|\nabla_\tau \psi|_{O-\Theta_1}$.

(H4) Every edge $e \in \Theta_1$ belongs to a circuit $\beta \subset \{e\} \cup (O - \Theta_1)$.

The assumption (H5) means that there exists a positive constant C such that for every $(\psi, 0) \in \Psi_{ad}(0)$ we have

$$|\psi|_{O-\Theta_1} \leq C|\nabla_\tau \psi|_O.$$

It leads to a kind of Poincaré inequality. Combined with the assumption (H4), it insures that the semi norm $|\nabla_\tau \psi|_O$ is a norm on $\Psi_{ad}(0)$.

(H5) In each connected component of $O - \Theta_1$ there is a vertex belonging to $\sigma_0 \cup \Gamma_-$.

The assumption (H6) is a compatibility condition between the various voltage sources (the amplifier's outputs are generally called active voltage source).

(H6) There is no circuit solely made up of edges belonging to $\Theta_4 \cup \Theta_0$.

The assumption (H7) is equivalent to the following assertion. For every $j \in \mathbb{P}^0(\Theta_4)$ there exists a function $(\psi, 0) \in \Psi_{ad}(0)$ such that $\nabla_\tau \psi = j$ on Θ_4.

Consider the circuits β included in O satisfying $\beta \cap \Theta_4 \neq \emptyset$. There exists a subset $O^* \subset O$ of edges such that the network $O - O^*$ does not contain such a circuit β. The set O^* is said to be minimal if for any $\Theta^{*1} \subset O^*$ ($\Theta^{*1} \neq O^*$), $O - \Theta^{*1}$ contains at least one circuit β satisfying $\beta \cap \Theta_4 \neq \emptyset$ (see [R1]).

(H7) There exists such a minimal set O^* verifying $O^* \cap (\Theta_0 \cup \Theta_4) = \emptyset$.

The assumptions (H1-H7) are assumed to hold. Now we are ready to state the theorem of existence and uniqueness.

Theorem 1 *(i) Under the assumptions made in section 2.1, 2.2 and 2.3, for every given data f^i, r^i, i_d and , the variational formulation (4. 6) has an unique solution. (ii) The triplet $(\mathbf{u}, \phi, i)$ is solution of the variational formulation (4. 6) if and only if there exists an extension $i \in P^0(\Theta)$ of the current $i \in P^0(\Theta_4)$ such that the triplet $(\mathbf{u}, \phi, i)$ is solution of the piezoelectricity equations coupled with the electric circuit equations.*

The proof of this theorem is an extension of the proof of theorem 1 in [L3].

3 A TWO-DIMENSIONAL MODEL FOR THE SHELL

In this section, we state the two-dimensional shell model related to the previous three-dimensional one, based on the Reisner-Mindlin kinematic.

1 Parameterization of the shell

The tools of differential geometry that are recalled here come from [B3, B4, B5, C1, D1 and G1]. Consider $\widehat{\omega}$ an open bounded subset of the euclidean plane $\mathbb{R}^2$ such that the middle surface $\bar{\omega}$ of the shell is the range in $\mathbb{R}^3$ of the set $\widehat{\omega}$ by a mapping $(\xi^1, \xi^2) \to \varphi(\xi^1, \xi^2)$. We assume that $\widehat{\varphi}$ and $\partial\widehat{\omega}$ are sufficiently smooth. In addition, we assume that the vectors $\mathbf{a}_\alpha = \partial\widehat{\varphi}/\partial\xi^\alpha$, are linearly independent for any $\xi = (\xi^1, \xi^2) \in \widehat{\omega}$. These two vectors define the tangent plane $T_{\mathbf{m}}(\bar{\omega})$ to the surface $\bar{\omega}$ at the point $\mathbf{m} = \widehat{\varphi}(\xi)$. The normal vector to the tangent plane is given by

$$\mathbf{N}(\mathbf{m}) = \mathbf{a}_3 = (\mathbf{a}_1 \times \mathbf{a}_2)/|\mathbf{a}_1 \times \mathbf{a}_2|.$$

The set $(\mathbf{a}_1, \mathbf{a}_2, \mathbf{a}_3)$ defines the local covariant basis at the generic point $\mathbf{m}$ of the middle surface. We denote by $a_{\alpha\beta}$ and $b_{\alpha\beta}$ the first and the second fundamental forms of the middle surface defined by

$$a_{\alpha\beta} = a_{\beta\alpha} = \mathbf{a}_\alpha.\mathbf{a}_\beta$$

and

$$b_{\alpha\beta} = b_{\beta\alpha} = -\mathbf{a}_\alpha.\mathbf{a}_{3,\beta} = \mathbf{a}_3.\mathbf{a}_{\alpha,\beta} = \mathbf{a}_3.\mathbf{a}_{\beta,\alpha}.$$

The contravariant basis $(\mathbf{a}^1, \mathbf{a}^2, \mathbf{a}^3)$, associated to the covariant one $(\mathbf{a}_1, \mathbf{a}_2, \mathbf{a}_3)$, is defined by $\mathbf{a}_\alpha.\mathbf{a}^\beta = \delta_\alpha^\beta$ and $\mathbf{a}^3 = \mathbf{a}_3$. The matrix $(a^{\alpha\beta})$ is the inverse of the matrix $(a_{\alpha\beta})$. Using these metric tensors $(a^{\alpha\beta})$ and $(a_{\alpha\beta})$, one defines mixed and contravariant components by

$$b_\alpha^\beta = a^{\beta\lambda} b_{\lambda\alpha}, b^{\alpha\beta} = a^{\alpha\lambda} a^{\beta\mu} b_{\lambda\mu}$$

and conversely

$$b_{\alpha\beta} = a_{\alpha\lambda} b_\beta^\lambda = a_{\alpha\lambda} a_{\beta\mu} b^{\lambda\mu}.$$

The thickness of the shell is an application defined by $h : (\xi^1, \xi^2) \in \hat{\omega} \longrightarrow \mathbb{R}^{*+}$. So the shell Ω is an open subset of $\mathbb{R}^3$ defined by

$$\Omega = \{\mathbf{M} \in \mathbb{R}^3 : \mathbf{OM} = \widehat{\varphi}(\xi^1, \xi^2) + x_3 a_3(\xi^1, \xi^2), (\xi^1, \xi^2) \in \hat{\omega}, \\ \text{and } -h(\xi^1, \xi^2) < x_3 < h(\xi^1, \xi^2)\}.$$

Consider $\mathbf{g}_\alpha = (\delta_\alpha^\mu - x_3 b_\alpha^\mu)\mathbf{a}_\mu$ and $\mathbf{g}_3 = \mathbf{a}_3$. The vectors $\mathbf{g}_1$ and $\mathbf{g}_2$ are parallel to the tangent plan $T_{\mathbf{m}}(\omega)$ and $\mathbf{g}_3$ is normal to this plan. It turns out, that $(\mathbf{M}, \mathbf{g}_1, \mathbf{g}_2, \mathbf{g}_3)$ is a local reference system at each point $\mathbf{M}$ of Ω. The volume element $d\Omega$ is given by

$$d\Omega = g^{\frac{1}{2}} d\xi^1 d\xi^2 dx_3 = (\frac{g}{\bar{a}})^{\frac{1}{2}} \widetilde{a}^{\frac{1}{2}} d\xi^1 d\xi^2\, dx_3 = (\frac{g}{\bar{a}})^{\frac{1}{2}} d\omega dx_3$$

with $\widetilde{a} = \det(a_{\alpha\beta})$. From this, we deduce that the surface element is $d\omega = \widetilde{a}^{\frac{1}{2}} d\xi^1 d\xi^2$. The boundary of ω is $\partial\omega = S_0 \cup S_1$, where S_0 is the clamped part of ω. We assume that the shell Ω has a constant thickness $2h$, where $h > 0$ and is small enough so that, for every point $\mathbf{M} \in \overline{\Omega}$ there is a unique pair $(\mathbf{m}, x_3) \in \bar{\omega} \times [-h, h]$, satisfying $\mathbf{OM} = \mathbf{Om} + x_3\mathbf{N}(\mathbf{m})$.

2 The two-dimensional shell model

Among the most popular thin elastic shell models, there is one which is taking into account the transverse shear effect. This model is derived using the Reisner-Mindlin kinematic. It is based on the three following assumptions. First, particles located along normal direction before deformation remain aligned after deformation along a line which is generally no longer to the deformed middle surface. Thus

$$\mathbf{u} = \overline{\mathbf{u}}(\mathbf{m}) + x_3\boldsymbol{\theta} \text{ where } \boldsymbol{\theta} = \theta_\alpha(\mathbf{m})\mathbf{a}^\alpha$$

is the rotation in the normal direction $\mathbf{a}_3$, and $\overline{\mathbf{u}} = \overline{u}_i\mathbf{a}^i$. Second, the stresses along the thickness are approximately plane. This means that $\boldsymbol{\sigma}^{33}$ can be neglected. Third, the shell is homogenous in its thickness, in the sense that the tensors R^{ijkl}, d^{kij} and c^{ij} are independent of x_3. A fourth assumption related to the electric potential is inspired by the plate model [C3] which was derived using an asymptotic method. We assume that the electric potential ϕ is affine with respect to x_3.

Let us define the strain tensors for the two-dimensional shell model :

$$\epsilon^0_{\alpha\beta}(\overline{\mathbf{u}}) = \frac{1}{2}(\overline{u}_{\alpha|\beta} + \overline{u}_{\beta|\alpha}) - b_{\alpha\beta}\overline{u}_3, \epsilon^1_{\alpha\beta}(\overline{\mathbf{u}},\boldsymbol{\theta}) = \frac{1}{2}(\theta_{\alpha|\beta} + \theta_{\beta|\alpha} - b^\lambda_\alpha d_{\lambda\beta}(\overline{\mathbf{u}}) - b^\lambda_\beta d_{\lambda\alpha}(\overline{\mathbf{u}})),$$

$$\epsilon_{\alpha 3}(\overline{\mathbf{u}},\boldsymbol{\theta}) = \frac{1}{2}(u_{3,\alpha} + b^\lambda_\alpha\overline{u}_\lambda + \theta_\alpha)$$

where $d_{\lambda\alpha}(\overline{\mathbf{u}}) = \overline{u}_{\lambda|\alpha} - b_{\lambda\alpha}\overline{u}_3$, and $\overline{u}_{\lambda|\alpha} = \overline{u}_{\alpha,\beta} - \Gamma^\rho_{\alpha\beta}\overline{u}_\rho$. The admissible function set is

$$W^{shell}_{ad}(v_d) = \{(\overline{\mathbf{v}},\boldsymbol{\beta},\psi,j) \in H^1_{S_0}(\omega)^5 \times \Psi^{shell}_{ad}(v_d)\}$$

where

$$\begin{aligned} \Psi^{shell}_{ad}(v_d) &= \{(\psi,j) \in H^1(\overset{o}{\overline{\Theta \cup \omega}}) \cap \mathbb{P}^1(\Theta) \cap \mathbb{P}^0(\omega) \times \mathbb{P}^0(\Theta_4), \\ \psi &= 0 \text{ on } \sigma_0 \text{ and } -L\nabla_{\boldsymbol{\tau}}\psi = v_d \text{ on } \Theta_0\}. \end{aligned}$$

Here $H^1_{S_0}(\omega) = \{w \in H^1(\omega) \text{ such that } w = 0 \text{ on } S_0\}$ and $\mathbb{P}^0(\omega)$ is the set of functions constant on ω. The rigidity and piezoelectricity tensors for the shell model are given by :

$$\begin{aligned} \widetilde{R}^{\alpha\beta\gamma\delta} &= R^{\alpha\beta\gamma\delta} + R^{\alpha\beta 33}l^{\gamma\delta 33}, \ \widetilde{R}^{\alpha\beta\gamma 3} = 2R^{\alpha\beta\gamma 3} + R^{\alpha\beta 33}t^{\gamma 333}, \\ \widetilde{R}^{\alpha 3\gamma\delta} &= 2R^{\alpha 3\gamma\delta} + 2R^{\alpha 333}l^{\gamma\delta 33}, \ \widetilde{R}^{\alpha 3\gamma 3} = 4R^{\alpha 3\gamma 3} + 2R^{\alpha 333}t^{\gamma 333}, \\ \widetilde{d}^{i\alpha\beta} &= d^{i\alpha\beta} + l^{\alpha\beta 33}d^{i33}, \ \widetilde{d}^{i\alpha 3} = 2d^{i\alpha 3} + t^{\alpha 333}d^{i33}, \end{aligned}$$

where $l^{\mu\eta 33} = -\frac{R^{33\mu\eta}}{R^{3333}}$ and $t^{\mu 333} = -\frac{2R^{33\mu 3}}{R^{3333}}$.
We assume that the forces $\mathbf{f} = (f^i)_i$ due to volume and surface forces are defined as $\mathbf{f} = \overline{\mathbf{f}} + x_3 g^\alpha\mathbf{a}^\alpha \in L^2(\omega)^3$, and the lateral forces $\mathbf{r} = (r^i)_i$ verify $\mathbf{r} = \overline{\mathbf{r}} + x_3\gamma_\alpha\mathbf{a}^\alpha$, where $(\overline{\mathbf{r}},(\gamma_\alpha)_\alpha) \in L^2(S_1)^5$.

Taking into account the kinematics assumptions, the two-dimensional shell variational formulation is stated as follows. Find $(\bar{\mathbf{u}},\theta,\phi,i) \in W_{ad}^{shell}(v_d)$ such that, for every $(\bar{\mathbf{v}},\beta,\psi,j) \in W_{ad}^{shell}(0)$:

$$\begin{aligned} a^{shell}((\bar{\mathbf{u}},\theta,\phi),(\bar{\mathbf{v}},\beta,\psi)) + b_1^{shell}(\psi,i) &= l^{shell}(\bar{\mathbf{v}},\beta,\psi) \\ b_2^{shell}(\phi,j) &= 0. \end{aligned} \tag{2. 9}$$

where

$$a^{shell}((\bar{\mathbf{u}},\theta,\phi),(\bar{\mathbf{v}},\beta,\psi)) = \int_\omega D(\bar{\mathbf{v}},\beta,\psi)\overline{\mathcal{K}}_h^{0t} D(\bar{\mathbf{u}},\theta,\phi)\, d\omega + \int_{\Theta_2} Lg\nabla_\tau\psi\nabla_\tau\phi dl(x),$$

$$b_1^{shell}(\psi,i) = \int_{\Theta_4} i\nabla_\tau\psi dl(x), b_2^{shell}(\phi,j) = \int_{\Theta_4} \nabla_\tau\phi j dl(\mathbf{x}) - \int_{\Theta_3} k\nabla_\tau\phi j dl(\mathbf{x}),$$

$$l^{shell}(\mathbf{v},\psi) = -\int_{\Theta_1} i_d\nabla_\tau\psi dl(\mathbf{x}) + \int_\omega \overline{f}^i\overline{v}_i d\omega + 2h^3/3\int_\omega g^\alpha\beta_\alpha d\omega + \int_{S_1}\{\overline{r}^i\overline{v}_i - M^\alpha\gamma_\alpha\}dS,$$

$$D(\bar{\mathbf{v}},\beta,\psi) = (\epsilon^0_{\alpha\beta}(\bar{\mathbf{v}}),\epsilon_{\alpha 3}(\bar{\mathbf{v}},\beta),\epsilon^1_{\alpha\beta}(\bar{\mathbf{v}},\beta),\psi),$$

and

$$\overline{\mathcal{K}}_h^0 = \begin{pmatrix} \widetilde{R}^{\alpha\beta\gamma\delta} & \widetilde{R}^{\alpha\beta\gamma 3} & 0 & \widetilde{d}^{3\alpha\beta}/2 \\ \widetilde{R}^{\alpha 3\gamma\delta} & \widetilde{R}^{\alpha 3\gamma 3} & 0 & \widetilde{d}^{3\alpha 3}/2 \\ 0 & 0 & \frac{h^2}{3}\widetilde{R}^{\alpha\beta\gamma\delta} & 0 \\ -\widetilde{d}^{3\gamma\delta}/2 & -\widetilde{d}^{3\gamma 3}/2 & 0 & c^{33}/4 \end{pmatrix}.$$

Theorem 2 *Assuming that assumptions of sections 2.1 ... 3.2 are fulfilled, then for any given forces* **f** *and* **r**, *any given current* i_d *and any given voltage* v_d, *the variational formulation (2. 9) of the two-dimensional shell model has an unique solution.*

The proof of this theorem is a combination of the proofs of theorem 1 in [L3] and existence and uniqueness proof for the two-dimensional shell model.

3 Model with collocated actuators and sensors

In view of vibration control application, one assume that the shell upper face metallization is divided in two insulated parts ω^{in} and ω^{out} having intricately shapes. The part ω^{in} is used as a sensors when ω^{out} is used as an actuator. This assumption of collocated actuator and sensor is now interpreted. The mean value of any function $f \in L^2(\omega)$ on ω^{in} (respectively on ω^{out}) is denoted by $\langle f\rangle^{in}$ (respectively $\langle f\rangle^{out}$). If an actuator and a sensor are collocated then $\langle\epsilon_{\alpha i}(\overline{\mathbf{v}})\rangle^{in} \simeq \langle\epsilon_{\alpha i}(\overline{\mathbf{v}})\rangle^{out}$ for $\alpha \in \{1,2\}$ and $i \in \{1,2,3\}$. Since these two mean values are assumed to be equal, the indices ^{in} and ^{out} can be removed and one can denote these mean values by $\langle\epsilon_{\alpha i}(\overline{\mathbf{v}})\rangle$. In addition, the sensor is assumed to be connected to the input of a voltage amplifier, when the actuator is connected to the output of a voltage to voltage amplifier. So, the relation between the electric potentials ϕ^{in} of the sensor and ϕ^{out} of the actuator is expressed using the transfer function defined by $\mathbf{K}$: $\phi^{out} = \mathbf{K}\ \phi^{in}$. By another way, since the current vanishes on the sensor ω^{in}, so ϕ^{in} is linearly dependent of $\left\langle \epsilon^0_{\alpha i}(\overline{\mathbf{v}})\right\rangle^{in}$. Finally, the two-dimensional shell model becomes : find $(\overline{\mathbf{u}}, \theta) \in W^{colloc}_{ad} = H^1_{S_0}(\omega)^5$ such that for every $(\overline{\mathbf{v}}, \beta) \in W^{colloc}_{ad}$

$$a^{colloc}((\bar{\mathbf{u}},\theta),(\bar{\mathbf{v}},\beta)) = l^{shell}(\bar{\mathbf{v}},\beta) \tag{3. 10}$$

where

$$\begin{aligned}
a^{colloc}((\bar{\mathbf{u}},\theta),(\bar{\mathbf{v}},\beta)) &= \int_\omega D(\bar{\mathbf{v}},\beta)\widetilde{\mathcal{K}}_h D(\bar{\mathbf{u}},\theta)\ d\omega + C(\overline{\mathbf{v}},\beta)\frac{\mathbf{Id}+\mathbf{K}}{c^{33}}C(\overline{\mathbf{u}},\theta),\\
D(\bar{\mathbf{v}},\beta) &= (\epsilon^0_{\alpha\beta}(\bar{\mathbf{v}}), \epsilon^0_{\alpha 3}(\bar{\mathbf{v}},\beta), \epsilon^1_{\alpha\beta}(\bar{\mathbf{v}},\beta)),\\
C(\overline{\mathbf{v}},\beta) &= \widetilde{d}^{3\alpha\beta}\left\langle \epsilon^0_{\alpha\beta}(\overline{\mathbf{v}})\right\rangle^{in} + \widetilde{d}^{3\alpha 3}\left\langle \epsilon^0_{\alpha 3}(\overline{\mathbf{v}},\beta)\right\rangle^{in}
\end{aligned}$$

and

$$\widetilde{\mathcal{K}}_h = \begin{pmatrix} \widetilde{R}^{\alpha\beta\gamma\delta} & \widetilde{R}^{\alpha\beta\gamma 3} & 0 \\ \widetilde{R}^{\alpha 3\gamma\delta} & \widetilde{R}^{\alpha 3\gamma 3} & 0 \\ 0 & 0 & \frac{h^2}{3}\widetilde{R}^{\alpha\beta\gamma\delta} \end{pmatrix}.$$

4 NUMERICAL SIMULATION

In this section, a numerical simulation of a vibrating three layers shell with co-located actuators and sensors is presented. Two simulations have been carried out using two different electric networks being designed in order to damp the shell vibrations.

1 Extension of the two-dimensional model

In the previous section a two-dimensional piezoelectric shell model has been presented. The shell was single layered and in equilibrium. The simulation is based on an extension of this model to the case of a vibrating three layers piezoelectric shell. More precisely, the shell is made of an elastic layer (indexed by $\zeta = 2$) and of two piezoelectric layers (indexed by $\zeta = 1, 3$) glued on each faces of the elastic layer. Each piezoelectric layer ζ is partitioned in a regular mesh of $n \times n$ piezoelectric transducers $(\omega^\zeta_{i,j})_{i,j\in\{1,..,n\}^2}$ having their sides parallel to the vectors $\mathbf{a}_1$ and $\mathbf{a}_2$. Each of the transducers is electrically insulated of its neighbors, and is a system of collocated sensor $\omega^{\zeta,in}_{i,j}$ and actuator $\omega^{\zeta,out}_{i,j}$. The electric network is distributed in each piezoelectric layer. It can connect any cell to their neighbors. So, in each piezoelectric layer indexed by ζ, the transfer function $\mathbf{K}_\zeta$ between the distributed fields $\phi^{\zeta,in}$ and $\phi^{\zeta,out}$ is no longer a scalar but a matrix of transfer functions. Now, two particular examples of circuits, that is, two examples of $\mathbf{K}^\zeta$ are detailed. In both cases, the circuits are identical on both layers. So, $\mathbf{K}^1 = \mathbf{K}^3 = \mathbf{K}$ and for the sake of simplicity, the indices ζ will be removed.

Example 1 (local circuit) : This circuit is constituted of a differential amplifier associated to each pair of collocated sensor and actuator : $(\omega_{i,j}^{in}, \omega_{i,j}^{out})$. The input and the output of the same amplifier being connected to $\omega_{i,j}^{in}$ and $\omega_{i,j}^{out}$. Let us denote by k the amplification coefficient, so $\mathbf{K} = -k\frac{d}{dt}$. So, the relation imposed by this network between ϕ^{in} and ϕ^{out} is $\phi_{i,j}^{out} = -k\frac{d\phi_{i,j}^{in}}{dt}$ for every $i, j \in \{1, ..n\}^2$. The equivalent circuit for a one dimensional problem is presented on the figure 4.

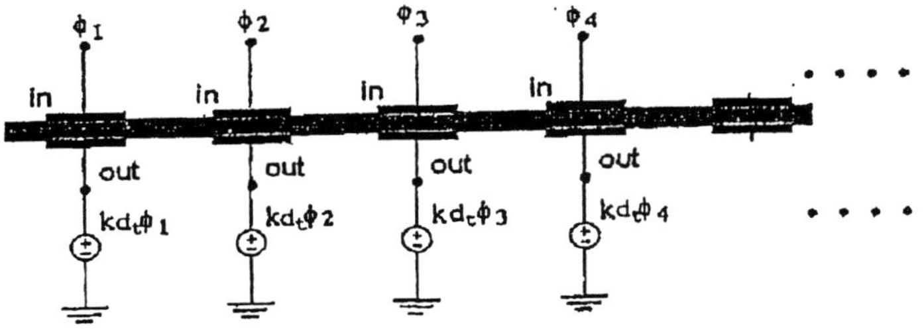

Figure 4: Local circuit for a one dimensional problem

Example 2 (non-local circuit) : The circuit is represented on the figure 5. The transfer matrix $\mathbf{K}$ is such that

$$\phi_{i,j}^{out} = (\mathbf{K}\phi^{in})_{i,j} = k\frac{d}{dt}(\phi_{i,j+1}^{in} + \phi_{i,j-1}^{in} + \phi_{i+1,j}^{in} + \phi_{i-1,j}^{in} - 4\phi_{i,j}^{in}).$$

The equivalent circuit for a one dimensional problem is presented on the figure 5, where $k = k_1 k_2$.

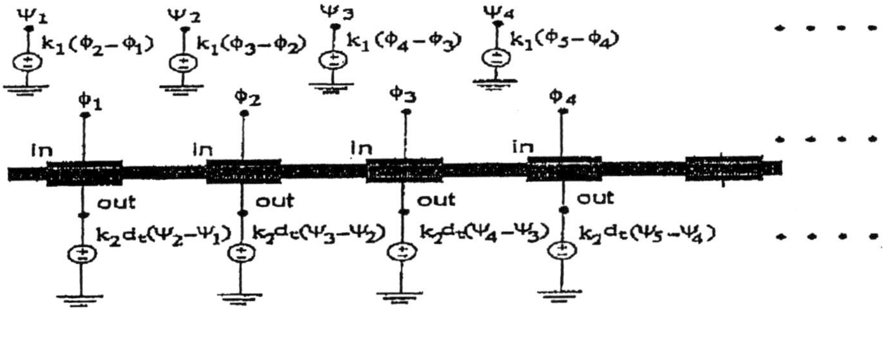

Figure 5: Non local circuit for a one dimensional problem

2 Description of the physical problem

In this section, we report our numerical simulation results. Let us quote that in this work, our focus is not to derive an optimized algorithm for vibrations damping.

We restrict our attention to the simulation of the whole coupled electro-mechanical system. Our choice of particular electronic circuits is only illustrative of our modelling.

We consider a quarter of a cylindrical elastic thin shell of radius 0.4 meter and thickness 5,8 mm. It is assumed to be clamped on its upper and lower extremities, where height = 0 and height = 1 meter, see figure 6. Other boundaries are assumed to be free. The thickness of piezoelectric transducers is equal to 2.7 mm. Material properties of elastic shell and of piezoelectric material are summarized in table 1. These values are copied from [L4]. A time periodic pressure force was applied uniformly on the shell surface. Its frequency was equal to 400 Hz. It was stopped after 4 periods.

The simulation is based on our formulation of a three-layered piezoelectric shell coupled with an electronic circuit. Actuators and sensors are still assumed to be collocated, but electric potentials are not eliminated as it was done in the model of section 3.3. The discretization of the mechanical fields is done using an extension to piezoelectric thin shells of the degenerate finite element introduced by S. Ahmad, B.M. Irons and O.C. Zienkiewicz [A2] for elastic thin shells, and improved by C. Gelin [G1] for the prevention of the locking effect. Electric potentials are discretized by a constant on each element. The finite element mesh is constituted by 64 regular rectangular elements. An insulated piezoelectric transducers is associated to each finite element. The two electric circuits previously mentioned have been tested. They are referred respectively as the local one and the non-local one.

Property	Elastic shell	Piezoelectric shell (VIBRIT)
d311	0 C/m^2	-5.4 C/m^2
d322	0 C/m^2	-5.4 C/m^2
d333	0 C/m^2	13.5 C/m^2
d312	0 C/m^2	11.7 C/m^2
Permittivity	10^{-9} F/m	7.210^{-9} F/m
Young's modulus	7.1 10^{10} N/m^2	6.4 10^{10} N/m^2
Poisson's ratio	0.33	0.31
Density	2.7 10^3 Kg/m^3	7.6 10^3 Kg/m^3

Table 1
Material properties of elastic and piezoelectric shell

3 Numerical results

Circumferential, longitudinal and transverse displacements of the center of the shell are plotted in figures 7 to 15 in the three following cases : open circuit, local circuit and non local circuit. The coefficient k for these particular circuits is taken equal to 0 (uncontrolled case), to 10 and to 100 (controlled case with local circuits) and 6.4 and 64 (controlled case with non local circuits).

Comparison of the effect of the local and of the non local controllers, in terms of decay rate of the solutions, show that for a lower amplification coefficient the non local controller is more efficient than the local one. This preliminary result is a first jusfication of the interest of considering control systems based on distributed circuits. A more detailled study of the effect of such a distributed control using a distributed circuit can be found in [M1].

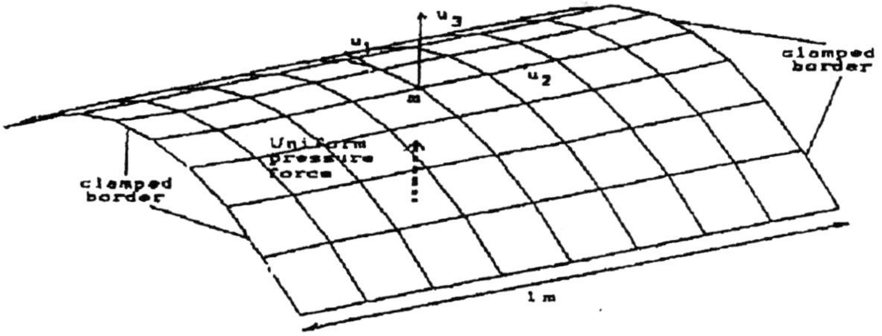

Figure 6: Mesh for a quarter of cylinder

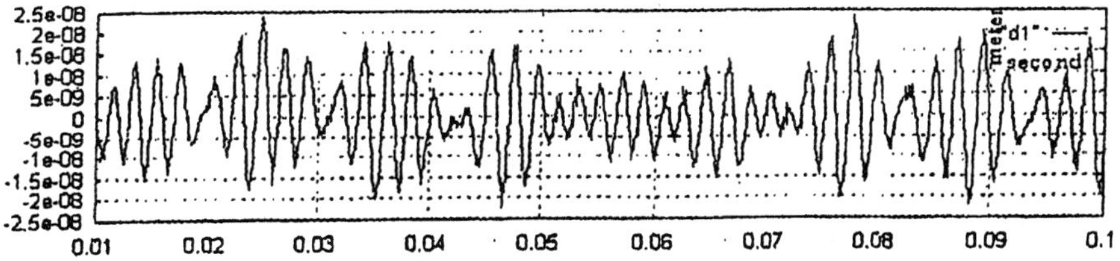

Figure 7 : Circumferential displacement : Open circuit

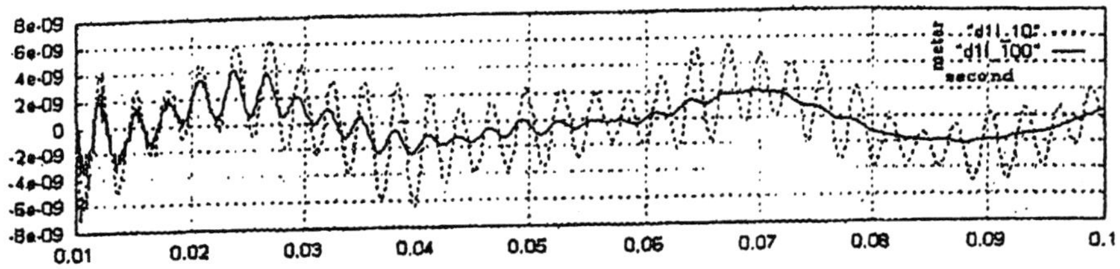

Figure 8 : Circumferential displacement : Effect of the local circuit

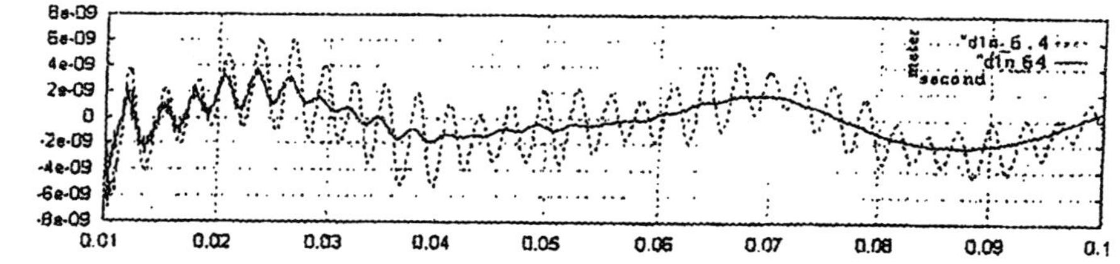

Figure 9 : Circumferential displacement : Effect of the non local circuit

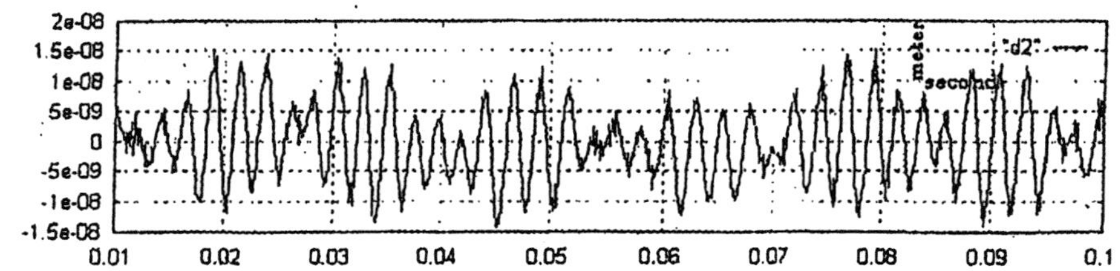

Figure 10 : Longitudinal displacement : Open circuit

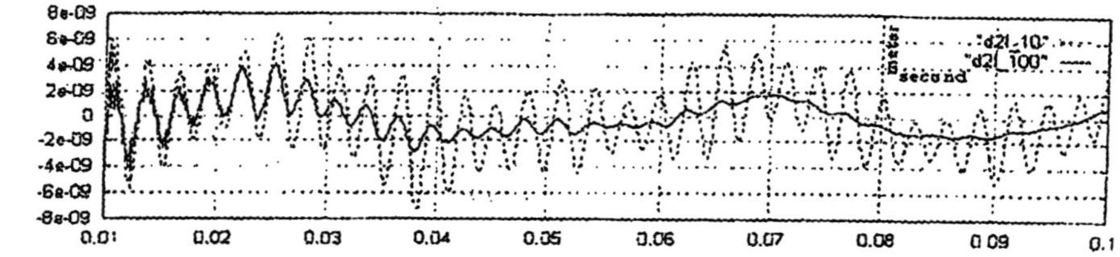

Figure 11 : Longitudinal displacement : Effect of the local circuit

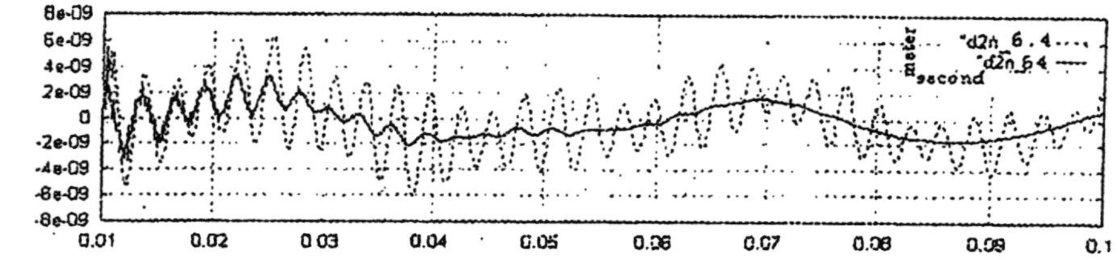

Figure 12 : Longitudinal displacement : effect of the non local circuit

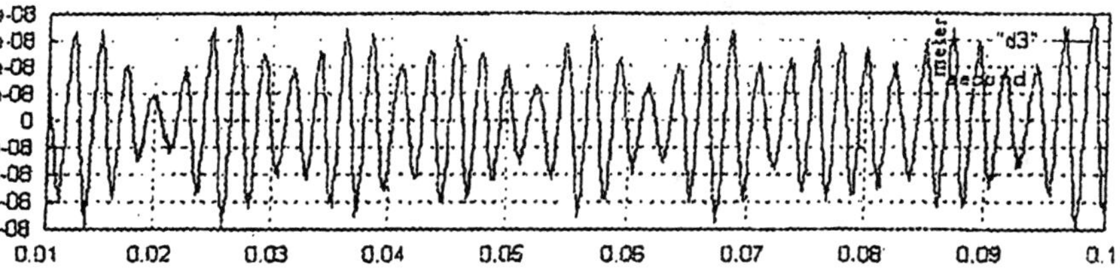

Figure 13 : Transverse displacement : Open circuit

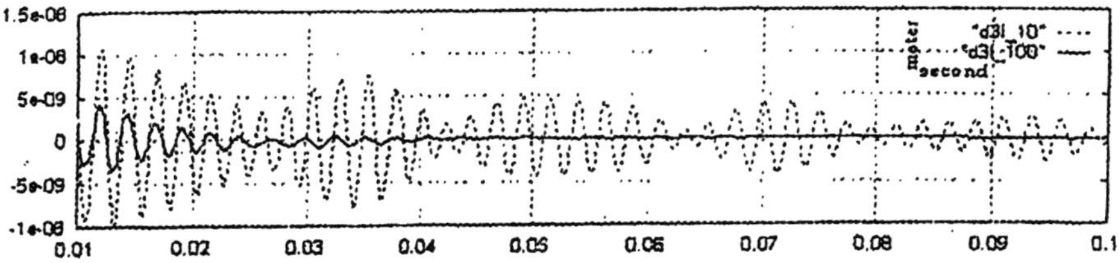

Figure 14 : Transverse displacement : Effect of the local circuit

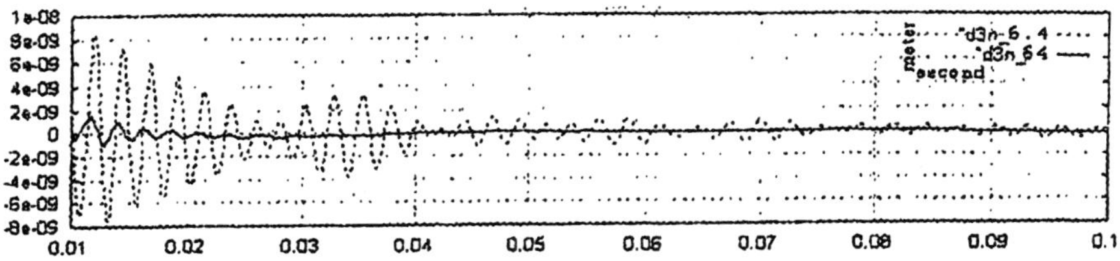

Figure 15 : Transverse displacement : Effect of the non local circuit

References

[A1] S. Ahmad, B. M. Irons, O. C. Zienkiewicz, *Analysis of thick and thin shell structures by curved finite elements*, Int. J. for Num.meth. in Eng. Vol.2, pp 419-451. (1970).

[A2] F. Ali Mehmeti, Non linear waves in networks, *Mathematical Research, Vol.80, Akademie Verlag, Berlin* (1994).

[B1] J. von Below, Parabolic Network Equations, Habilitation thesis, Eberhard-Karls-Universität Tübingen (1993).

[B2] C. Bernardi, C. Canuto, Y. Maday, *Generalized inf-sup conditions for Chebyshev spectral approximation of the Stokes Problem.* SIAM J. Numer. Anal. Vol. 25, n° 6. (1988).

[B3] M. Bernadou, Finite element methods for thin shell problems. Ed. John Willey & Sons. (1994).

[B4] M. Bernadou, P. G. Ciarlet, and B. Miara, *Existence theorems for two-dimensional linear shell theories,* J. of Elasticity, vol. 34, n° 2, pp 111-138, (1984)

[B5] M. Bernadou and D. Haenel, *Some Remarks about piezoelectric shells,* Civil-Comp Ltd, Edinburgh, Scotland, (1994)

[C1] P. G. Ciarlet, *Modèles bi-dimensionnels de coques : Analyse asymptotique et théorèmes d'existence, Boundary Value problems for Partial Differential Equations and Applications,* J. L. Lions and C. Biacchi, Editors, pp. 61-80, Masson, Paris, (1993).
[C2] P. G. Ciarlet, H. Le Dret, R. Nzengwa, *Junction between three dimensional and two-dimensional linearly elastic structures,* J. Math. Pures et Appl., n° 68, pp. 261-295, (1989).
[C3] E. Canon, M. Lenczner, *Models of elastic plates with piezoelectric inclusions. Part I : Models without homogenization.* Math. Comput. Modelling, Vol. 26, n°. 5, pp. 79-106, (1997).
[D1] Ph. Destuynder, Thèse d'état, Université P. M. Curie, Paris 6. (1980).
[G1] C. Gelin, Rapport du Laboratoire de Mécanique Appliquées de Besançon. (1992).
[H1] D. Haenel, Thèse de mathématiques appliquées, Université P. M. Curie, Paris 6. (1998).
[L1] J. Lagnese, J.E. Leugering, G. Schmidt, E.J.P.G.., Modeling Analysis and Control of Dynamic Elastic Multi-Link Structures, *Birkhauser, Boston, Basel, Berlin* (1994).
[L2] N.N. Le khan'chau, The theory of piezoelectric shells, *Prikl. Matem. Mekhan. USSR, Vol.50, n° 1, pp. 98-105. (1986).*
[L3] M. Lenczner and G. Senouci-Bereksi, *Homogenization of electric networks including voltage to voltage amplifiers,* Mathematical Models and Methods in Applied Sciences, Vol. 9, n° 6, pp 899-932 (1999).
[M1] M. Kader., Lenczner M. et Mrcarica Z. *Approximation d'un contrôle optimal distribué par un circuit électronique réparti : Application au contrôle de vibrations,* to appear in C.R. Acad. Sci. Paris, Série II b, (2000).
[N1] S. Nicaise, Polygonal interface problems. *Methoden und Verfahren der mathematische Physik, Band 39, Peter Lang, Frankfurt a.P., Berlin, Bern, New York, Paris, Wien* (1993).
[N2] S. Nicaise, *Contrôlabilité exacte d'un problème couplé pluri-dimensionel,* C.R. Acad. Sci. Paris, Série I, Math. tome 311, n° 1, pp. 19-22, (1999).
[L3] Reinhard Lerch, *Simulation of piezoelectric devices by two and three-dimensional finite elements,* IEEE vol 37 n° 2, (1990).
[R1] A. Recski, Matroid Theory and its Applications in Electric Network theory and in Statics. *Algorithms and Combinatorics 6, Springer Verlag. (1989).*
[R2] N. N. Rogacheva, *Classification of free piezoceramic shell vibrations,* J. Appl. Math. Mech., Vol. 50, pp 106-111, (1986).
[R3] N. N. Rogacheva. *Equations of state of piezoceramic shells,* J. Appl. Math. Mech., Vol.45, pp 677-684, (1982).
[S1] A. Saidi, Thèse Mathématiques appliquées, Université P.M. Curie, Paris 6. (1997).
[T1] H. S. Tzou, Piezoelectric shells : Distributed sensing and control of continua, *Kluwer Academic publishers, Dordrecht, The Netherlands. (1993).*
[V1] Vlach, J., Singhal, K, Computer methods for circuit Analysis and Design, *Van Nostrand Reinhold Company, New York. (1983).*

Aknowlegment : The authors thanks the referee who read the manuscript carefully and suggested many improvements.